BROWN'S BOOK OF CARBURETORS

COMPILED BY MICHAEL H. BROWN

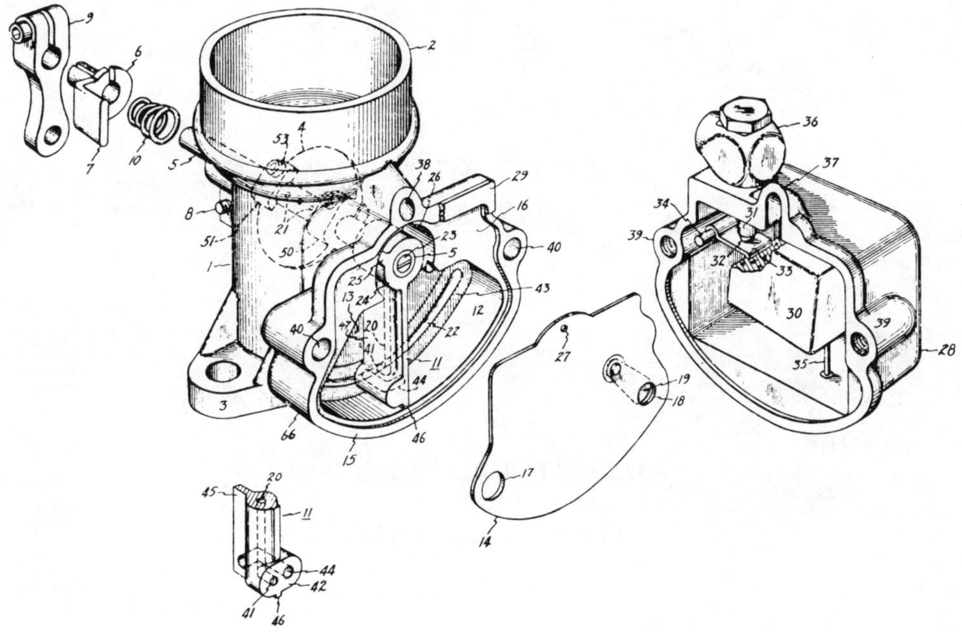

Alternate Energy Ideas
FROM
DESERT PUBLICATIONS

BROWN'S BOOK OF CARBURETORS

Compiled By Michael H. Brown

ISBN: 0 - 87947 - 306 -1

DESERT PUBLICATIONS

Cornville, Arizona 86325

Table Of Contents

Introduction

Basically, the types of carburetors Detroit refuses to produce fall into two categories: those that use a variable venturi or similar principle -- that is, altering the percentage of air flow to coincide with the amount of gasoline sucked into the engine, and those that pre-heat the fuel to a complete vapor thereby utilizing all the fuel and preventing carbon formation in the engine.

Either way you get better gas mileage.

Why won't Detroit produce these devices?

It boils down to two reasons:

First, if fuel is turned into a complete vapor before combustion there are few (if any) modifications needed to run the vehicle on alcohol (easily homemade), olive oil, kerosene, or whatever else you can get your hands on that has a "flash point". The oil barons own plenty of auto manufacturing stock and have a vested interest in the *status quo*.

Second, carbon formation from incomplete combustion is the number one cause of engine wear. Observe the insides of a propane-powered engine after a quarter million miles: clean as a pin. And no wear to speak of. If the average auto engine lasted 250,000 miles instead of the present 100,000 miles the present extremely lucrative spare parts business -- on engine parts alone -- would suffer at least a 50% drop in sales.

Lastly, you've heard all the stories of how General Motors or Ford bought up this or that patent and then "sat" on it. We think you will find this book interesting.

A couple of these goodies may already be in production by the time this hits print. Write them (not us) to make sure. If any of you enterprising machinists or mechanics manage to assemble some of these items and can maintain some semblance of production drop us a line and we'll help you sell them. We are only interested in completed projects, we simply don't have time to answer all the "I think I can do it" letters.

If you want extra patents listed on the back of these patents write the U.S. Patent Office (again, not us) and enclose 50 cents for each patent you want by number (e.g., 1,997,497 Pogue) and name. Be willing to wait a couple of months -- the Patent Office is no ball of fire.

The Mystery Of The 200 M.P.G. Pogue Carburetor

Reprinted from the September, 1953 issue of CARS Magazine

In Montreal today, a tired, resigned man goes about his work, trying to forget the past. He hopes to brush from his memory the fact that 16 years ago he panicked the Toronto Stock Exchange, threw a fright into major oil companies, upset manufacturers of carburetors and captured the immediate, unflagging interest of every motorist in the world. His name is Pogue. Charles Nelson Pogue, automotive engineer. His moment of glory, his claim to fame lies in the invention of an automobile carburetor. Its principle was supposed to be a delayed mixing process that produced a more gaseous charge of fuel and delivered upwards of 200 miles per gallon of standard gasoline.

Or so some people say. And this is the year the world may discover the truth. For this is the year of the Pogue carburetor — the year the last of the Pogue patents runs out.

Shrouded in the vagueness of time and forgotten documents, Pogue and his carburetor have become a world-wide legend. Whenever automotive minded people get together, whether they be technical brains or garden-variety motorists, the talk sooner or later turns to "that man from Canada who invented the 200-mpg carburetor". Sunday newspaper supplements and syndicated columnists perenially rediscover the item and wonder about rumors that Pogue was kidnapped, murdered, beaten, bought off.

Wonder no longer.

For the first time in 14 years Charles Nelson Pogue spoke to a reporter. I am that reporter and this is his story.

In the opening months of 1936, stock exchange offices and brokers were swamped with orders to dump all oil stock immediately. Imperial Oil of Canada, for instance, suffered a six-point drop. Why? Well, during June and July of the previous year, it was rumored that a small eight-cylinder coupe fitted with a mysterious carburetor had been driven from Winnipeg, Maintoba to Vancouver, British Columbia — a distance of 1,879½ miles — on 14½ gallons of gasoline for an overall average of around 130 miles per gallon!

This was not the first time, however, that the carburetor had made news. Since Pogue had started on the invention some 17 years before (about the end of World War I) and there had been earlier impractical models, Pogue had created whispers throughout the North American continent. Oil men snickered at the idea and most motorists laughed off news of the invention as just another illogical contraption.

However, in 1936 the manager of one of Winnipeg's largest automobile dealerships was reported in the authoritative *Canadian Automotive Trade* to have said: "Today I drove a coupe with the Pogue carburetor 216.8 miles on an imperial gallon of gasoline (roughly the equivalent of a gallon and a fifth in the United States). The car performed per-

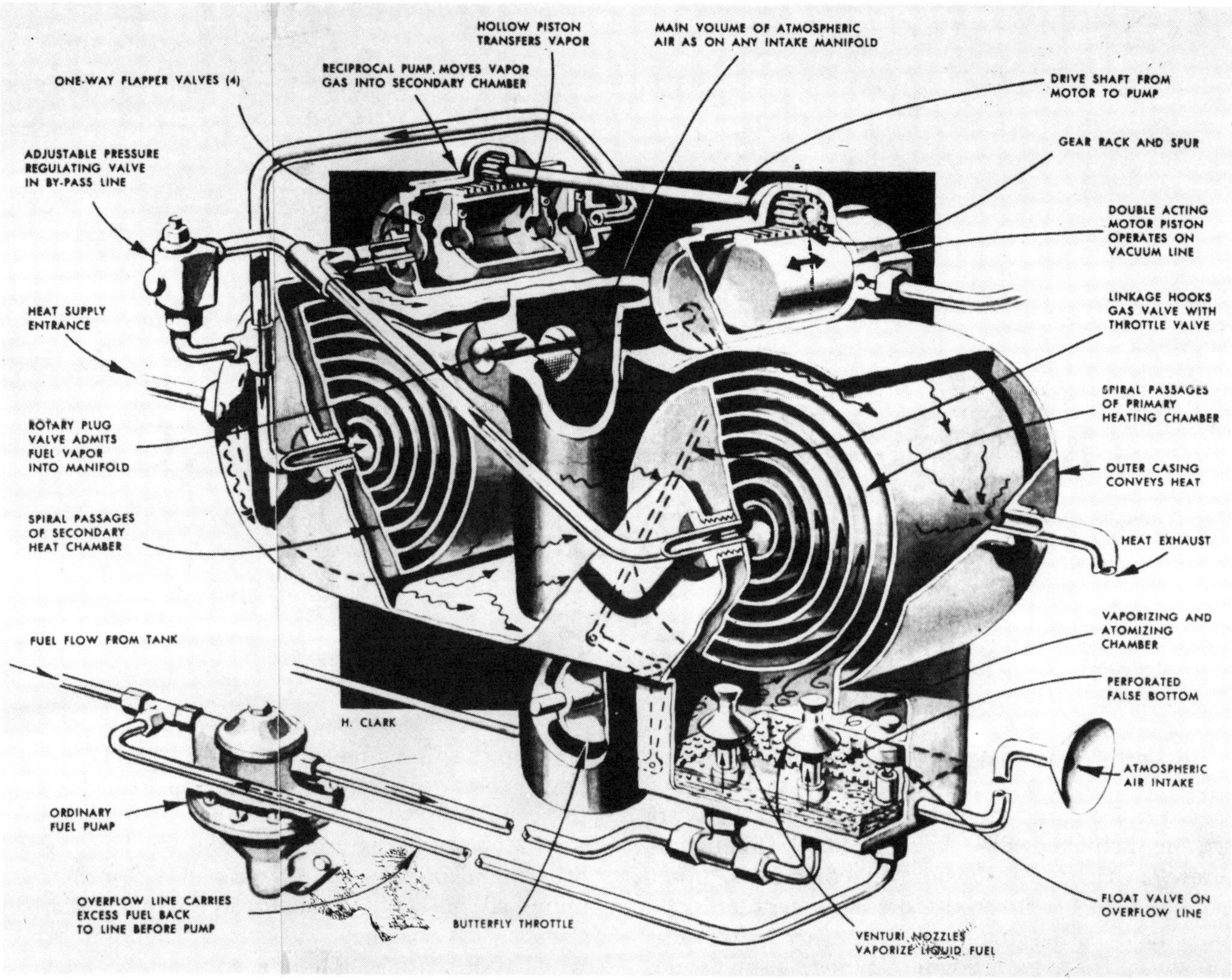

THE POGUE CARBURETOR

fectly in every respect . . . I was able to throttle it down to one mile an hour and the car pulled along smoothly and steadily."

The president of a leading automobile agency in the same area (which was Pogue's original home and birthplace of the invention) was quoted by the *Schrader Service News* of Brooklyn, N.Y., for Summer 1936 as saying: "I made a test today of the Pogue carburetor installed on an eight cylinder coupe. The speedometer showed that the car had run over 8,000 miles. I drove the car 26.2 miles on one pint of gasoline. The performance was 100 per cent in every way."

An engineer in a Canadian automobile plant was quoted as saying: "At the time of testing we covered 25.7 miles on one pint of gasoline."

In addition to these personal testimonials, other articles appeared in the daily press and automotive trade journals about better idling, faster pickup and in one case, a test car which was kept in an unheated garage during a cold Canadian winter offered no starting difficulties during the entire period.

Overwhelming controversy met these testimonials. The December 12, 1936 issue of *Automotive Industries* said: "It is stated that on the announcement of these results (216 miles per imperial gallon) the stocks on the Toronto Exchange took a tumble, that men of great wealth or powerful political connections are financing the development, and that the inventor's laboratories have been broken into in two different instances and working models stolen . . . it is added that the stolen models were incomplete . . . if the inventor could substantiate the claims, the consequences would be most important . . . fails to show any features hitherto unknown in carburetor practice and absolutely gives no warrant for crediting the remarkable

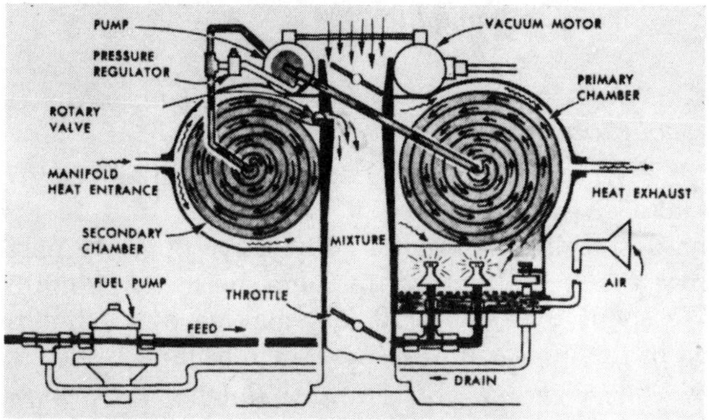

ENGINE MANIFOLD

Operation of Pogue carburetor is relatively simple. Fuel pump, lower left of diagram above, forces gasoline through fuel line and out, as fine mist, through two nozzles at bottom of primary heating chamber. Only pure vapor is drawn up through spiral passages. When vapor reaches center of primary chamber, a vacuum-operated reciprocating pump forces it through conduit into secondary heating chamber, where vapor goes through second set of spiral passages. Vapor then enters downdraft air horn where it mixes with air and flows into intake manifold of engine. Gasoline which is not vaporized in the primary chamber condenses, flows back through drain to fuel line and into pump.

results claimed . . . we shall remain exceedingly skeptical . . . "

Other articles at that time said that the carburetor was too impractical for use in the ordinary course of driving because it was too dangerous. These writings pointed out that the Pogue carburetor was liable to explode at any time.

Scores of carburetion experts and oil company engineers made public pronouncements on the impossibility of the sevenfold increase claimed by the Pogue invention. One oil company engineer even pointed out that gasoline did not, nor never could, contain that high a percentage of explosive power.

But the fact remains that Charles Nelson Pogue was the holder of some five American and seven Canadian patents issued to him between 1928-1936. The Canadian Patent Office lists the first patent for an improvement in carburetion as filed on April 3, 1928 and the last on June 23, 1936. The first United States patent was issued on March 11, 1930 from an application filed August 20, 1927, and the last was issued on January 7, 1936. Since the life span of a patent is 17 years, Pogue's last legal claim to his controversial carburetor runs out his year.

Columnist E. V. Durling wrote in a 1949 column, "Some organizations make a practice of waiting 17 years for a patent to expire and then start to manufacture the patented article . . . that brings to mind that there was patented in the United States in the early 1930s a carburetor said to make it possible to travel from 150 to 200 miles on one gallon of gasoline . . . it has never reached the market. However, the patent on it expires in a few years. I can hardly wait to see what happens then."

Neither can Charles Nelson Pogue.

Now the general manager and part owner of a dingy second floor machine shop in Montreal which specializes in the production of oil filters and filter elements, Pogue is anxious to see if anyone will succeed where he did not.

Each time his story is "rediscovered" Pogue is swamped with mail from all 48 states, all provinces of Canada and at least 29 foreign countries. His fame is so widespread throughout the Dominion that a letter addressed to "Pogue, Canada" and mailed in San Antonio, Tex., once reached him in three days.

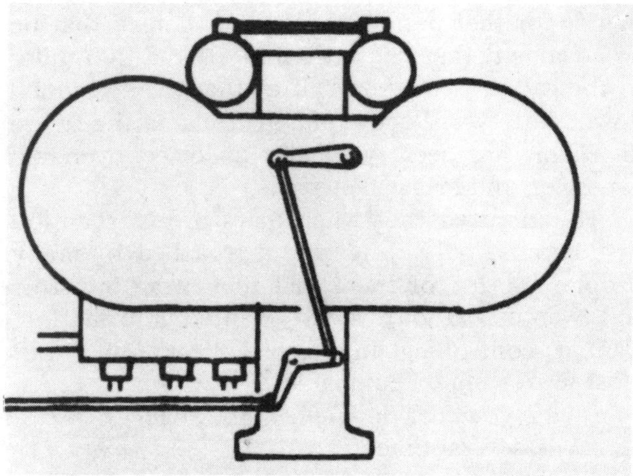

THROTTLE LINKAGE

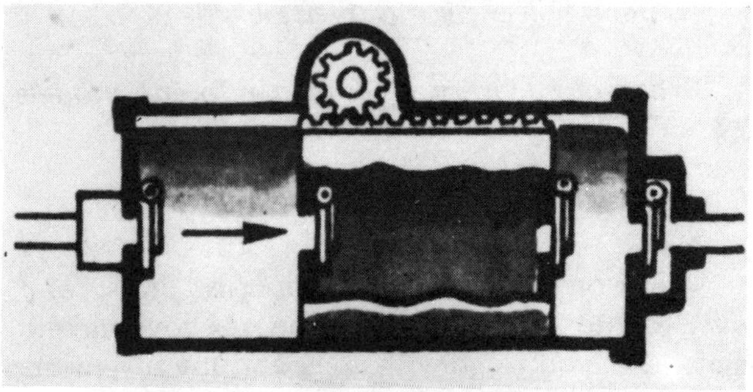

SECTION THROUGH RECIPROCAL PUMP

Although he is not in hiding (the Montreal telephone directory lists his home address and telephone number), Pogue has a healthy dread of the mail influxes which arrive at his home or office. Each mention of the carburetor seems to drive deeper his painful memory of the invention and his life with it.

Pogue is reluctant to talk about his invention. He spoke to this reporter "only because you're from New York and you've come this distance. I haven't seen a reporter or talked for publication in 14 years."

When reminded that he was the central figure in a controversy which had never been settled and further, that he was now a legend to a new generation, Pogue said: "I don't want to hear about that. I have nothing to gain by telling my story. If I knew I were to die tomorrow or if I were leaving the country, I would tell it . . . and it is quite a story. I'm alive and this is my home . . . " He waved a tired hand in the general direction of downtown Montreal.

But, then, with prodding he admitted that his own automobile ran for ten years on one of his carburetors but he refused to divulge performance figures for that period. Further, he denied that he ever claimed that the invention offered 200 miles to the gallon -- or even half of that. When pinned down, however, Pogue refused to name the figure he claims has been "violently distorted by newspapermen and magazine writers."

He admitted that when news of his invention first became public, he was approached by many people. "A lot of them said they were investors and wanted to buy stock, wanted information, wanted controlling interests. I later found out most were from oil companies."

"Were you ever threatened, Mr. Pogue?"

"Yes, several times."

"How and by whom?"

"I won't answer that."

"Were you ever physically attacked?"

"No."

"Was your workshop broken into and models stolen?"

"Several times."

"Are any of your carburetors still in existence?"

"No."

"Were you ever the victim of political pressure?"

"What do you think? Say you're a government official and you've got oil stock and my invention came along. What would you do? I had pressure from both Canadian and American politicians. One American is now a big shot in Washington."

"Will you name him?"

"No."

"Does automatic failure face all inventors of revolutionary new products?"

"No, providing they invent something easy to make. Now you take a carburetor, one of the hardest things to make. I lost $100,000. My partner, that's his picture up there, he lost $100,000. We spent over $20,000 just making dies. They're all in the next room. We use them as junk."

"Why didn't you produce the Pogue carburetor?"

"First, we couldn't get materials, then distribution, then the war came along, then . . ."

"There's a story which says you were bought off. Is this true?"

"No."

"The carburetor is still yours then until all patents expire?"

"Yes."

"Did you try to sell the rights to your invention?"

"Where? The companies buy from their own suppliers to make that double profit. What company would give this carburetor away? Change a little bit of chrome here and there, that's all they do for competitive purposes."

"Don't you think the public would have bought your carburetor for installation on standard cars?"

"I don't want to talk about it any more."

And that was the unsatisfying end of Charles Nelson Pogue's first interview in 14 years, perhaps the last time Pogue will speak about his invention to anyone. Earlier in the two-and-a-half hour interview, this information came to light:

Charles Nelson Pogue, tired and frightened, is hiding a great deal of information for no apparent reason that I can fathom. Although visibly moved when confronted with the picture of young inventors who might be aided by his experiences, he would not tell his story. Even when confronted with the fact that many readers would consider his refusal to speak a certain indication of fraud, he merely shrugged. He refused even to pose for pictures and threatened prosecution for invasion of privacy if photographs were taken against his will.

One emotional moment came when he was asked if his invention was dangerous. He leaped up, towered over his gadget-strewn desk and waggled his forefinger: "That's nonsense. Gasoline will not explode until it has oxygen. In my invention the gasoline was heated before it was mixed with air.

It's entirely possible many people misunderstood that, but my invention would never explode. Never!"

He also disclosed that the first rash of publicity brought letters of inquiry from the governments of Italy and Canada and personal visits from representatives of Nazi Germany and Communist Russia. He hinted darkly at financial and political considerations offered to him by these governments but, when pressed for details, refused to continue.

Taking another tack, Pogue pointed out that the United States and Canadian patent offices hold thousands of carburetor patents which are superior to those in use. And yet they are never investigated by the manufacturers, according to him. In the same vein, he described the steam automobile as "the most practical vehicle ever designed" and asked why it has been discarded; He believes that if as much research had gone into steam propulsion as into the gasoline engine, all early objections to that form of locomotion could have been eliminated.

"Why?" he said. "Ask the companies why . . ."

He claimed he received a letter of invitation to submit his carburetor to the United States government during World War II. Pogue wrapped up his invention and traveled to Washington, checking into a D.C. hotel. Although he flashed the letter, government officials refused to see him, finally ended by sending him to a testing laboratory in a government building. There, Pogue found the man in charge to be a dollar-a-year engineer. His momentary puzzlement over the familiarity of the man was soon evaporated when he recalled that this individual had approached him in 1936 as an investor.

"Naturally," Pogue commented on his Washington trip, "they kept me waiting around two weeks in my hotel room and ended by turning down the carburetor."

However, a letter written by a firm of Canadian attorneys to Alan Gooden, a young New Jersey resident who actually makes a living selling information about the Pogue carburetor, pointed out that the Canadian Army had run tests on the Pogue carburetor during World War II and had found that although mileage per gallon had been increased, that increase did not offset the tremendous loss of engine power which they encountered. Official Canadian sources, however, claim to know nothing of such tests.

Basically, the Pogue carburetor created no new principles of carburetion. Similar to an ordinary perfume-bottle atomizer, any carburetor mixes a high percentage of air with a small percentage of gasoline to create an explosive mixture. Generally the proportion runs 13 to 17 parts of air to one part of fuel. When mixed with the air, the fuel forms a vapor which, in the case of gasoline, falls instead of rises like other gases. Strangely enough, this fact was overlooked for years by manufacturers and all production carburetors were of the updraft type; It wasn't until the early '30s that downdraft carburetors were introduced.

During the late '20s and early '30s so many fuel-saving devices were patented and placed on the market in the United States and Canada that the National Bureau of Standards in Washington created a special department to handle the testing of these devices. For a fee of $35, the Bureau tested the devices and informed their sponsors as to whether the claims made for the inventions could be substantiated. The average claim for gasoline savings with this influx of gadgets was 25 per cent. Although Pogue's claim made the others seem negligible by comparison, it still excited no undue comment until the first "personal testimonials" came out of Winnipeg in 1936.

Judging by today's standards of what constitutes good mileage, the Pogue carburetor claims indicated a fantastic increase in one fell swoop. Is it possible?

There is a certain maximum economy mixture of gasoline and air (generally conceded to be about 16 to 1). If a carburetor delivers this with the air, this is accepted as the best it can do. Actually, "the best" is possible only under ideal conditions. That is, a one-cylinder engine operating under a constant load — fluctuations in power would upset the entire mixing process.

The gas-air mixtures going into the multi-cylinder engine are not even. To give maximum gasoline mileage, any carburetor should be adjusted down to the leanest (least gas) possible mixture. But, if this is done, the mixture may become so lean in one or more of the cylinders far away from the carburetor, that those cylinders won't fire at all. A richer mixture is generally offered for everyday use to make sure that every cylinder is receiving a proper amount of gasoline. In addition, a richer mixture offers better acceleration and slightly more power.

The matter of gasoline engine efficiency is one which has stumped automotive engineers since the advent of the horseless carriage. Because of high power losses in the carburetor, in the exhaust gases, and the low power factor at which engines work, gasoline has never delivered its theoretical potential.

What, then, did the Pogue carburetor do to overcome these power losses and offer the mileage claimed?

Basically, the Pogue carburetor heated the gasoline before mixing it with air. With this system, the inventor claimed that the imperfect mixing of gasoline in air (which results in small unvaporized droplets of gasoline which do not explode, which burn slowly, unevenly and increase internal heat — all bad for performance and mileage) was eliminated and every possible energy unit in the gasoline was being utilized.

To provide a perfect gas-air mixture for the engine, Pogue believed that a heating chamber was needed to thoroughly vaporize the gasoline before it joined the air stream and entered the cylinders.

Although automobiles do have a heating chamber built into the intake manifold, this chamber heats the gasoline and air mixture only *after* it leaves the carburetor. Although this helps the automobile to start easier and warm up faster, it also creates an expansion of air and consequently, a smaller charge of the air-gas mixture is drawn into the cylinder. Pogue decided that the only way to produce a real gasoline vapor was to heat the gasoline and not the air.

His first vaporizing carburetor (1930) had a single chamber with hollow walls which allowed hot exhaust gases to circulate through the carburetor and warm the gasoline in the chamber. When heated, the gasoline should turn into vapor. But there were no great claims made for this particular model.

Later, in 1935, Pogue turned up with an improved carburetor which featured two vaporizing chambers. This time, the inventor claimed thorough vaporization of the gasoline. And this is the carburetor which started the oil stock tumble.

Pogue admitted to a little difficulty in starting the car however, for the gasoline was not warmed until the exhaust gases reached it. So the first charge of gasoline going into the engine was far rawer than those later when it had been heated.

But with the exception of this one minor flaw, Pogue still believes his carburetor is completely practical.

So the mystery of the Pogue carburetor lies not in the device itself, but in the semi-secret story surrounding its invention, development and eventual abandonment.

In a further effort to clear up some of the most controversial aspects of the case, CARS contacted R. Pepper, now editor of the *Canadian Automotive Trade*, the publication which originally printed some of the most sensational testimonials.

First, could CARS have a copy of the original article?

"I am sorry but there are no back issues or photostats of the 1936 article on the Pogue carburetor available."

What about those original reports which were printed? Were they true?

"The carburetor never was a practical success and the original report was entirely misleading as there was no expert ever tried it to prove that it did give the mileage suggested in the article."

And then, as an added bonus, Pepper volunteered: "It might interest you to know — if you do not already know it — that it is impossible to get more than 45 imperial miles per gallon on a 3,000 pound standard production vehicle with a 100-hp engine under ordinary driving circumstances."

Impossible? Not according to Charles Nelson Pogue.

Although he will not reveal his exact age, Pogue is not yet an old man. His face shows the strain of years and the backs of his hands are latticed with a network of fine lines and veins which are trademarks of the man who dreams with his hands. The grease-rimmed fingers drum incessantly against the scarred, battered desk top behind which he sits. He squints constantly against the hard sun which cuts through the one smeared window, lies softly across the splintered, greasy wooden floor.

There has probably been more conjecture and less fact written about this man and his invention than any legend of modern time. His case has been utilized to spearhead attacks against big business, against the government itself.

And yet, men of fine reputation and standing indicate that Pogue is a fraud, his invention a failure.

But the legend goes on, losing no momentum through time, gathering about it the darkness of ignorance.

And the tired, bitter inspiration for that legend sits today in that bare office in Montreal, musing quitely, trying desperately not to remember the past — for he has assigned it to a place completely secret all to himself.

Section I.
Variable Venturi Type
Carburetors

July 24, 1956 R. R. DIETZ 2,755,818

CARBURETOR ADJUSTING MECHANISM

Filed Dec. 18, 1953 2 Sheets-Sheet 1

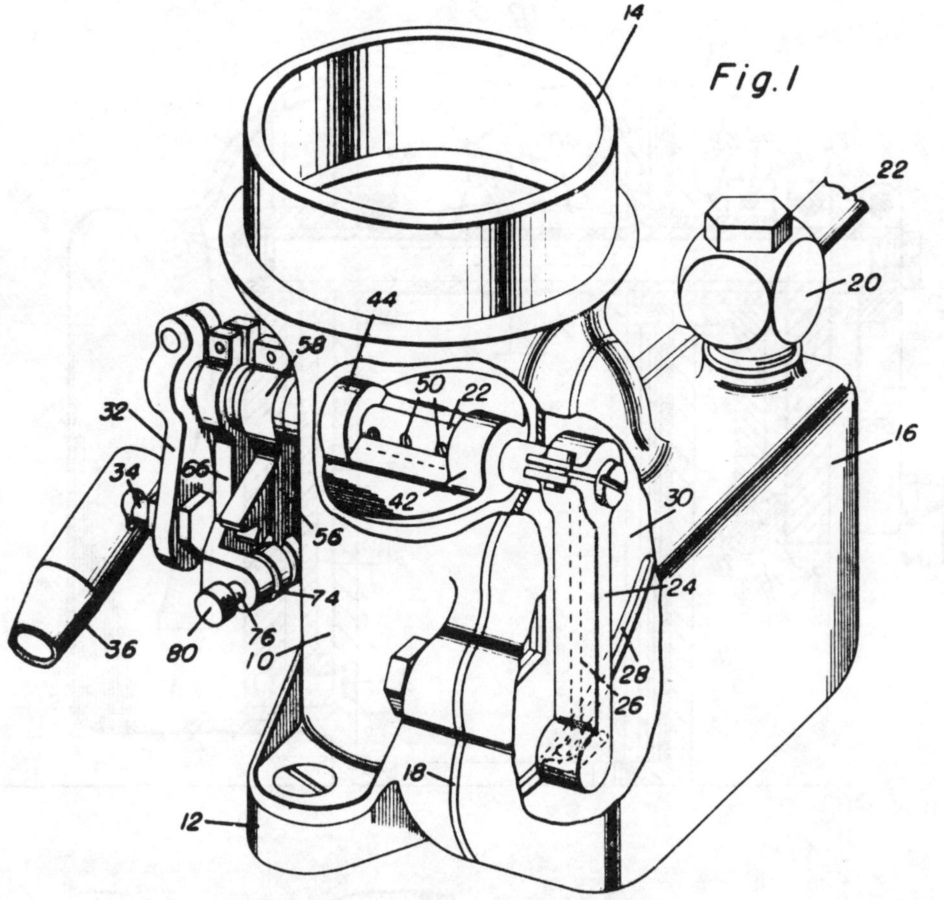

Fig. 1

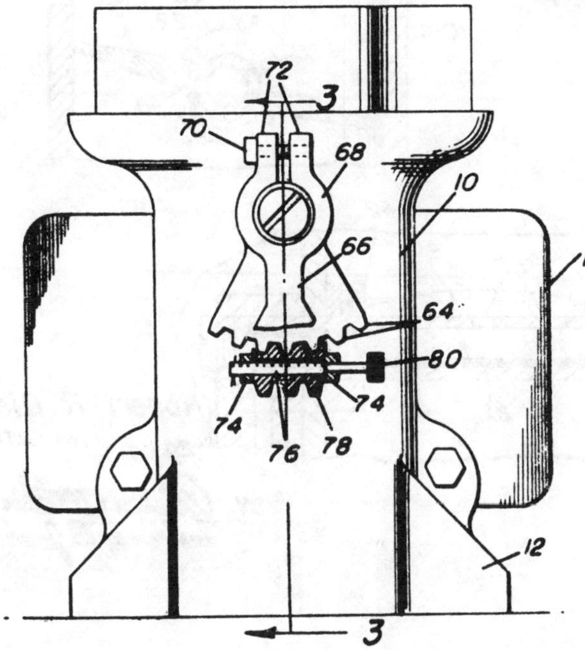

Fig. 2

Robert R. Dietz
INVENTOR.

BY Clarence A. O'Brien and Harvey B. Jacobson
Attorneys

July 24, 1956 R. R. DIETZ **2,755,818**

CARBURETOR ADJUSTING MECHANISM

Filed Dec. 18, 1953 2 Sheets-Sheet 2

Fig. 3

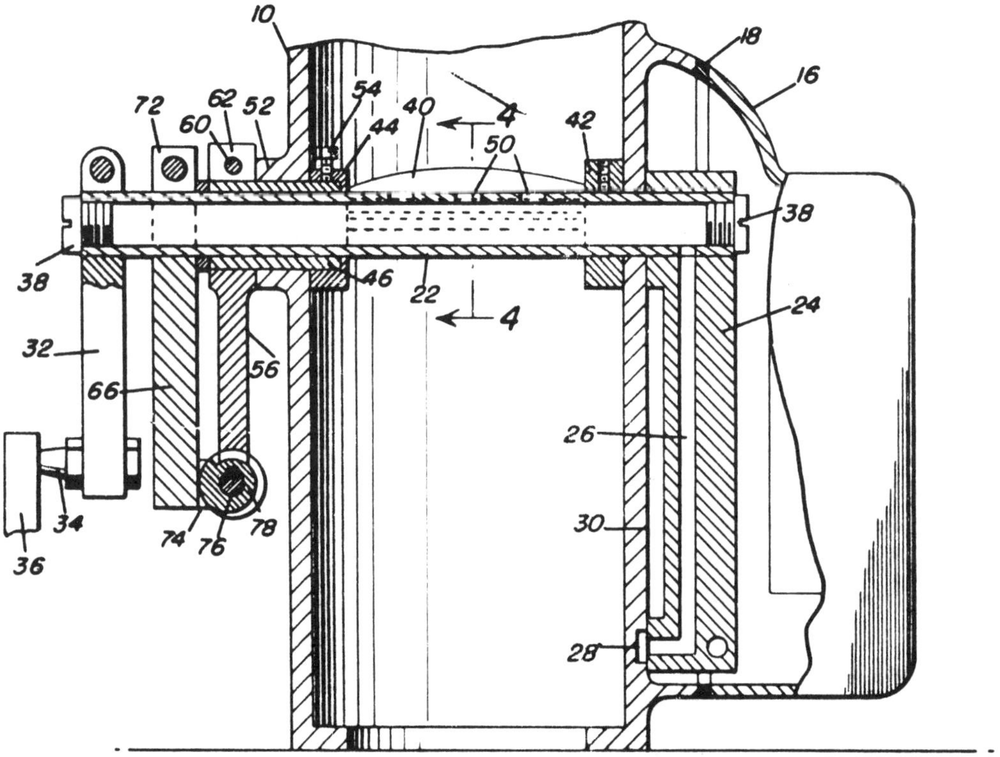

Fig. 4

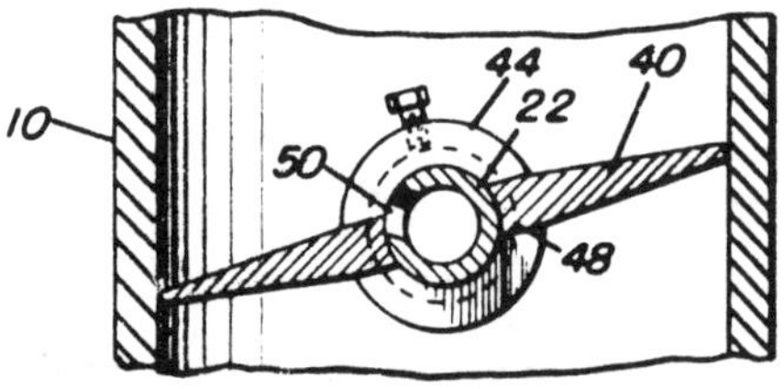

Fig. 5

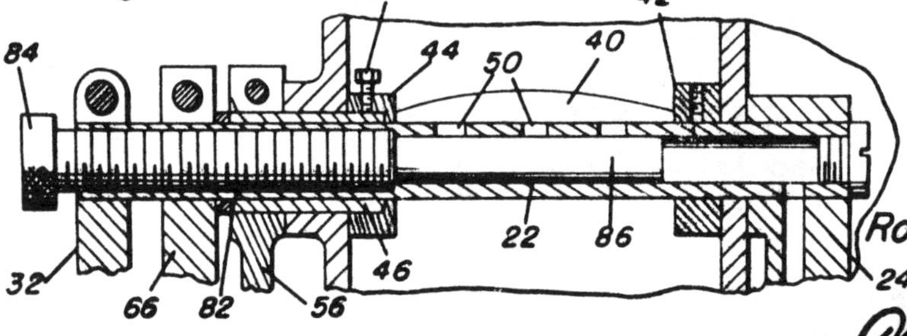

Robert R. Dietz
INVENTOR.

BY Clarence A. O'Brien
and Harvey B. Jacobson
Attorneys

United States Patent Office

2,755,818

Patented July 24, 1956

1

2,755,818

CARBURETOR ADJUSTING MECHANISM

Robert R. Dietz, Havertown, Pa.

Application December 18, 1953, Serial No. 398,994

3 Claims. (Cl. 137—637.3)

This invention relates generally to carburetors, and pertains more particularly to mechanism for adjusting the metering system of carburetors of the type shown in the patents to John R. Fish, Nos. 2,214,273 and 2,236,595, issued September 10, 1940, and April 1, 1941, respectively.

A primary object of this invention is to provide an improved form of metering adjustment mechanism for use in conjunction with carburetors of the type illustrated in the above mentioned patents, wherein the metering adjustment can be accomplished with a minimum of effort by merely manipulating mechanism disposed exteriorly of the carburetor assembly.

These together with other objects and advantages which will become subsequently apparent reside in the details of construction and operation as more fully hereinafter described and claimed, reference being had to the accompanying drawings forming a part hereof, wherein like numerals refer to like parts throughout, and in which:

Figure 1 is a perspective view of a carburetor with portions thereof broken away to show the operating mechanism;

Figure 2 is a side elevational view of a carburetor having the improved attachment positioned thereof, portions of the attachment being broken away to show the relationship of parts;

Figure 3 is a vertical section taken through the carburetor assembly shown in Figure 1 illustrating details of the attachment;

Figure 4 is a vertical section taken substantially along the plane of section line 4—4 in Figure 3; and

Figure 5 is a sectional view showing a modified form of adjusting mechanism.

Referring now more particularly to the drawings, the carburetor illustrated includes a barrel portion 10 having a mounting flange portion 12 at its lower end for mounting the carburetor on an intake manifold of an internal combustion engine. The upper end of the barrel is provided with a collar 14 to which an air cleaner mechanism is conventionally secured. A portion 16 of the carburetor body is removable from the remainder of the carburetor, a gasket 18 being provided therebetween, and this body portion forms the gasoline reservoir for the carburetor, suitable float and float valve mechanism being associated therewith, as is conventional. A fitting 20 is provided on the upper surface of the float chamber body, and a gasoline inlet line 22 is connected therewith in a manner well understood in the art.

Referring now more particularly to Figures 1 and 3, it will be noted that the barrel 10 has a pair of diametrically opposed openings therein receiving the throttle shaft member 22, one end portion of which projects into the float chamber and has rigidly affixed thereto an arm member 24 provided with a fuel passageway 26 opening at its lower end into the fuel feed groove 28 on the side wall 30 of the carburetor. The opposite end of the throttle shaft extends outwardly from the opposite side of the barrel 10 and is provided thereon with an actuating arm member 32 connected by a suitable fitting member 34 to throttle linkage member 36 by means of which rotation of the throttle shaft is effected by manipulation of the conventional accelerator pedal. The throttle shaft is of tubular configuration and has its opposite ends closed by the plug members 38. A butterfly valve 40 is disposed within the barrel 10, as is conventional, and this valve has diametrically opposed hub portions 42 and 44 received on a portion of the throttle shaft 22 and a sleeve member 46, respectively. Each of the hub portions 42 and 44 is provided with a radial aperture internally threaded to receive a setscrew member. Normally, in this type of carburetor, the sleeve member 46 is non-existent, this sleeve forming a part of the present invention, and the hub portions of the butterfly valve are rigidly affixed to the throttle shaft 22 by means of setscrews received in the threaded apertures in the hub portions.

In the conventional carburetor structure, it is to be noted that the diameter of the throttle shaft 22 is greater than the width of the central web porion 48 of the butterfly valve 40 and that the throttle shaft is provided with a series of longitudinally spaced openings 50 which communicate with the upstream side of the barrel 10 in the manner shown most clearly in Figures 1 and 4. In the conventional structure, when it is necessary to change the metering adjustment of the apertures 50 to be more or less uncovered by the central web portion 48 of the butterfly valve, it is necessary to remove the air cleaner on the carburetor, loosen the setscrews and manually rotate the butterfly valve. This method of adjusting the metering of the carburetor is cumbersome, since it necessitates the removal o. the air cleaner and manipulation of the setscrew members in the hub portions 42 and 44 of the throttle valve, and also manipulation of the throttle valve with respect to the throttle shaft to effect the desired fuel setting. Additionally, this method does not lend itself readily to rapid adjustment since it is impractical to make the adjustment while the engine is running, so that it makes it take several trial adjustments of the assembly to effect the desired end result.

To overcome this cumbersome method of adjusting the metering system of the carburetor, the boss 52 on the carburetor barrel 10 which is normally apertured to journal the throttle shaft 22 is reamed out or otherwise enlarged to receive the previously mentioned sleeve 46 and the hub 44 of the throttle valve 40 is also reamed out to receive an inner end portion of this sleeve in the manner shown most clearly in Figure 3. One of the setscrew members 54 is then replaced in the hub 44 to rigidly affix the throttle valve 40 to the sleeve 46, the other setscrew being omitted in the hub portion 42 such that the butterfly valve is, in effect, journaled on the throttle shaft 22. The sleeve 46 projects outwardly through the boss 52 and its outer end receives an arm member 56 whose split upper end 58 surrounds the sleeve and is clampingly engaged thereon by virtue of the clamping screw 60 extending through the ears 62 at the upper end of the arm 56. The lower end of the arm is formed as a gear segment having the teeth 64 thereon, in the manner shown most clearly in Figures 1 and 2. A second arm member 66 has its split upper end 68 surrounding the throttle shaft 22 and clampingly engaged thereon by virtue of the clamping screw 70 engaging through the ears 72 in a manner similar to the first mentioned arm. The lower end of this second arm is provided with the laterally projecting, spaced parallel leg members 74 which are positioned below the gear segment teeth 64 and these legs are apertured to receive a shaft 76 to which the worm gear 78 is suitably feathered. One end portion of the shaft 76 is provided with a hand wheel portion 80, and it is to be understood that the legs 74 normally frictionally engage on the shaft 76 to such an extent that the unit is self-locking. The worm gear 78, of course, meshes with the gear segment teeth 64 and it

2,755,818

3

will be manifest that by manipulation of the shaft 76 and consequently the worm gear 78, relative rotation between the throttle shaft 22 and the butterfly valve 40 will be effected such as to change the amount the apertures 50 are covered or uncovered by the central portion 48 of the butterfly valve. Accordingly, it will be manifest that metering adjustment of the carburetor is thus effected without removal of the air cleaner or any other associated carburetor mechanism, and that the adjustment may be effected very rapidly and while the engine is running to produce the most accurate setting of the metering holes 50.

Referring now more particularly to Figure 5, wherein a modified form of adjusting mechanism is shown, it will be seen that in addition to the previously described adjusting mechanism, one of the end plugs 38 has been removed from the throttle shaft 22 and this end portion of the throttle shaft enlarged and internally threaded to receive the metering needle 82. The outer end of the metering needle is provided with a hand wheel portion 84 and the inner end 86 is of reduced diameter corresponding to the internal diameter of the throttle shaft such that manipulation of the metering needle 82 may effect the blocking off of one or more of the metering holes 50 in the throttle shaft, Figure 5 illustrating a position in which all of the metering holes have been blocked off. In this manner, a finer adjustment of the metering mechanism may be effected than is otherwise produced by the provision of the previously described assembly alone.

From the foregoing, the construction and operation of the device will be readily understood and further explanation is believed to be unnecessary. However, since numerous modifications and changes will readily occur to those skilled in the art, it is not desired to limit the invention to the exact construction shown and described, and accordingly, all suitable modifications and equivalents may be resorted to, falling within the scope of the appended claims.

What is claimed as new is as follows:

1. In a carburetor, a barrel, a throttle shaft projecting diametrically through said barrel and having one end journaled therein, a sleeve rotatably received on said throttle shaft and journaling an intermediate portion thereof in said barrel, a butterfly valve disposed within said barrel having diametrically opposed hub portions receiving a portion of said shaft and an inner end portion of said sleeve respectively, said butterfly valve being fixed to said sleeve, said throttle shaft being hollow and having a plurality of apertures between the hub portions of said

4

butterfly valve, a first arm fixed to said sleeve externally of said barrel, a second arm fixed to said throttle shaft adjacent the first arm, a gear segment formed on the free end of said first arm, a worm gear rotatably carried by the free end of said second arm in mesh with said segment, and means for rotating said worm shaft to effect relative rotation between said throttle shaft and butterfly valve.

2. In a carburetor, a barrel, a throttle shaft projecting diametrically through said barrel and having one end journaled therein, a sleeve rotatably received on said throttle shaft and journaling an intermediate portion thereof in said barrel, a butterfly valve disposed within said barrel having diametrically opposed hub portions receiving a portion of said shaft and an inner end portion of said sleeve respectively, said butterfly valve being fixed to said sleeve, said throttle shaft being hollow and having a plurality of apertures between the hub portions of said butterfly valve, a first arm fixed to said sleeve externally of said barrel, a gear segment formed on the free end of said first arm, a second arm fixed to said throttle shaft adjacent the first arm, a pair of laterally extending spaced legs on the free end of said second arm, a worm gear rotatably received between said legs in mesh with said segment, and means for rotating said worm shaft to effect relative rotation between said throttle shaft and butterfly valve.

3. In a carburetor, a barrel, a throttle shaft projecting diametrically through said barrel and having one end journaled therein, a sleeve rotatably received on said throttle shaft and journaling an intermediate portion thereof in said barrel, a butterfly valve disposed within said barrel having diametrically opposed hub portions receiving a portion of said shaft and an inner end portion of said sleeve respectively, said butterfly valve being fixed to said sleeve, said throttle shaft being hollow and having a plurality of apertures between the hub portions of said butterfly valve, a first arm fixed to said sleeve externally of said barrel, a second arm fixed to said throttle shaft adjacent the first arm, and means connecting said arms for effecting relative rotation between said throttle shaft and butterfly valve.

References Cited in the file of this patent

UNITED STATES PATENTS

1,574,435 Martin ---------------- Feb. 23, 1926

FOREIGN PATENTS

444,625 France ------------------------ 1912

July 30, 1957 J. R. FISH **2,801,086**

CARBURETOR

Filed Jan. 14, 1954 2 Sheets-Sheet 1

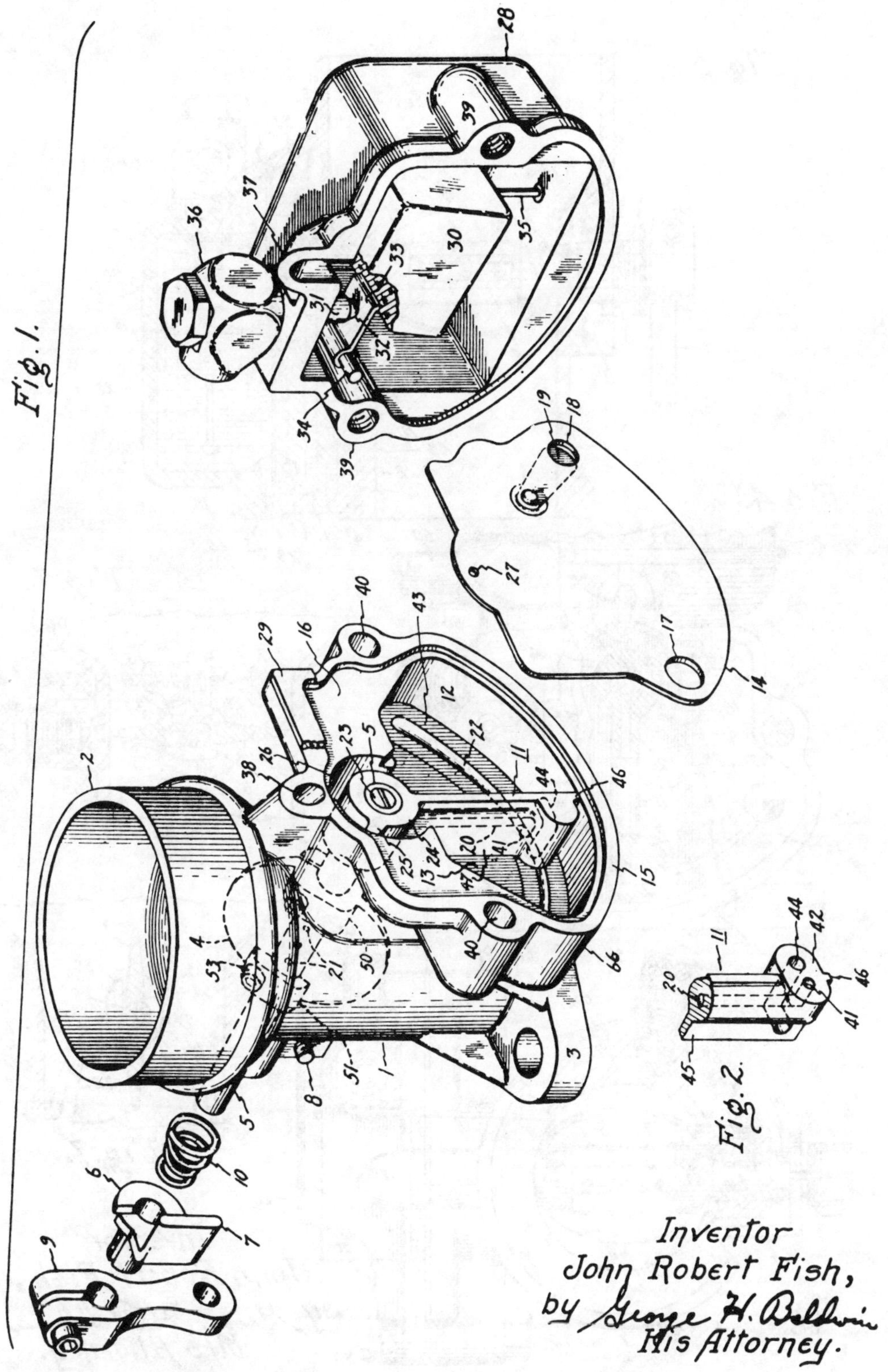

Inventor
John Robert Fish,
by George H. Baldwin
His Attorney.

July 30, 1957 | J. R. FISH | **2,801,086**

CARBURETOR

Filed Jan. 14, 1954 | 2 Sheets-Sheet 2

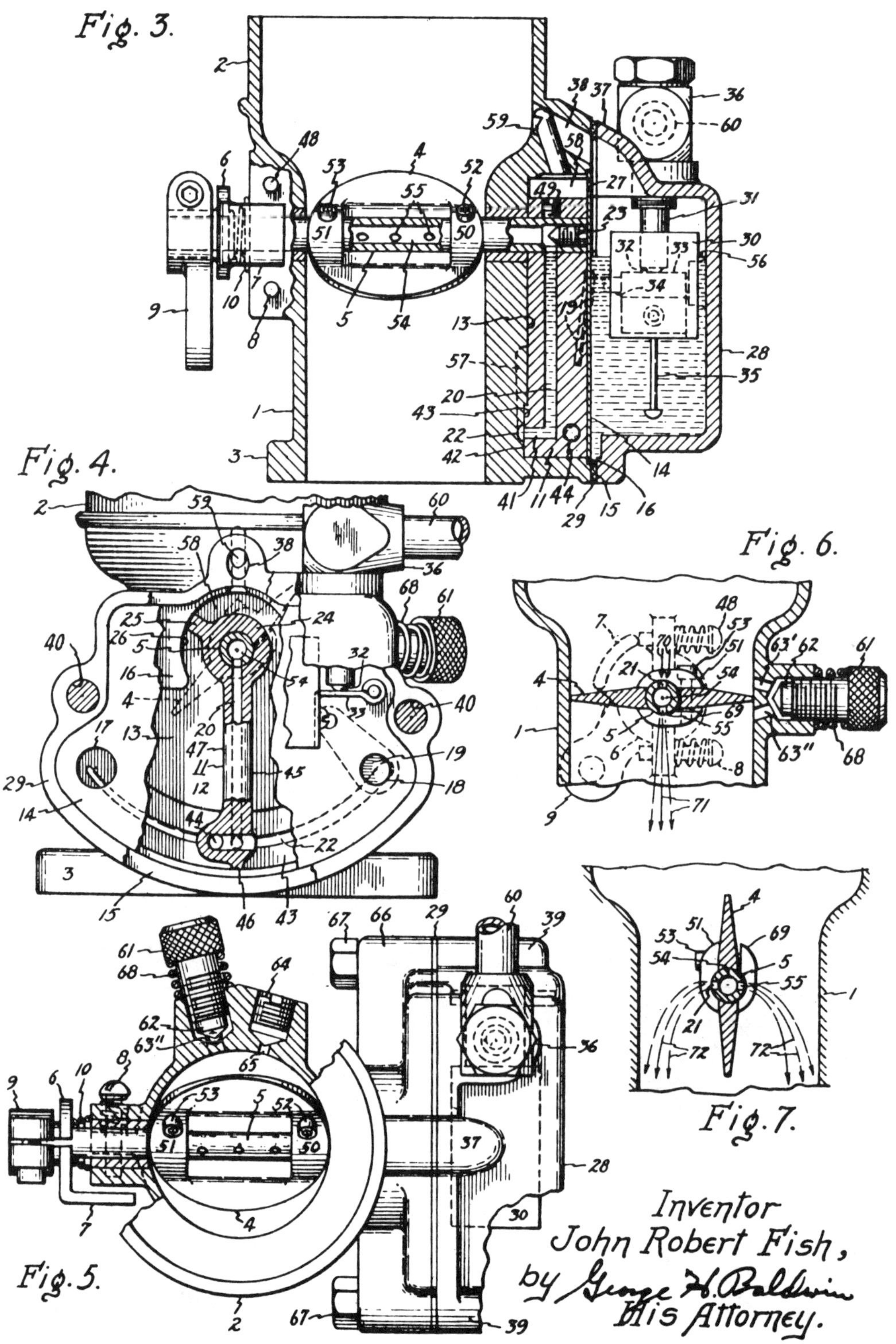

Inventor
John Robert Fish,
by George H. Baldwin
His Attorney.

United States Patent Office

2,801,086

Patented July 30, 1957

1

2,801,086

CARBURETOR

John Robert Fish, Daytona Beach, Fla., assignor to Fish Carburetor Corporation, a corporation of Massachusetts

Application January 14, 1954, Serial No. 403,979

12 Claims. (Cl. 261—34)

My invention relates to carburetors particularly suitable for providing gasoline in metered amounts to the intake manifold of an internal combustion engine.

A general object of my invention is to improve the reliability and accuracy of fuel and air metering of internal combustion engine carburetors.

A further object of my invention is to provide a carburetor having a minimum of simplified adjustment arrangements but which may be readily adjusted for operation in connection with engines of a very wide range of displacement and horsepower characteristics, and a carburetor which, when adjusted for any engine within the range, will perform at high efficiency.

An additional and more specific object of my invention is to improve the operation and simplify the construction of carburetors of the type comprising a throttle located at the throat of the main air passage and wherein fuel is supplied through the throttle shaft and introduced into the air stream at the throttle.

The novel features which I believe to be characteristic of my invention are set forth with particularity in the appended claims. My invention itself, however, both as to its organization and method of operation, together with further objects and advantages thereof, may best be understood by reference to the following description taken in connection with the accompanying drawings, in which:

Fig. 1 is an exploded perspective view of a carburetor in accord with my invention;

Fig. 2 is a fragmentary broken away view of a lower end portion of the fuel supply arm of the carburetor;

Fig. 3 is a partially cut away side view of the assembled carburetor of Fig. 1;

Fig. 4 is a partially cut away end view of the carburetor showing details of the fuel supply arm and chamber;

Fig. 5 is a top view of the carburetor partially broken away to show details of the throttle and the throttle control arrangement, and details of a vacuum spark control arrangement and of a low speed adjustable air bypass arrangement;

Fig. 6 is a fragmentary sectional view taken through the throttle and showing the throttle in closed or idling position; and

Fig. 7 is a fragmentary sectional view showing the throttle in fully open position.

As seen in Fig. 1, the carburetor comprises a main body member 1 shaped internally to form a venturi passage between an upper air inlet portion 2 and a lower outlet portion arranged for connection, by means of flange 3, to the intake manifold of the engine. Within the main air passage of the body, at a venturi or constricted throat portion, a throttle 4 is disposed, the throttle being carried on a throttle shaft 5 to be rotated between an approximately horizontal idling position and an approximately vertical full throttle position. A stop member 6, having a protrusion 7 adapted to meet an idling adjustment screw 8 mounted on the carburetor body, is arranged to be slipped over shaft 5 and to be locked in position thereon by a throttle link 9. Spring 10 is interposed between member 6 and the carburetor body to urge link 9 and member 6 away from the body and thus to urge shaft 5 in the direction toward the left in Fig. 1.

2

The end of shaft 5 opposite to that at which the link 9 is attached carries a fuel supply arm 11. The arm 11 operates in a fuel supply chamber 12 formed with a substantially flat vertical wall 13 against and end of the body 1. The wall 13 is approximately the shape of a sector of a circle, and arm 11 sweeps arcuately back and forth across the surface 13 with opening and closing throttle and throttle shaft movements. A wall plate 14 is adapted to fit within the edges 15 of the walls of chamber 12 against a recessed surface 16, and the wall plate 14 seals against surface 16 and seals, with sufficient clearance for a running fit, against arm 11 and completes the closed fuel supply chamber 12. It will be understood that the clearance between arm 11 and wall 14 is sufficient to permit free swinging, yet sufficiently small to cause the arm to function as a vane pump in the chamber 12. Fuel, such as gasoline, is admitted through wall plate 14 into the chamber 12 through ports 17 and 18, the latter port having a flap or feather valve member 19 arranged to permit pressure to be built up by swinging of arm 11 and to prevent the flow of gasoline from the respective portion of the supply chamber when sufficient pressure is produced to seat the valve member 19 to close port 18.

As so arranged, the arm 11 divides sectorial chamber 12 into two separate sectorial compartments having respective volumes which vary inversely, in accord with the angular position of arm 11, one such compartment lying in what will be herein considered to be in advance of or ahead of the arm 11 and the other lying behind the arm. The forward compartment formed ahead of the arm serves primarily to supply an increased fuel charge to the intake air stream during acceleration and has two communications with the carburetor barrel, one through a passage 20 in the fuel supply arm and the other through an acceleration nozzle hereinafter described and shown best in Fig. 3. The rearward compartment behind the arm 11, together with the forward compartment, normally, during constant power periods, supplies fuel to the air stream through the passage 20 of the arm only. The fuel passing through passage 20 enters the throttle shaft 5 and is expelled into the air stream through vaporizing ports, such as the ports 21, shown in dotted lines, these ports being drilled in the throttle shaft to intersect the central opening of the throttle shaft.

Metering of the fuel from chamber 12 into the air stream is accomplished by a combination of a metering groove 22 formed into the wall 13 of the chamber and of a needle restriction valve controlled by an adjusting screw 23, and in other ways as hereinafter more particularly described. The groove 22 is seen to increase gradually in size, preferably both as to width and depth, from its left hand end as seen in Fig. 1 to its right hand end. Gasoline entering channel 20 of the arm must pass through groove 22, and the flow is more restricted when arm 11 is at the position corresponding to idling throttle position than when the arm is at full throttle position. The groove 22 extends at least throughout the full distance traversed by the arm 11 and, preferably, extends at its larger end beyond the extreme full throttle position of the arm. The groove may also extend beyond the extreme closed throttle position of the arm 11.

The hub 24 formed at the upper end of arm 11 serves to attach the arm to the end of hollow throttle shaft 5 and a projecting vane 25 is carried by the hub to engage, in a running fit, against the cylindrical arcuate surface 26 which surrounds the hub through about 260 degrees. Vane 25 and the hub 24 form, with the balance of arm 11, a seal, within the limits of the running fit, between the forward and rear compartments of chamber **12**. The small cylindrical extension of the chamber bounded by walls **26** is desirable to permit communication between the accelerating nozzle and the chamber at a point above the normal fuel level. The cylindrical extension of the chamber further permits a pressure equalizing connection between the fuel supply chamber and the float reservoir, above the normal fuel level in each, through a small aperture **27** provided in plate **14**.

A float reservoir **28** is arranged to fit against edges **15** of the fuel supply chamber body, a sealing gasket **29** being interposed. Float **30** within the reservoir controls the fuel level therein by means of a float valve **31** preferably comprising a caged ball **32** to seal against a seat within the valve body. The ball **32** is arranged to rest against a generally horizontal portion **33** of a hinge member which pivotally supports float **29** from a stationary pin **34**. A post **35** extends from the bottom of float **29** and limits downward pivoting of the float about pin **34**, thereby to retain ball **32** in the cage at the lower end of the body of valve **31**. A seat for the ball **32** is provided above the ball and within the body of the valve, and raising of the float on its pivot on pin **34** serves to raise the ball, thereby to tend to close the valve and to limit or stop the flow of fuel into the float reservoir from a supply line connecting with the valve **31** through a suitable coupling **36**.

An upstanding hollow hood **37** at the top of the reservoir body **28** aligns with a channel **38** of the carburetor body, and channel **38** completes a conduit to the main air passage of the carburetor above the throat. The very slightly less than atmospheric pressure existing in the main air passage above the throat is thus applied to the space above the fuel within the float reservoir. Internally threaded openings in bosses **39** of body **28** cooperate with drilled openings **40** of the carburetor body **1** to permit bolting of body **28** to the carburetor body.

As seen in the fragmentary view of Fig. 2, the arm **11** is drilled longitudinally to provide a fuel supply passage **20** which terminates at the lower portion of the arm in a short transverse passage **41** opening through the inner face **42** of the arm. The inner face of the arm in formed to lie closely against the inner wall of the fuel supply chamber, shown as wall **13** in Fig. 1, and if the wall **13** has a recessed portion, as at **43** in Fig. 1 along the fuel metering groove **22**, the lower portion of face **42** of the arm should protrude into such recessed portion to provide engagement with running fit between the face **42** and the wall **13** from one end to the other of arm **11**. The fit of the arm against the recessed surface **43** is preferably a very close running fit.

The point at which passage **41** opens through face **42** is positioned to face and communicate with metering groove **22**, and with swinging of arm **11**, this point traverses the groove. A second channel **44** is provided in the lower end of arm **11** to open through face **42** at a second point also facing groove **22**, and channel **44** communicates with the groove at face **42** at a point just behind or to the rear of the point of communication of passage **41**, and furnishes a connection in the nature of a bypass connection between this rearward point on face **42** and the forward face **45** of arm **11**. The provision of a small downward projection **46** on the lower end of arm **11** permits the arm substantially completely to seal, as permitted by the running fit, between the forward compartment of chamber **12** lying ahead of forward face **45** of the arm and the rearward compartment lying behind rear face **47** of the arm, the only communication, normally, between the chambers being through the short length of the groove **22** which is covered by face **42**, and also through a short portion of this covered length and through bypass channel **44** in sequence.

Fig. 3 is a side view of the assembled carburetor almost entirely in section. Link **9** is affixed at the control end of throttle shaft **5** and clamps stop member **6** to the

shaft. Adjustable screws **8** and **48** on the carburetor body **1** are arranged to cooperate with protrusion **7** of the stop member to adjust the extreme idling and full throttle positions of shaft **5**, and, accordingly, the extreme positions of the throttle and arm.

Fuel supply arm **11** is affixed, preferably in a pressed fit, at the end of throttle shaft **5** opposite to link **9** and by means of a suitable set screw **49**, and compression spring **10** urges shaft **5** longitudinally in the direction to retain inner surface **42** of arm **11** in a running fit against the inner wall **13** of the fuel supply chamber.

Throttle **4** is mounted to hollow shaft **5** within the carburetor throat. The configuration of throttle **4** may be seen, by a comparison of the several figures, to comprise an element circular in plan aspect of less maximum thickness than the diameter of the shaft **5**, except for collars **50** and **51**, the plate being of gradually reduced thickness with distance from the shaft to provide relatively thin swinging edges remote from the shaft. The shaft **5** passes along a diameter of the circular throttle element and so divides the element that portions of the shaft surface are exposed through each face of the element. Split collars **50** and **51** are formed as a part of the throttle element at opposite ends of the diameter along which shaft **5** passes, and these collars surround and clamp to the shaft to retain the throttle in position thereon. Clamping bolts **52** and **53** draw the collars **50** and **51**, respectively, tight on the shaft, and the bolts are operable by a wrench which may be introduced at air inlet portion **2** to loosen the bolts for adjustment of the rotative position of the throttle on the shaft, the bolts being thereafter tightened to lock the throttle in its desired adjusted position.

The throttle shaft **5** is seen to be hollow through the throttle element **4** and to the end of the shaft to which arm **11** is attached. The end of the shaft at link **9** is closed. A plug, not shown, may be employed to close the link end of the shaft, or the shaft may be formed from a solid rod longitudinally drilled from the arm end to form an internal passage into and through the position of the throttle. The passage **54** thus formed in shaft **5** supplies fuel to atomizing or distributing ports **55** opening from the passage **54** into the air stream below the throttle **4** in its idling or partially open position, the supply of fuel being restricted by a needle valve adjustable by screw **23**. Passage **54** communicates with fuel supply passage **20** of arm **11**, and it will be seen that, in the absence of air flow through the carburetor, the fuel stands at the level **56** in passage **20**, as controlled by float **30**. Passage **20** is in communication with groove **22** through its lower horizontal portion **41**, and the groove **22** will be seen to be of tapering or gradually increasing depth in accord with the broken line **57** defining the portion of the groove extending from the lowermost portion shown in Fig. 3. A normally empty portion **58** of the fuel supply chamber exists above hub **24** of the fuel supply chamber and this normally empty portion communicates through a small orifice **27** in plate **14** with the upper portion, within hood **37**, of float reservoir **28**. During acceleration, upon forward motion of arm **11**, the portion **58** of the fuel supply chamber will be filled with fuel under appreciable pressure and a suitable portion of this fuel, but obviously only a small portion, will be ejected back into the float reservoir through orifice **27**, while the major flow from the upper part of the chamber will be through an accelerating nozzle **59** into the air stream. The size of orifice **27** is proportioned to control the amount of the accelerating charge which will be delivered, and if too great a charge is found to be delivered, the orifice **27** may be reamed to a larger size. Nozzle **59** is in communication with the portion **58** of the fuel supply chamber, and the member comprising the nozzle extends upwardly into channel **38** formed in the carburetor body **1**. As heretofore mentioned, channel **38** communicates between the air passage of the carburetor, at its air inlet portion 2, and the reservoir 28. The nearly atmospheric air pressure at the intake portion 2, accordingly, is transferred to the volume above the fuel level in reservoir 28 and from reservoir 28 through orifice 27 to the volume above the fuel level in the fuel supply chamber. As heretofore explained, fuel is introduced into reservoir 28 through a suitable fuel line coupling 36 and through a control valve 31, the valve being controlled by the position of valve control ball 32 as regulated automatically by the position of float 30. Ball 32 is prevented from falling out of the lower end of valve body 31 by the provision of post 34 which contacts the bottom wall of reservoir 28 to prevent dropping of the float 30 below a predetermined position. The location of flap or feather valve member 19 in open position is indicated in Fig. 3, and it will be understood that communication between reservoir 28 and the fuel supply chamber is through the port controlled by member 19 and through a second open port 17, each of which has been heretofore described. Plate 14 is held against recessed surface 16 by portions of reservoir member 28 which extend inwardly over the edges of the plate. Gasket 29 also overlies edge portions of plate 14, since the gasket conforms to the flange of reservoir member 28.

Fig. 4 is an end view of the carburetor taken from the fuel reservoir end, parts being broken away to disclose certain features of the internal construction. Fuel inlet line 60 provides fuel through coupling 36 into reservoir 28 under the control of ball 32 which engages portion 33 of the supporting hinge for float 30. Fuel from the float reservoir may pass freely through the port 17 in the plate 14 into the rearward compartment of fuel supply chamber 12, which lies behind rear face 47 of arm 11. Fuel may also pass through port 18 under the control of feather valve member 19 into the forward compartment of chamber 12 ahead of face 45 of arm 11. The fuel is furnished from the supply chamber into passage 54 of shaft 5 through longitudinal passage 20 of arm 11, and access of the fuel to the passage 20 is through the portions of channel 22 lying immediately ahead of an behind the opening of passage portion 41. Channel 44 formed near the lower end of arm 11 communicates between the chamber in advance of forward face 45 of the arm and a portion of the groove 22 immediately behind passage 41. During constant speed operation when arm 11 and the throttle member 4 are stationary, fuel flows into channel 22 from the compartments both ahead of and behind arm 11. The restriction imposed by the groove on the flow into passage 41 is considerably greater when the throttle is near idling position and when the arm 11 is adjacent port 17 than when the throttle is further opened and arm 11 is near port 18, since groove 22 varies both in width and depth from a narrow and shallow size at its end near port 17 to a wider and deeper size adjacent the port 18 end.

Accelerating forward motion of arm 11 causes the fuel in the compartment in advance of forward face 45 of arm 11 to be increased in pressure, while the pressure within the compartment behind rear face 47 will remain almost constant or decrease very slightly, because port 17 is open at all times. Except for the provision of channel 44, accelerating conditions would tend to cause a flow of fuel through groove 22 from the forward compartment to the rearward compartment past the opening of passage 41. The aspirating effect of such flow would stop or reduce the flow of fuel into passage 41. The provision of channel 44 causes the introduction of a desirable flow of fuel under pressure into channel 22 immediately behind the opening of passage 41 during acceleration, and the result is that fuel under more than normal pressure then flows into channel 41 from groove 22 from both directions. During acceleration, accordingly, channel 44 eliminates the tendency to drain channel 22 which would otherwise exist. Channel 44 serves an additional purpose during deceleration. Pressure within groove 22

might tend to build up as the arm swings toward port 17 under deceleration conditions, but bypass channel 44 reduces the pressure in the channel in the area immediately adjacent the opening of passage 41 toward which the opening is moving.

Fuel flows freely through port 18 into the forward compartment of chamber 12 during constant power and deceleration periods, but upon acceleration, feather valve flap member 19 seals off port 18 to permit the fuel ahead of arm 11 to be increased in pressure. This increase of pressure forces fuel up into the normally empty upper portion 58 of the fuel supply chamber and through accelerating nozzle 59 into the air stream. The vane 25 extending from hub 24 of arm 11 seals in a running fit against the walls 26 of the chamber portion 58 and permits fuel pressure in the forward compartment and in portion 58 to build up substantially above the approximately atmospheric pressure within the rearward compartment behind face 47. The fuel under pressure in chamber portion 58 provides a suitable accelerating charge of fuel through nozzle 59 into the air stream. It will be noted in Fig. 4 that channel 38 is only partially filled by nozzle 59, and that free air communication through channel 38 is not appreciably affected by the provision of the nozzle.

The knob 61 of a vacuum control plug is also seen in Fig. 4, the purpose of which will be more particularly described hereinafter in connection with Fig. 6.

Fig. 5 is a top view of the carburetor, partially broken away, and disclosing the position of the throttle 4 and ports 21 as presented to the air entering upper portion 2 of the carburetor air passage. Control knob 61 is provided to adjust valve member 62 which controls the idling and low speed air-fuel ratio by bypassing air around the throttle through passages 63. In the idling range, the valve member 62 provides much of the control. The groove 22 is shallow and varies little in the low speed range, whereas the effect of passages 63′, 63″ changes substantially with a small change in throttle position in the low speed range. Plug 64 is shown as closing a single passage 65 communicating with the carburetor throat just above the position of the throttle at full idling position, and the passage 65 may be used, by removing the plug 64, to provide an automatic spark control vacuum line connection if desired. The arrangement of link 9 to clamp stop member 6 to shaft 5, and the arrangement of protrusion 7 of the stop member, which may engage full throttle adjustment screw 8, are clearly seen in Fig. 5. The relative positions of reservoir 28 and of the fuel supply chamber within portion 66 of the main carburetor body, as held together by bolts 67, with interposed gasket 29, are shown, and it will be noted that hood portion 37 of the float reservoir 28 aligns with a projecting portion of the carburetor body into which the accelerating nozzle extends, and through which equalizing communication between the approximately atmospheric air pressure in portion 2 of the carburetor air passage and the upper portion of reservoir 28 is established.

Certain members of the carburetor which have been previously fully described are identified in Fig. 5 for purposes of comparison, and it will be understood that elsewhere in the drawings symbols have been applied to parts shown in certain figures without specific description in connection with the particular figure to assist in comparing the several views. Throughout the drawings, like symbols have been employed to designate like elements or portions.

Figs. 6 and 7 are sectional end views taken substantially transversely of throttle shaft 5 at the center of the carburetor air passage, Fig. 7 being fragmentary and diagrammatic in part.

Fig. 6 discloses conditions which may exist when protrusion 7 of the stop member is in contact with idling adjustment screw 48. At this time, throttle 4 may be horizontal, or substantially horizontal, and air may be passing into upper ports **21** and into channel **54** of the throttle shaft, to force a mixture of vaporized fuel and air downwardly through lower fuel delivery or atomizing ports **55**. Additional air may be supplied by a bypassing channel system, through **63'** communicating with the air passage above throttle **4** and out through passage **63''** into the air passage below throttle **4**. The amount of air supplied through bypassing passages **63'**, **63''** is controllable in accord with the position of vacuum control plug **62** as adjusted by knob **61**. Spring **68** is provided to prevent creeping of plug **62** from the manually adjusted position of the plug. The connection of throttle **4** to shaft **5** by means of collar **51**, tightened by bolt **53**, which pulls collar end **69** toward the lower face of throttle plate **4**, is apparent in Fig. 6.

The conditions existing with closed throttle as shown in Fig. 6, wherein air indicated by arrows **70** flows into upper ports **21** and flows downwardly toward the intake manifold carrying atomized fuel through lower ports **55**, as indicated by arrows **71**, continue to exist as the throttle is moved into partially open position. As the throttle opens, however, air may pass around the edges of the throttle plate remote from shaft **5**, and air flow around these edges, of course, increases as the throttle is further opened. At about three-quarters throttle, or shortly theretofore, fuel begins to be discharged from both ports or orifices **21** and **55**, as indicated by the arrows **72** in Fig. 7. Fig. 7 discloses the throttle position at full throttle, and the stream of fuel from ports **21** and **55**, under full throttle condition, will be substantially equal and, of course, relatively large amounts of fuel will be supplied. The tapered shape of throttle **4**, as seen in Fig. 7, provides a decreased throat area at the level of the shaft **5**, thereby to increase the venturi action at this point. The orifices **21** and **55**, while shown to be all of equal size, may be of different sizes. For example, if it is desired to provide a richer stream toward one side of the throat than toward the other side, the ports toward the corresponding respective one end of the shaft **5** within the throat may be made larger than the ports toward the other end of the shaft.

Since fuel is supplied through the shaft for introduction into the air stream exactly at the throttle level, the turbulence created by the throttle in the air stream appreciably assists in forming a homogeneous and thorough dispersal of the fuel in the air stream.

Under idling conditions, with the throttle **4** horizontal, or nearly horizontal, and substantially closing the main air passage, air enters the upper vaporizing ports **21**, as seen in Fig. 6, and carries fuel with it in passing on through the lower vaporizing ports **55** into the intake manifold. A pressure differential of about 20 or 21 inches of mercury will normally exist between the nearly atmospheric pressure in the intake portion **2** above the throttle and the partially evacuated passage below the throttle which connects with the intake manifold. The position of the arm in respect to the groove at idling throttle position is alternatively adjustable by changing the idling position of the throttle shaft, conveniently by adjustments to idling adjustment stop screw **48**. Such adjustments to the stop screw **48**, of course, also changes the idling position of the throttle **4**, and thereby affect the amount of air flowing around the throttle. Separate adjustments are provided for altering the position of the throttle on the throttle shaft, and this is accomplished by means of bolts **52** and **53** which may be loosened to permit the throttle to be manually adjusted to a desired position on the shaft by inserting a suitable tool down through the air intake portion **2**. After such adjustment, the throttle is again locked on the shaft by tightening bolts **52** and **53**. The amount of air passing the throttle through low speed air bypass passages **63'** and **63''** is also adjustable, and knob **61** and valve screw **62** are provided for this purpose. By a combination of adjustments to idling adjustment stop screw **48**, adjustments to the position of the

throttle **4** on its shaft and adjustments to air bypass valve screw **62**, a wide range of adjustment of the air-fuel ratio at idling and low speed is permitted.

Since fuel, during idling, is freely admitted through ports **17** and **18** from the float reservoir into the fuel supply chamber, under the atmospheric or nearly atmospheric pressure which is maintained in the fuel supply chamber and fuel reservoir through channel **38** from the intake portion **2** of the main air passage, and since the gasoline level with constant throttle is normally just below the level of the throttle shaft, nearly 21 inches of mercury of pressure difference is applicable to force gasoline from the ports **55**. Actual measurement indicates that the pressure differential is such that the pressure within the hollow interior of the throttle shaft and through passage **20** is about 16 inches of mercury below atmospheric pressure. Accordingly, it will be seen that the principal metering of the gasoline, or other fuel, occurs at the groove **22**.

It is an important feature of the carburetor of this invention that the pressure differential which causes the fuel to flow from the supply chamber into the air stream is very much greater than in prior art carburetors. The metering of fuel does not depend primarily on small pressure changes, and the rate of fuel flow is not appreciably altered by atmospheric pressure changes, for example. Under constant power, or steady state, conditions, the size of the groove **22** at the arm **11** is controlling for lower power settings, while for higher power settings, the maximum fuel supply rate may be limited by the adjustable screw controlled valve **23**, or by the size of groove **22**, or by the size of the passage **41**. In the preferred arrangement, groove **22** is made larger than passage **41**, and the size of passage **41** limits the maximum fuel supply rate. If, in operation, the maximum rate is found to be too great for the particular engine with which the carburetor is being used, screw **23** is adjusted to limit the maximum rate by imposing addition restriction at screw **23**. It is to be noted that all of the ports **21** and **55**, on respective opposite sides of the throttle shaft, come into fuel supplying operation when the throttle is opened beyond about three-quarters throttle position. Accordingly, the total area of all of the ports **21** and **55** becomes available for the supplying of fuel at the maximum rate and at no time, preferably, is the rate of flow metered by the available area of the vaporizing ports.

Upon rapid movement of the throttle link **9** in an accelerating direction, important changes take place. The arm **11** moves toward the wider and deeper end of groove **22**, thereby providing less restriction to the flow of fuel into the throttle shaft for vaporizing into the air stream through ports **21** and **55**, feather valve **19** closes to seal port **18**, and the pressure within the fuel supply chamber **12** ahead of arm **11** increases. The increase in pressure forces fuel into the upper portion **58** of the fuel supply chamber and fuel is thus ejected from nozzle **59** into the air stream. Since the rear compartment of the fuel supply chamber **12** behind arm **11** is in communication with the float reservoir through relatively large open part **17**, the fuel pressure behind arm **11** remains nearly at atmospheric pressure, and the net effect is that fuel is forced into passage **41** and passage **20** of the arm at more than atmospheric pressure. During the forward accelerating motion of arm **11**, some fuel will be forced back through the small aperture **27** in wall plate **14** into the float reservoir, but aperture **27** is sufficiently small to pass, ordinarily, only a relatively small amount of fuel during the relatively brief period of acceleration. The aperture **27**, therefore, serves as a bypass which is proportioned in size to provide a desired amount of accelerating fuel. Alteration of the size of aperture **27** will, of course, alter the amount of fuel supplied in any given accelerating movement of the throttle and arm.

During the forward movement of arm **11** in the fuel supply chamber, the short bypass channel **44** formed from the forward face **45** of arm **11** and communicating with the groove **22** immediately behind opening **41** at the lower end of passage **20** performs a useful function in supplying fuel from the forward compartment at substantially the pressure existing therein to the groove **22**, whereby opening **41** receives fuel through groove **22** from both directions under the increased pressure of the forward chamber during acceleration periods. In the absence of channel **44**, the forward movement of the arm and sucking of fuel from passage **20** would tend to drain channel **22** and substantially reduce the fuel pressure in groove **22** behind opening **41**. It will be understood that channel **44** is primarily effective during relatively rapid motion of arm **11**.

Channel **44** provides another and different function during rapid accelerating motion of arm **11** in preventing the build up of pressure in groove **22** in the portion of the groove toward which the arm is moving, by bypassing fuel from the point of communication of channel **44** with the groove out through the face **45** of arm **11**. Thus channel **44** serves to increase the amount of fuel supplied through channel **20** to ports **21** and **55** during rapid accelerating movement of the arm and to overcome any tendency to supply increased fuel through passage **20** upon rapid decelerating movement of the arm. Gradual acceleration does not require an extra accelerating charge and none is provided, since gradual accelerating motion of arm **11** is insufficient to cause feather valve **19** to close and no accelerating charge is forced through nozzle **59** or through passage **20**.

As the throttle is opened toward full throttle position, the streamlined form of the throttle element **4** causes the throat of the air passage more nearly to approach venturi form. The action of the ports **21** in supplying gasoline to the air stream changes at approximately three-quarter throttle position from the admission of air into the hollow throttle shaft to the ejection of fuel into the air stream. Thus, at full throttle position, with the throttle member **4** subsantially vertical, approximately equal streams **72** of fuel will issued from ports **55** and **21** on opposite sides of the throttle shaft.

When the throttle is at other than full throttle position, at least the larger portion of the fuel will be introduced through the ports **55** into the air stream below the throttle. Turbulence in the area below the throttle is sufficient to cause thorough dispersion of the fuel particles into the air stream and the action of the throttle is, accordingly, such as to provide improved fuel mixing under all conditions, it being understood that the spray produced by air passing into ports **21** and, mixed with fuel, out of ports **55**, during idling conditions, provides good dispersion, and that the six sprays of fuel emerging from the throttle in fully open position similarly provides even distribution of the fuel throughout the air stream. If other than even distribution of fuel is desired, the ports **21** and **55** may be of different sizes.

Particular advantages of the carburetor of this invention, in accord with the above description include the following features. Simplicity and sturdiness of the parts are achieved, together with relatively large tolerance, as will be apparent from a consideration of the drawings. The seal or fit of fuel supply arm **11** against the opposite flat walls of the sectoral fuel supply chamber is not critical in that leakage around the arm, between the forward portion and rearward portion of the chamber as divided by the arm, has no effect under constant power conditions and very little effect during decreasing power operation of the throttle. Upon accelerating forward movement of the arm **11**, an accelerating charge is furnished through accelerating nozzle **59**, as well as through the ports **21** and **55**, and even substantial leakage past the arm decreases the amount of the charge very little. While leakage around the arm is not desired, and while the running fit of the arm against the walls is made a reasonably close fit to provide as complete a seal as possible within the tolerances of inexpensive manufacture, satisfactory operation of the carburetor is obtainable even though the fit of the arm is so loose as to permit considerable leakage during accelerating motion of the arm.

No choke is required in the carburetor since a mere pushing down of the accelerator delivers a starting mixture of increased richness through the accelerating nozzle and because of the highly effective atomization and mixing of fuel with incoming air which results particularly from the arrangement of the ports **21** and **55** at the throttle. It will be seen, also, that no separate idling system is necessary; the fuel during idling enters the air stream through the passages and ports which are used in power operation.

It is not necessary to provide delayed throttle closing to prevent stalling of the engine with the present carburetor, since there is no loading of the carburetor under any operating conditions or sequence of operations. When the throttle closes, both air an fuel are cut off and only the idling mixture can enter the intake manifold, the fuel being metered through the small end of groove **22** and the air being metered through ports **21** and **55** and bypass passage **63′, 63″**.

During closing movement of the throttle, feather valve member **19** provides some restriction to the flow of fuel into the forward compartment of the fuel chamber, and the port **18** at the feather valve being somewhat smaller than the open port **17** into the rearward compartment, the pressure behind the arm **11** remains substantially atmospheric while the pressure in the forward compartment drops slightly below atmospheric. The net effect is that fuel flow into the air stream is reduced as the throttle closes, and no pumping of fuel into the air stream occurs with swinging of the arm **11** toward the closed throttle position. Centrifugal force on the fuel in metering groove **22** upon very rapid closing of the throttle further tends to prevent pumping upon rapid deceleration.

The carburetor clears itself of water and obstructing particles with ease and rapidity. The ports **17** and **18** into the fuel supply chamber are large enough freely to pass foreign matter. The pumping action associated with swinging of arm **11** tends to clear matter from the groove **22** and from about the opening into passage **41**, and further causes flow back and forth through channel **44** to clear the channel. The pumping action is similarly effective to free other parts of the system.

The carburetor is not sensitive to fluctuations in float reservoir fuel level resulting from bumps, tilting, or the like, nor is it subject to surges from negotiating curves in an automobile, for example. The arrangement is such that the fuel, being confined in relatively small chambers, does not spill over from one part of the carburetor to another in any manner which will adversely effect operation. The connection provided between the upper portions of the chambers and the air intake portion of the barrel, together with the large pressure difference which causes the fuel to flow into the air stream, greatly reduce or eliminate offside enrichment of the mixture and malfunctioning from changes in altitude or barometric pressure.

The groove **22** is preferably proportioned to be small throughout approximately one-quarter of the full arm movement distance at the low-speed end of the groove, and gradually to increase in size toward its full throttle end. The first one-quarter of throttle motion, accordingly, changes the fuel supply rate primarily as a result of motion of the throttle member with respect to bypass passage **63, 63″**. Thereafter, the gradually increasing size of groove **22** meters the gradually increasing fuel rate, while the opening throttle increases the amount of air passing into the intake manifold.

While I have shown and described only certain preferred embodiments of my invention by way of illustra-

tion, many modifications will occur to those skilled in the art, and I therefore wish to have it understood that I intend, in the appended claims, to cover all such modifications as fall within the true spirit and scope of my invention.

What I claim as new and desire to secure by Letters Patent of the United States is:

1. In a downdraft carburetor, a fuel supply nozzle, a throttle, said throttle having a major face disposed toward the engine intake manifold at idling settings and said nozzle opening through said face, a closed fuel supply chamber having a wall, an arm lying against said wall, pivotal mounting means to swing said arm across said wall, said arm having an opening thereinto from a portion toward said wall, said opening tracing an arcuate path across said wall with pivoting of said arm, said wall being grooved along said path, the size of said wall groove increasing gradually from a small size at one end to a relatively larger size at the other end, said arm and said means comprising a fuel conduit communicating between said arm opening and said nozzle, said mounting means being connected to open said throttle wtih swinging of said arm in the direction to dispose said arm opening toward said other end of said groove.

2. A downdraft carburetor having a main air passage comprising a constricted throat portion, a throttle disposed in said portion and having a fuel supplying aperture, a sectorial fuel supply chamber comprising a sectorial wall, an arm connected to said throttle and disposed against said wall and movable swingingly in said chamber and across said wall about the axis of said throttle, said arm having an opening toward said wall and remote from said axis, whereby said opening traces an arcuate path along said wall with swinging of said arm, said arm having an internal passage communicating between said opening and said aperture, said wall having a surface groove along siad path of a size gradually increasing with distance from one end thereof to the other, means to supply fluid fuel to said chamber for metering through said groove into said opening and arm passage, said arm dividing said chamber into a forward compartment and a rear compartment of respective volumes which vary inversely to one another with swinging of said arm, said forward compartment lying toward said other end of said groove and said rear compartment lying toward said one end of said groove, a check valve, said fuel supply means comprising a fuel reservoir communicating freely with said rear compartment and communicating through said check valve into said forward compartment, and an accelerating nozzle having a closed conduit connecting directly with said forward compartment and operative to supply an accelerating charge of fuel to said air passage in response to increased fuel pressure in said forward compartment produced by rapid swinging of said arm in the direction to decrease the volume of said forward compartment.

3. In a downdraft carburetor, a main air passage having a constricted throat portion, a throttle at said throat portion, a conduit into said throttle and having a port to admit fuel from said conduit into the air stream at said throttle, said port opening from an under portion of said throttle subject to at least substantially one-half of the full intake manifold vacuum when said throttle is in idling position, a fuel supply chamber comprising a wall having an elongated groove of a size gradually increasing from one end to the other of said groove, a fuel receiving member having an internal passage communicating at one end with said conduit, said receiving member having an opening into the other end of said passage, means to move said member to sweep said opening along said groove, said last means being responsive to throttle movements, whereby said opening is positioned at points on said groove corresponding with respective settings of said throttle.

4. A downdraft carburetor comprisnig an air passage having a throat area, a throttle in said area having upper and lower major faces disposed generally horizontally across said throat at idling settings, a hollow throttle shaft mounting said throttle and having apertures into said air passage from the hollow interior of said shaft through said lower major face, means to rotate said shaft and throttle between different power settings, a fuel supply arm mounted externally of said air passage on said shaft and swingable with rotation of said shaft, said arm having a fuel supply passage communicating with the hollow interior of said shaft, wall means defining a closed fuel supply chamber housing said arm and comprising a wall traversed by said arm in accord with throttle shaft rotation, said arm having an opening through a portion thereof which lies against said traversed wall, said opening communicating with said fuel supply passage of said arm, said wall having an elongated groove extending along the path traversed by said opening with said swinging of said arm, said groove being of a size gradually increasing with distance from one end to the other of said path.

5. In a downdraft carburetor comprising a sectorial fuel supply chamber, a hollow throttle shaft having ports to expel fuel from the hollow interior of said shaft into the main air stream, and an arm disposed in said chamber and connected to swing with rotation of said shaft and having a longitudinal passage communicating with said hollow interior of said shaft, a side of said arm being arranged to contact and sweep across an inner face of said chamber and having an opening toward said face communicating with said passage, said opening being positioned to travel an arcuate path across said face in response to said shaft rotation, the provision in said face of a groove along the full extent of said path tapering in cross-sectional area from an area much smaller than that of said opening at one end of said groove gradually to an increased area of at least substantially as great as the area of said opening at the other end of said groove, the portion of said arm contacting said face having a back edge spaced from said opening toward one end of said groove and having a forward surface extending outwardly from said wall and facing generally toward said other end of said groove, and a bypass passageway opening at one end through said portion of said arm, said one end opening being located between said edge and said first opening and being directed and disposed to open into said groove and said bypass passageway communicating between said first end opening and said second end opening which is located to open through said forward surface.

6. In a downdraft carburetor having a constricted throat air passage, a throttle member, a rotatable control shaft for said throttle member, a fuel nozzle opening into said air passage at said throat and below said throttle when said throttle is in idling position, a fuel supply system including a fuel reservoir and channeled means communicating between said reservoir and said nozzle, said means comprising fuel flow restricting means, said restricting means comprising a movable member connected to said shaft for movement therewith between respective limit positions corresponding respectively to full throttle and idling throttle positions, said restricting means further comprising relatively movable orifice and orifice covering means, a first of said orifice and orifice covering means having a gradual taper and at least one of said orifice and orifice covering means being connected with said movable member for movement therewith, said taper being proportioned gradually to increase the restriction to the flow of fuel to said nozzle in response to throttle closing rotation of said shaft and gradually to decrease said restriction in response to throttle opening rotation of said shaft.

7. A downdraft carburetor comprising a main air passage having a constricted throat portion, a throttle member in said throat portion, a plurality of nozzles enter-

2,801,086

13

ing at said throat portion subject at idling settings of said throttle to substantially full intake manifold vacuum, a fuel supply chamber having an upper opening communicating with a source of at least substantially atmospheric pressure, and means to furnish fuel from said chamber to said nozzles, said means comprising a continuously and gradually variable restrictor valve controlling a restriction to the flow of fuel through said means, said restrictor valve being connected to said throttle member for operation therewith and being arranged gradually to reduce the restriction to fluel flow with gradually increasing power settings of said throttle member and gradually to increase the restriction with gradually decreasing power settings of said throttle member.

8. A downdraft carburetor comprising a main air passage having a constricted throat portion, a throttle member in said throat portion, a plurality of nozzles entering at said throat portion subject at idling settings of said throttle to substantially full intake manifold vacuum, a fuel supply chamber having an upper opening communicating with a source of at least substantially atmospheric pressure, means to furnish fuel from said chamber to said nozzle, said means comprising a continuously and gradually variable restrictor valve controlling a restriction to the flow of fuel through said means, said restrictor valve being connected to said throttle member for operation therewith and being arranged gradually to reduce the restriction to fuel flow with gradually increasing power settings of said throttle member and gradually to increase the restriction with gradually decreasing power settings of said throttle member, said valve comprising a member movable in said chamber, with opening movements of said throttle member, toward an end of said chamber to tend to increase the pressure of the fuel in the portion of said chamber toward said end, said chamber having an outlet accelerating charge conduit communicating from said end portion to said air passage, said end portion having an inlet and pressure equalizing port and a normally open check valve member arranged to seal said port upon rapid increase in pressure in said end portion in response to rapid motion of said throttle member and of said member movable with said throttle member toward said end, said port being effective to maintain normal pressure in said end portion during slowly increasing throttle movements and during decreasing movements and during stationary periods of said throttle member.

9. A downdraft carburetor comprising a main air passage having a constricted throat portion, a throttle in said throat portion, means to rotate said throttle between an idling position and a full throttle position, said throttle having an upper major face disposed toward the inlet of said air passage at idling throttle position and at full throttle facing generally toward a side of said air passage, said throttle having a lower major face opposite said upper face, said means comprising a hollow shaft extending into said throttle, said throttle having an opening extending therethrough from said upper face to said lower face and communicating with the hollow interior of said shaft, and means to supply fuel to the hollow interior of said shaft, said last means comprising a continuously and gradually variable fuel flow restricting mechanism connected to said first means and operable gradually to increase the restriction to the flow of fuel in response to gradual rotation of said throttle toward said idling position and gradually to decrease the restriction in response to gradual opening of said throttle.

10. In a downdraft carburetor, a fuel supply chamber, a movable piston member in said chamber, said chamber having a fuel inlet port in a wall thereof, said piston

14

member being arranged to increase the pressure of fuel in a portion of said chamber which includes said port in response to motion of said member in a predetermined direction, a feather valve member disposed to cover said port in response to said increased pressure and to uncover said port for the admission of fuel to said chamber in the absence of such increased pressure, fuel metering means connected with said piston member variably to restrict the flow of fuel from said chamber in accord with the position of said piston member in said chamber, means forming a main air passage having a constricted throat portion, an accelerating nozzle in said air passage connected to said portion of said chamber, a second nozzle at said throat portion, a throttle in said air passage, means to supply fuel through said metering means to said second nozzle, and means to operate said throttle between greater and less power positions and concurrently and correlatively to move said piston in said chamber, said second nozzle opening into said throat below the idling position of said throttle.

11. In a downdraft carburetor having an air passage, a throttle in said passage, a fuel supply nozzle opening into said passage on the intake manifold side of said throttle, operating means for said throttle, a chamber, means to supply fuel to said chamber, a movable member in said chamber connected to said throttle operating means for movement therewith, and a fuel conduit connecting between said nozzle and said chamber and terminating in a fuel receiving orifice in said chamber, said chamber and said member comprising cooperating means variably to restrict the flow of fuel into said orifice in accord with the position of said member in said chamber.

12. In a downdraft carburetor, an air passage having a constricted throat, a nozzle positioned to supply fuel into said air passage at said throat, a throttle control member, a fuel chamber having a wall, a piston member connected to said throttle control member for movement in said chamber along said wall thereof and having a passage communicating at one end with said nozzle and at its other end opening from said piston member toward said wall, whereby said passage opening traces a path along said wall with said movement of said piston member, said piston having a forward face and being arranged to increase pressure at said face within a forward portion of said chamber in response to movement of said piston member in a predetermined forward direction, said wall having a groove tapering along said path cooperating with said opening, and a bypass channel in said piston member connecting between a first opening of said channel through said forward face and a second opening of said channel toward said wall into said groove, said second opening of said channel being spaced from said forward face by a greater distance than said first mentioned passage opening, whereby said channel opens into said groove behind said passage opening.

References Cited in the file of this patent

UNITED STATES PATENTS

1,126,127	Swan	Jan. 26, 1915
1,325,688	Burr	Dec. 23, 1919
1,383,044	Weiland	June 28, 1921
1,801,684	Oelerich	Apr. 21, 1931
2,214,273	Fish	Sept. 10, 1940
2,236,595	Fish	Apr. 1, 1941
2,463,535	Holley	Mar. 29, 1949
2,617,638	Udale	Nov. 11, 1952

FOREIGN PATENTS

611,784	France	July 17, 1926

June 4, 1968 J. B. YORK, JR **3,386,710**

FUEL SYSTEM

Filed Sept. 16, 1965

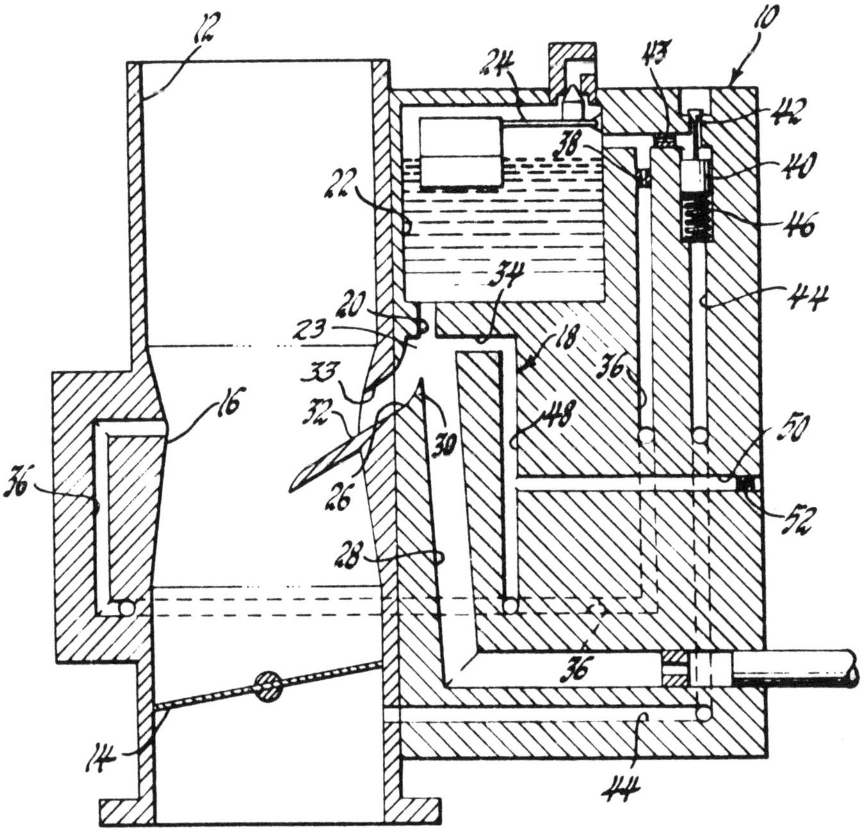

INVENTOR.

Jerome B. York, Jr.

BY

J. L. Carpenter

ATTORNEY

United States Patent Office

3,386,710

Patented June 4, 1968

1

2

3,386,710

FUEL SYSTEM

Jerome B. York, Jr., Royal Oak, Mich., assignor to General Motors Corporation, Detroit, Mich., a corporation of Delaware

Filed Sept. 16, 1965, Ser. No. 487,801
5 Claims. (Cl. 261—36)

This invention relates to fuel systems and is particularly directed to fuel systems utilizing fluid amplifier technology.

Because spark ignition internal cumbustion engines are quite sensitive to the proportions of the air-fuel mixture in the combustion chamber, innumerable systems have been developed in attempts to supply the engine with an air-fuel mixture proper for the various operating conditions. The systems in common use, however, include a venturi located in the air inlet to create a pressure signal which varies with the rate of air flow to the engine; this pressure signal determines the rate at which fuel is delivered to the engine.

Unfortunately however, the venturi signal does not vary linearly with air flow. When utilizing a large venturi to avoid restricting air flow at high air flow rates, the venturi signal is weak at low air flow rates. This weak signal would result in insufficient fuel delivery to the engine. On the other hand, a small venturi which would provide an adequate fuel metering signal at low air flow rates would also restrict air flow at high air flow rates. For these reasons, a compromise is generally adopted in which a separate idle system provides fuel at the lowest air flow rates, a large venturi and one or two small boost venturi provide a fuel metering signal over the part throttle and wide-open throttle operating ranges, and off-idle discharge ports smooth the transition from the idle system to the venturi metering system as the throttle opens.

Fuel systems have recently been developed which utilize fluid amplifiers to amplify the venturi pressure signal and deliver fuel to the engine at a rate determined by the amplified signal. Such an arrangement makes a complex system unnecessary.

As more extensively discussed in recent literature, Control Engineering of January 1963, for example, a fluid amplifier uses a relatively weak control signal to deflect a fluid jet. Because relatively large deflections in the jet are obtained with weak control signals, a more sensitive control function is obtained by utilizing the deflection of the jet rather than by directly utilizing the control signal.

This invention, directed to such a fluid amplifier controlled fuel system, provides means to vary the rate of flow through the amplifier in response to engine operating conditions. Thus, this invention causes the mixture to be enriched whenever acceleration or maximum power are required of the engine.

In addition, this invention provides means for controlling the gain of a fluid amplifier so that the desired amplification of a control signal such as venturi pressure may be achieved.

The details as well as other objects and advantages of this invention are disclosed in the following description and in the accompanying drawing which illustrates a fuel system carburetor embodying this invention.

Referring to the drawing, a carburetor **10** has an air inlet **12** controlled by a throttle **14** in the customary manner. A venturi **16** provides a restriction within air inlet **12** which reduces the pressure of the air flow therethrough to create a control signal related to the rate of air flow.

A fluid amplifier, indicated generally at **18**, has an inlet **20** opening from the carburetor fuel bowl **22** into an interaction region **23**. The fuel level within bowl **22** is maintained at a substantially constant level by a conventional float and inlet valve mechanism **24**. Fuel flow from bowl **22** provides a stream of fuel through amplifier **18** which is issued along a predetermined path within interaction region **23** and which has a predetermined energy state. The fuel stream is discharged from interaction region **23** through outlets **26** and **28**. With the amplifier construction so far described, the proportion of fuel discharged from outlet **26** is determined by the location of the splitter **30**.

As air flows through the carburetor inlet **12** and the venturi **16**, a pressure drop is produced to create a signal which is transferred through a fuel inlet nozzle **32**, a passage **33**, and amplifier outlet **26** to amplifier **18**. This pressure signal has an energy state much lower than that of the air stream issued from inlet **20** but deflects the fuel stream to change the proportion of fuel discharged through outlet **26**. As air flow through the carburetor inlet **12** increases, the pressure in venturi **16** drops further; this reduced pressure sensed through amplifier outlet **26** deflects more of the fuel flow through amplifier **18** to outlet **26**.

In addition to the control by venturi pressure, the pressure at a bias port **34** provides a low energy signal which affects fuel flow through amplifier **18**. As the pressure at bias port **34** increases, more of the fuel flow from inlet **20** will be deflected to outlet **26**. The pressure at bias port **34** is controlled by means to be described below.

An amplifier which operates as described is known as a "proportional" amplifier since the proportion of fuel discharged through outlet **26** varies with the pressure differential across the fuel stream, i.e., with the pressure differential between bias port **34** and venturi **16**.

The rate of fuel flow through inlet **20** is determined by the difference in the pressure on the free surface of the fuel and the pressure in amplifier inlet **20**. As described above the pressure in venturi **16** is sensed within amplifier **18** adjacent inlet **20**, thereby causing the pressure in inlet **20** to vary with the venturi signal. A passage **36** connects the bowl chamber **22** above the fuel level with venturi **16** to vary the pressure above the fuel level with the venturi signal. Thus, the difference in the pressure on the free surface of the fuel and the pressure in amplifier inlet **20** is held substantially constant so that fuel flows through amplifier **18** at a predetermined rate. Amplifier **18** is constructed so that the fuel flow from inlet **20** is divided and a portion directed through fuel inlet **32** to form an air-fuel mixture suitable for economy operation of the engine.

A restriction **38** in passage **36** prevents a sudden change in the rate of air flow through carburetor air inlet **12** and the concomitant pressure change in venturi **16** from suddenly changing the pressure above the fuel in bowl **22**. Therefore, upon a sudden throttle opening and the sudden increase in the air flow rate, the pressure above the fuel will be slowly reduced until it is balanced with venturi pressure. During this period of gradual fuel bowl pressure reduction, the fuel flow through amplifier inlet **20** and thus the fuel flow discharged through amplifier outlet **26** and fuel inlet **32** will temporarily be greater than that which would provide an economy air-fuel mixture. This increased fuel flow provides an enriched mixture to accelerate the engine.

A pressure responsive piston **40** controls a valve **42** to admit air through a restriction **43** to the fuel bowl chamber **22** above the fuel level. Intake manifold vacuum which exists in carburetor inlet **12** below throttle **14** is applied through a passage **44** against piston **40** in opposition to the bias of a spring **46**. During normal engine operation, the manifold vacuum will be sufficient to cause piston **40** to close valve **42**. When manifold vacuum reaches a value indicative of the engine's requirement for

3,386,710

3

an enriched mixture for maximum power operation, spring 46 overcomes the vacuum to raise the piston 40 and open valve 42. The air admitted to fuel bowl chamber 22 increases the pressure above the fuel level and causes an increased rate of fuel flow through amplifier inlet 20. As discussed above, this increased rate of fuel flow causes fuel to be delivered through fuel inlet 32 at a rate greater than that which would form an economy air-fuel mixture. The increased fuel flow thus provides an enriched mixture for power operation of the engine.

In some fluid amplifier constructions, the gain of the control signal causes too great a flow of fuel through the outlet 26. In order to reduce the sensitivity or gain of the fluid amplifier, this invention provides means to vary the pressure at bias port 34 in accordance with the control signal. A passage 48 connects bias port 34 to passage 36 and thus to venturi 16. A passage 50 connects passage 48 to atmosphere through a restriction 52. The pressure at bias port 34 is thus related to the venturi pressure signal but has an actual value intermediate the vacuum signal and the atmospheric pressure. Since the delivery of fuel through outlet 26 is determined by the pressure differential between bias port 34 and venturi 16, a reduction in pressure at bias port 34 reduces fuel delivery through outlet 26 and fuel inlet 32. By properly proportioning the restriction 52 within the atmospheric passage 50, the gain of amplifier 18 may be reduced to deliver fuel at any desired rate.

I claim:

1. An internal combustion engine fuel system comprising
- a nozzle adapted to discharge fuel for mixture with air,
- fluid amplifier means including
  - an interaction region
  - and an inlet port opening into said interaction region,
- a source of liquid fuel connected to said inlet port for issuing a liquid fuel stream along a predetermined path within said interaction region, said fuel stream having a certain energy state,
- an air inlet for air flow to the engine,
- a venturi in said air inlet connected to said interaction region for supplying a pressure signal indicative of the rate of air flow through said air inlet and having an energy state substantially lower than the energy state of the liquid fuel stream and variations of which cause the path of the fuel stream to vary within said interaction region in proportion to the pressure signal variations,
- said amplifier means further including outlet means opening from said interaction region for receiving the liquid fuel stream, said outlet means having
  - a pair of outlets
  - and a splitter for dividing the liquid fuel stream between said outlets in accordance with variations in the path of the fuel stream and for continuously varying the division of the fuel stream over the range of variations in the path of the fuel stream and for directing an increased flow of fuel to one of said outlets upon a change in the pressure signal indicative of increased air flow through said air inlet,
- passage means connecting said nozzle to said one outlet for directing liquid fuel from said one outlet to said nozzle,
- and regulating means for controlling the rate of fuel flow through said inlet port to compensate for the variations in pressure within said interaction region caused by the venturi pressure signal.

2. The fuel system of claim 1 wherein said source of fuel comprises a fuel bowl having means for maintaining fuel at a substantially constant level therein and wherein said regulating means comprises a regulating passage extending from said venturi to the top of said fuel bowl

4

whereby the pressure on the free surface of the fuel in said fuel bowl may vary in accordance with the venturi pressure signal.

3. The fuel system of claim 2 wherein said regulating passage includes a restriction for delaying a reduction in the pressure in said fuel bowl to provide an enriched mixture for acceleration.

4. The fuel system of claim 2 which further includes a vent passage opening into the top of said fuel bowl to supply air thereto and vent valve means responsive to an engine demand for an enriched mixture for opening said vent valve means whereby the pressure in said fuel bowl is increased.

5. An internal combustion engine fuel system comprising
- a nozzle adapted to discharge fuel for mixture with air,
- fluid amplifier means including
  - an interaction region
  - and an inlet port opening into said interaction region,
- a source of liquid fuel connected to said inlet port for issuing a liquid fuel stream along a predetermined path within said interaction region, said fuel stream having a certain energy state,
- an air inlet for air flow to the engine,
- a venturi in said air inlet connected to said interaction region for supplying a pressure signal indicative of the rate of air flow through said air inlet and having an energy state substantially lower than the energy state of the liquid fuel stream and variations of which cause the path of the fuel stream to vary in a certain plane within said interaction region in proportion to the pressure signal variations,
- said amplifier means further including outlet means opening from said interaction region for receiving the liquid fuel stream, said outlet means having
  - a pair of outlets
  - and a splitter for dividing the liquid fuel stream between said outlets in accordance with variations in the path of the fuel stream and for continuously varying the division of the fuel stream over the range of variations in the path of the fuel stream and for directing an increased flow of fuel to one of said outlets upon a change in the pressure signal indicative of increased air flow through said air inlet,
- passage means connecting said nozzle to said one outlet for directing liquid fuel from said one outlet to said nozzle,
- said amplifier means still further including a regulating port opening into said interaction region in said certain plane,
- and passage means connecting said venturi to said regulating port for transmitting another pressure signal to said interaction region in opposition to the first-mentioned pressure signal to reduce the variations in the path of the fuel stream caused by the first-mentioned pressure signal

References Cited

UNITED STATES PATENTS

2,675,217	4/1954	Slason	158—36
2,796,243	6/1957	McDuffie	261—72 X
3,001,539	9/1961	Hurvitz	137—81.5
3,258,023	6/1966	Bowles	137—81.5

FOREIGN PATENTS

1,257,050	2/1961	France.
694,387	7/1953	Great Britain
289,804	3/1954	Switzerland

HARRY B. THORNTON, *Primary Examiner.*

T. R. MILES, *Examiner.*

United States Patent [19] | [11] **3,747,901**
Coverston | [45] **July 24, 1973**

[54] **CARBURETOR APPARATUS**

[76] Inventor: **George C. Coverston,** 76 S. Russell St., Fallon, Nev. 89406

[22] Filed: **Apr. 21, 1972**

[21] Appl. No.: **246,258**

[52] **U.S. Cl.** **261/1,** 261/DIG. 39, 239/718, 261/116
[51] **Int. Cl.** **F02m 19/04**
[58] **Field of Search** 261/DIG. 39, 116, 261/1, 19; 239/118

[56] **References Cited**

UNITED STATES PATENTS

1,360,265	11/1920	Capell	239/118
1,399,655	12/1921	Rector	239/118
1,460,667	7/1923	Good	261/DIG. 39
2,093,444	9/1937	Hubbard	239/118
2,127,444	8/1938	Emerson	261/DIG. 39
3,325,975	6/1967	Coverston	55/225
3,567,124	3/1971	Jones	239/118

Primary Examiner—Tim R. Miles
Attorney—Paul D. Flehr, Robert B. Block et al.

[57] **ABSTRACT**

Carburetor apparatus for vaporizing and mixing a liquid fuel with a stream of gas. An incoming stream of gas is directed through an annular venturi. The outlet ends of a plurality of circumferential spaced-apart fuel jet tubes project into the throat portion of the venturi. The fuel jet tubes have their inlet ends in communication with a supply of fuel confined within a reservoir. Elongate inserts mounted within the tubes are radially spaced from the tube inner surfaces to define relatively small cross sectional area capillary fuel flow passageways. The inserts are adapted for both vertical and lateral displacement to dislodge any foreign matter which forms or deposits within the tube passageways. The end of each insert which projects into the fuel reservoir is reversely bent to form an enlarged portion serving to limit the vertical displacement of the insert.

2 Claims, 2 Drawing Figures

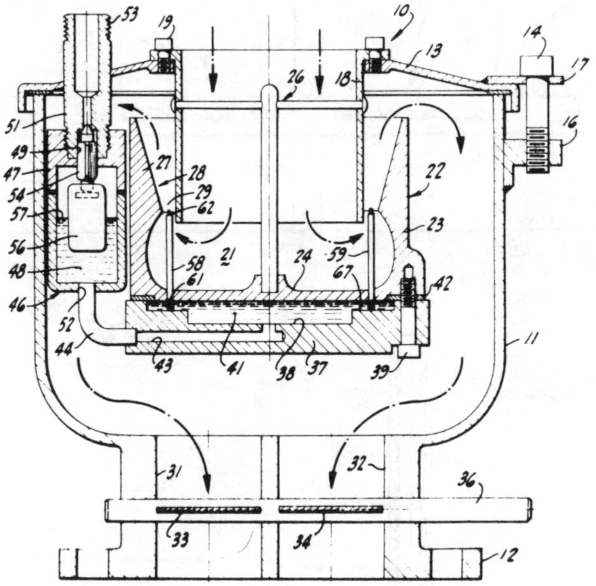

PATENTED JUL 24 1973

3,747,901

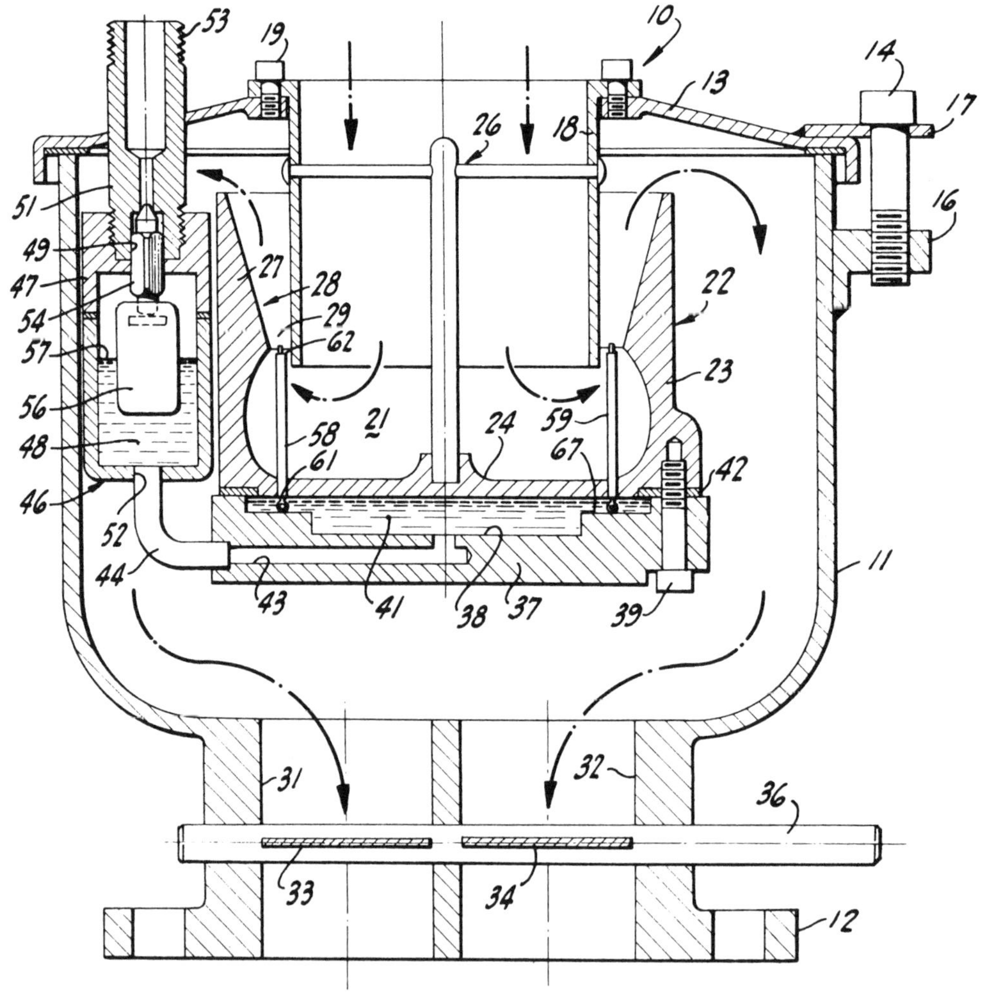

FIG_1

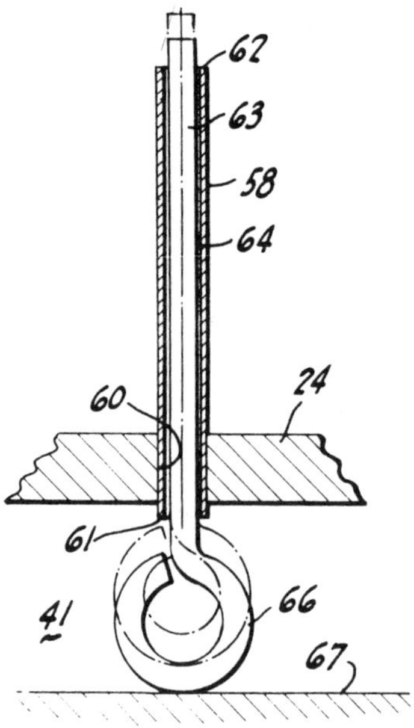

FIG_2

3,747,901

1

CARBURETOR APPARATUS

BACKGROUND OF THE INVENTION

Carburetors of known construction utilize various expedients to introduce and vaporize a liquid fuel into an incoming flow of air to form a combustible charge for induction into an internal combustion engine. In these known carburetors, a venturi flow passage is formed with metering fuel jets discharging into the air stream at the throat portion of the venturi.

In U.S. Pat. No. 3,325,975 issued to the present applicant, a carburetor is provided which creates a high degree of fuel vaporization and has provision for removal of fuel droplets through centrifugal action. The apparatus of the patent includes means forming an annular venturi with a plurality of circumferentially spaced-apart tubes or fuel nozzles arranged with their inlet ends disposed in the fuel supply and with their outlet ends disposed in the venturi throat. The mixture of vaporized fuel and air is directed by baffle means through a reverse flow path arranged to remove fuel droplets by centrifugal action so that a substantially droplet free combustion charge is inducted into the engine.

The tubular fuel jets utilized in known carburetors of the type described are subject to failure or reduced effectiveness for reasons which include the fact that foreign matter and contaminants in the fuels tend to lodge in or form deposits on the surfaces of the fuel tube or nozzles. Among the foreign matter which may be found in fuels of this nature are solid particulate matter such as dirt which may not be removed by a fuel filter, and dissolved contaminants which may gradually adhere to and form a gum or a varnish-like layer or deposit within the inner surfaces of the tube. The problem of foreign matter deposit becomes highly critical in carburetors employing relatively small diameter, capillary action fuel tubes of the type disclosed in Applicant's issued Patent where a high degree of fuel atomization is achieved. The flow area of such capillary tubes is so small that the likelihood of malfunction due to the presence of foreign matter is increased substantially.

OBJECTS AND SUMMARY OF THE INVENTION

This invention relates in general to improvements in apparatus for forming a mixture of gas and a vaporized liquid, and in particular relates to improvements in carburetor apparatus for atomizing or vaporizing liquid fuel for mixture with a stream of air to form a combustible charge for induction into an internal combustion engine.

It is a general object of the invention to provide carburetor apparatus which is effective to provide a high degree of liquid fuel atomization in a stream of gas for more complete combustion of the inducted charge, and thereby reduced pollutants in the exhaust.

Another object of the invention is to provide carburetor apparatus which efficiently atomizes and injects fuel into an air stream while at the same time avoids functional failure of carburetor operation which could otherwise occur as a result of any foreign matter contained in the fuel.

Another object of the invention is to provide carburetor apparatus of the character described utilizing fuel jet tubes in which insert members are carried within the tubes to define capillary flow passages therewith, with the inserts adapted to undergo limited vertical and axial

2

displacement to dislodge any foreign matter which may form or deposit within the tubes.

The invention includes a carburetor apparatus in which an air stream is directed through an inlet conduit into the throat of an annular venturi. A fuel supply reservoir is provided within the carburetor with a plurality of circumferential spaced-apart fuel jet tubes having inlet ends in communication with fuel in the reservoir and having outlet ends disposed in the venturi throat. Elongate insert members mounted in each of the tubes are radially spaced from the inner surfaces thereof to define relatively small diameter fuel flow passages effective to provide capillary action for withdrawing fuel from the reservoir and to inject fuel into the air stream for relative complete fuel atomization so that more complete combustion is achieved. The elongate inserts are mounted in the tubes for both axial and lateral displacement to dislodge any foreign matter which may form or deposit within the tubes. The ends of the inserts projecting into the reservoir are reversely bent to form enlarged portions for limiting the extent of axial displacement. The gaseous mixture of vaporized fuel and air is directed by the carburetor housing through throttle means for induction into an engine.

Additional objects and features of the invention will appear from the following description in which the preferred embodiments of the invention has been set forth in detail in conjunction with the drawings.

BRIEF DESCRIPTION OF THE DRAWINGS

FIG. 1 is a cross-sectional view in elevation of carburetor apparatus according to the invention;

FIG. 2 is a cross-sectional view to an enlarged scale of a fuel jet tube and insert of the apparatus of FIG. 1.

DESCRIPTION OF THE PREFERRED EMBODIMENT

In the drawing FIG. 1 illustrates generally at **10** carburetor apparatus constructed in accordance with the invention. The preferred form of carburetor apparatus **10** comprises a downdraft type carburetor including a cylindrical housing **11** extending from mounting flange **12** which in turn is secured by suitable fasteners such as bolts to an intake manifold, not shown, incorporated in the associated internal combustion engine. An annular cover **13** is secured to the upwardly open end of housing **11** by means such as a plurality of fasteners **14** extending between brackets **16** secured to the side of the housing and arms **17** secured to the cover. An air inlet conduit **18** is mounted by fasteners **19** within a central opening formed in cover **13**, and a suitable air cleaner, not shown, may be provided upstream of the inlet conduit.

Air inlet conduit **18** directs an incoming stream of air vertically downwardly into a chamber **21** formed within an upwardly open cylindrical structure **22** comprising an outer cylindrical wall **23** and flat bottom wall **24**. Cylindrical structure **22** is mounted by means of a bracket assembly **26** depending from within conduit **18**. The inner surface of cylindrical wall **23** is formed into an annular cusp-shaped configuration radially spaced about the outer periphery of inlet conduit **18** and with an upper portion **27** flaring outwardly from the conduit such that upwardly directed venturi means **28** is formed having an annular throat portion **29** adjacent the cusp of the cylindrical wall. The stream of air flowing from chamber **21** and exiting from the venturi means is reversed in direction by cover **13** and housing **11** for downward flow through the pair of passageways **31, 32** leading to the intake manifold. Throttle means is provided which includes a pair of butterfly-type valves **33, 34** mounted on a throttle shaft **36** for pivotal movement by means of a suitable throttle linkage, not shown.

A plate **37** having a circular recess **38** is mounted below bottom wall **24** of the cylindrical structure through suitable fasteners **39** to define a fuel supply reservoir **41.** A suitable annular gasket **42** is provided to form a fluid-tight seal between the peripheral rim of the plate and the bottom wall. A liquid hydrocarbon fuel is supplied to the reservoir through a passageway **43** formed in the plate and connected through tubing **44** with a float chamber assembly **46.** The float chamber assembly comprises a housing **47** defining a fuel chamber **48** having an inlet **49** in communication with fuel supply fitting **51** and an outlet **52** in communication with tubing **44.** The float chamber housing is threadably mounted to the downwardly projecting end of the fitting which in turn is mounted through an opening in cover **13** and connected at its upwardly projecting end **53** with a fuel supply line, not shown. A longitudinally grooved needle valve **54** is slidably mounted within housing inlet **49** for movement to and from a valve seat formed within fitting **51.** The needle valve directs fuel into the float chamber responsive to movement of a float **56** carried within the chamber for maintaining the fuel supply substantially at the illustrated level **57.**

Fuel jet means is provided to meter the fuel with a high degree of atomization into the incoming stream of air. This fuel jet means comprises a plurality of upright, circumferentially spaced-apart hollow capillary tubes **58, 59** mounted within openings **60** formed within bottom wall **24** in vertical alignment below venturi throat **29.** Each of the tubes has an inlet end **61** in fluid communication with the fuel supply reservoir and an outlet end **62** disposed in the venturi throat. The level **57** of the fuel within float chamber assembly **46** is established at an elevation which is below the level of the tube outlet ends.

As best illustrated in FIG. **2** an elongate rod-like insert **63** is mounted within each of the tubes. Preferably both the tubes and inserts are circular in cross-section, although the invention contemplates that other cross-sectional configurations may be employed. The insert **63** is radially spaced from the inner cylindrical surface of the tube to define an annular fuel flow passageway **64** of a relatively small cross-sectional area which is effective to create a capillary effect for withdrawing fuel from the reservoir into the tube, and at the same time to create a high degree of liquid subdivision as the fuel is sprayed into the air stream for a relatively greater atomization and vaporization of the fuel.

The inserts are mounted within each tube for both axial and lateral displacement to dislodge any foreign matter which may form or deposit within the small clearance between the insert and tube. Movement of the insert in this manner effects both abrasive and mechanical action against the foreign matter to prevent obstruction of passageway **64** and to ensure proper operation of the carburetor. Engine vibration and movement is carried through to the carburetor to generate lateral displacement of the inserts. Axial displacement of the inserts is generated by changes in the static pressure differential between the tube inlet and outlet ends due to variations in air stream flow rates through the venturi. Thus, an increased flow rate reduces the static pressure at the venturi throat tending to both aspirate added fuel through the tubes and raise the insert to the broken line position of FIG. **2.** A decrease in the air flow rate reduces the static pressure differential so that the insert is moved by force of gravity downwardly to the illustrated solid line position. Suitable means is provided to limit the extent of axial displacement of the inserts, and this means preferably comprises an enlarged extension **66** of the insert disposed within the reservoir and which is formed by reversely bending the lower end of the insert. This extension engages at its upper extremity with tube inlet end **61** to limit upward travel of the insert, and engages at its lower extremity with the surface of annular shoulder **67** formed within the recess of plate **37**, as best illustrated in FIG. **1.** The insert is sized to be of sufficient axial length so that it completely projects through the tube throughout the range of its axial displacement.

In the operation of carburetor **10** when connected with the intake manifold of an internal combustion engine, float chamber assembly **46** feeds sufficient fuel through tubing **44** to completely fill supply reservoir **41,** with the fuel moving upwardly by capillary action into each of the fuel jet tubes **58, 59.** Operation of the engine with the pumping action of its pistons inducts a stream of air downwardly through inlet conduit **18** where it reverses for upward flow through venturi means **28** and then again reverses for downward flow within housing **11** and is directed into the engine's intake manifold under control of throttle valves **33, 34.** The increased flow velocity of the air at venturi throat **29** creates a reduced static pressure which acts to aspirate and meter fuel with a high degree of atomization from the outlet ends of fuel jet tubes **58, 59.** At the same time the static pressure differential acts on the tube inserts moving them upwardly which, together with lateral displacement of the inserts as limited by the radial clearance within the tubes, serves to dislodge any foreign matter which may form or deposit within the tubes. Fuel which is withdrawn from reservoir **41** is replenished from float chamber assembly **46** which in turn is replenished by operation of float **56** and needle valve **54.**

While the embodiment herein is at present considered to be preferred it is understood that numerous variations and modifications may be made therein by those skilled in the art, and it is intended to cover in the claims all such variations and modifications as fall within the true spirit and scope of the invention.

I claim:

1. In a carburetor for mixing a liquid fuel with a gas, the combination of a conduit positioned to direct said gas into an incoming stream, means defining an annular venturi about said conduit, said venturi having a throat portion for reducing the static pressure of said stream of gas, a reservoir spaced below said throat portion for containing a supply of fuel, a plurality of vertically elongate hollow tubes each having an inlet end in fluid communication with fuel contained in said reservoir and having an outlet end in discharge relationship with said venturi throat portion, a plurality of elongate tube inserts, each insert being mounted in a respective one of said tubes and being radially spaced from the inner surface thereof to define a capillary fuel flow passageway, each insert being mounted for vertical and lateral displacement with respect to the tube inner surface associated therewith, and means to limit the vertical displacement of each insert between an upper position assumed under the influence of reduced static pressure of gas in said venturi throat portion and a lower position assumed under influence of gravity acting on said insert.

2. In a carburetor apparatus, the combination comprising means forming a reservoir containing a liquid fuel, means forming a continuous passageway for directing a stream of gas through said carburetor, said last mentioned means including means forming a venturi having a throat portion, fuel jet means for directing fuel from said reservoir into said stream of gas, said fuel jet means including one or more substantially vertically extending elongate tubes, each tube being formed with an inner cylindrical surface having a length extending between an inlet disposed in said reservoir in communication with said contained fuel and an outlet disposed in said venturi throat portion, an elongate insert carried within each of said tubes and being radially spaced

from said inner cylindrical surface to define a fuel flow passageway therewith, means forming upper and lower spaced stop surfaces in substantially vertical register and disposed below said tubes, weight means formed on the lowermost ends of each insert and disposed between said stop surfaces, and means mounting each insert for movement axially within its associated tube between extreme positions in which said weight means contacts, respectively, said upper and lower surfaces responsive to the resultant force acting on said insert from the force of gravity and the pressure force differential acting on the insert as a combination of static gas pressure within said venturi and fuel pressure within said reservoir whereby said movement of each insert dislodges any foreign matter disposed in said flow passageway, each insert being sized in length sufficiently greater than said length of the associated tube inner cylindrical surface whereby the cross sectional area of the flow passageway therebetween is independent of the axial position assumed by the inserts with respect to the tubes.

* * * * *

United States Patent [19] | [11] **3,752,451**
Kendig | [45] **Aug. 14, 1973**

[54] **FUEL METERING DEVICE FOR INTERNAL COMBUSTION ENGINE**

[75] Inventor: **Willard Z. Kendig,** Yettem, Calif.

[73] Assignee: **Pollution Control Industries, Inc.,** Torrance, Calif.

[22] Filed: **Jan. 21, 1972**

[21] Appl. No.: **212,687**

Related U.S. Application Data

[63] Continuation of Ser. No. 45,037, June 10, 1970, abandoned, Continuation-in-part of Ser. No. 865,239, Oct. 10, 1969, abandoned.

[52] **U.S. Cl.** ...... **261/39 B,** 261/DIG. 58, 261/44 R, 261/50 A, 261/50 AA, 261/71

[51] **Int. Cl.** **F02m 9/08**

[58] **Field of Search** ...... 261/50 A, 50 AA, 261/44 R, 39 B, 71, DIG. 39, 65, 41 B, DIG. 56, DIG. 58, DIG. 59

[56] **References Cited**

UNITED STATES PATENTS

1,120,183	12/1914	Duff	261/41 B
1,143,779	6/1915	Pembroke	261/50 A
1,429,534	9/1922	Renner	261/DIG. 58
1,510,366	9/1924	Whiteman	261/DIG. 58
1,893,920	1/1933	Winfield	261/DIG. 39
2,229,819	1/1941	Reid	261/DIG. 39
2,361,993	11/1944	Chandler	261/65
2,598,450	5/1952	Shaff	261/50 AA
2,600,368	6/1952	Winkler	261/50 AA
2,801,086	7/1957	Fish	261/34 A
2,957,463	10/1960	Schnabel	261/DIG. 19
3,249,346	5/1966	Bickhaus et al.	261/50 AA
3,282,573	11/1966	Phelps	261/51
3,350,073	10/1967	Hill	261/50 A

Primary Examiner—Tim R. Miles
Attorney—Miketta, Glenny, Poms & Smith

[57] **ABSTRACT**

A downdraft carburetor of the air valve type having a mixing passage provided at its upstream end with two pivoted plates or air valves forming a variable venturi, and provided with two pivoted throttle plates adjacent its downstream end. A fuel spray bar in the mixing passage between the venturi plates and the throttle plates discharges fuel into the mixing passage and is supplied with fuel by a pivoted pickup arm movable in a fuel chamber over a calibrated metering ramp which cooperates with the free end of the pickup arm to form a variable fuel metering clearance therebetween. The fuel chamber may be pressurized, or the level therein may be controlled by a float. The venturi plates are biased closed by springs, one of which is a bimetallic spring coming into operation only when the engine is cold to provide a choke.

8 Claims, 17 Drawing Figures

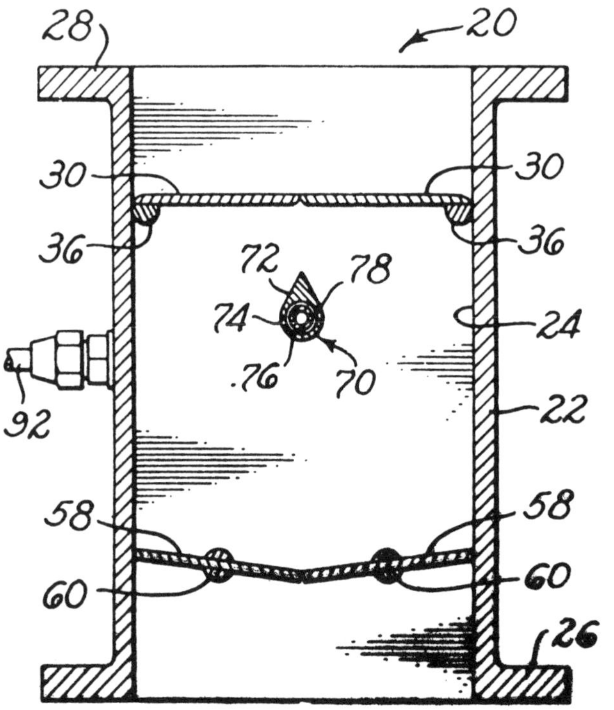

PATENTED AUG 14 1973 — 3,752,451

SHEET 1 OF 3

Fig. 1.

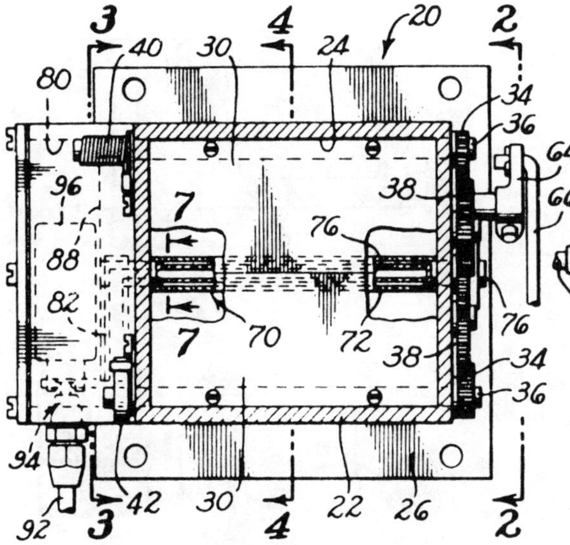

Fig. 2.

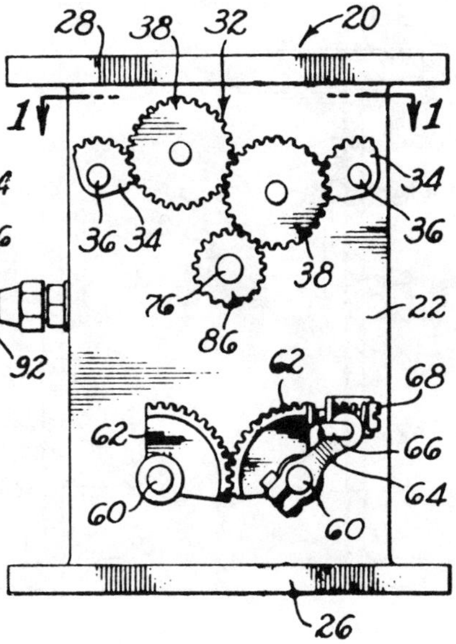

Fig. 3.

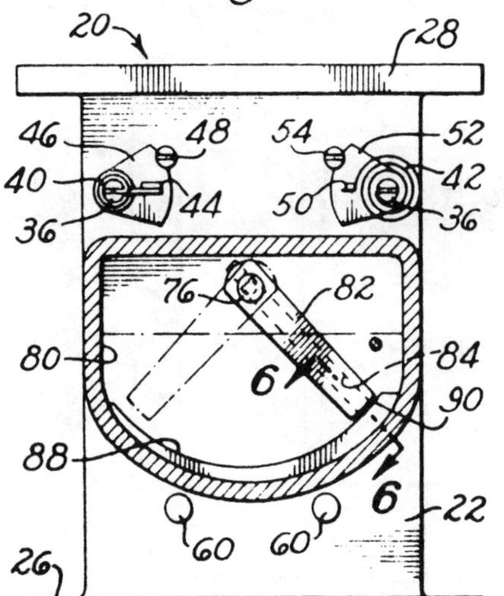

Fig. 4.

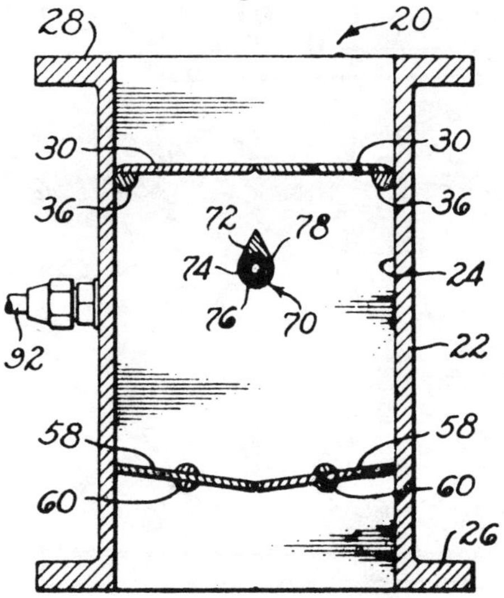

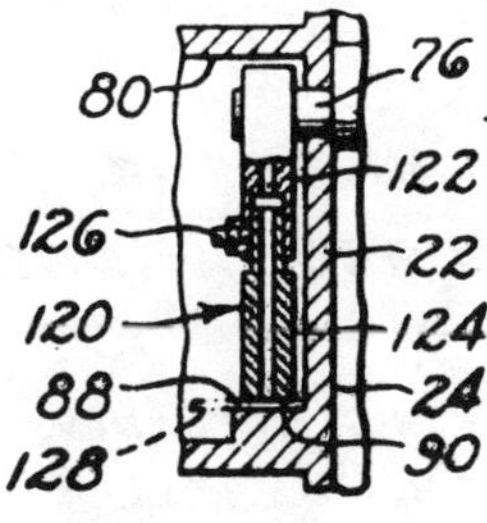

Fig. 12.

INVENTOR.
WILLARD Z. KENDIG

BY HIS ATTORNEYS.
HARRIS, KIECH, RUSSELL & KERN

PATENTED AUG 14 1973 3,752,451

SHEET 2 OF 3

Fig. 5.

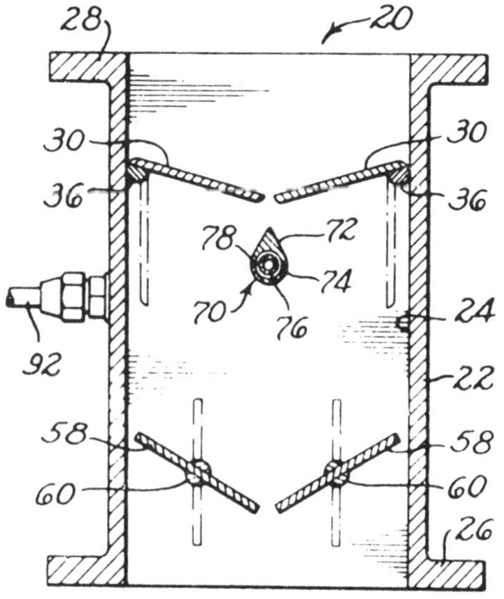

Fig. 6.

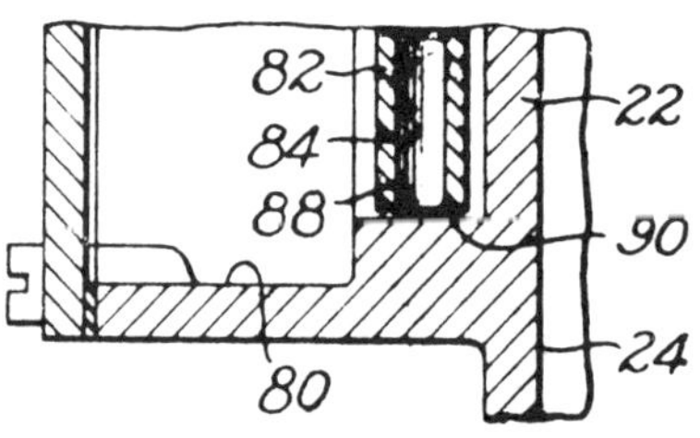

Fig. 7.

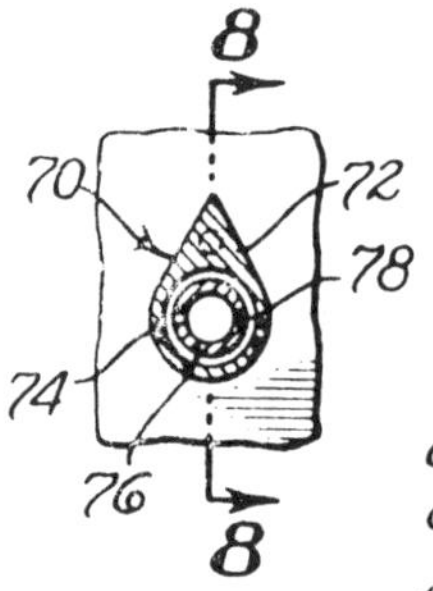

Fig. 8.

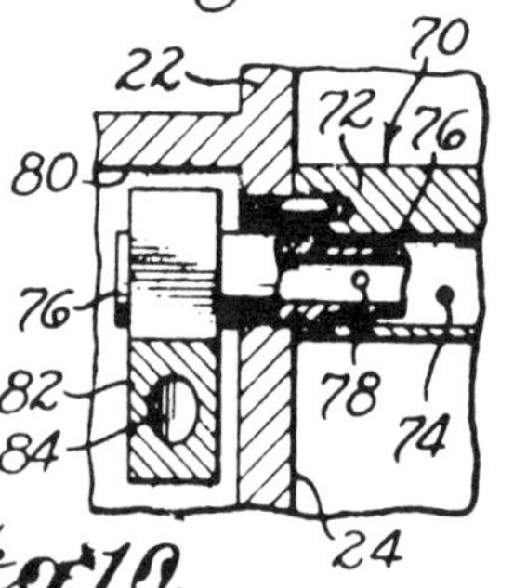

Fig. 9.

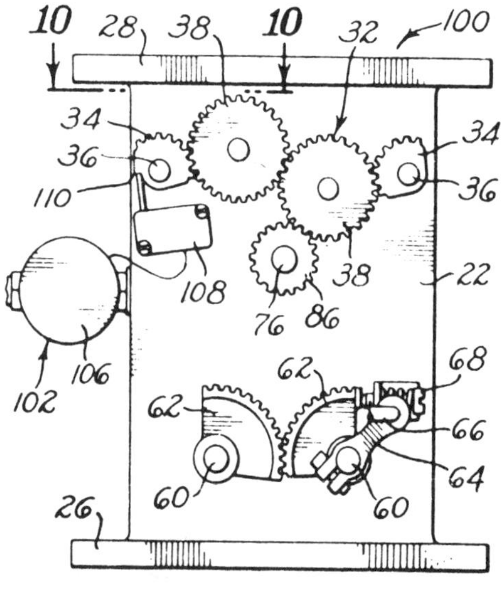

Fig. 10.

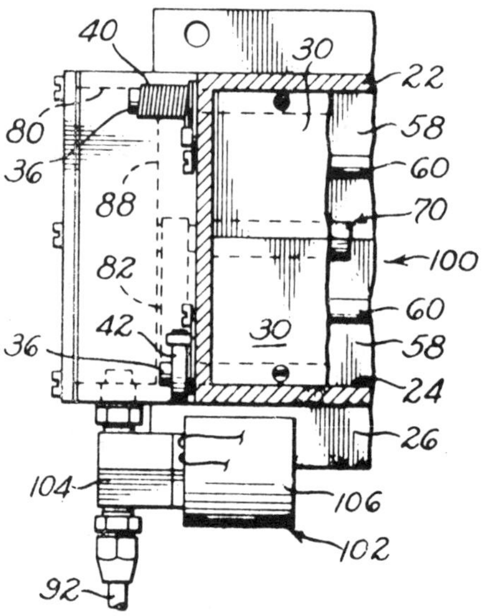

Fig. 11.

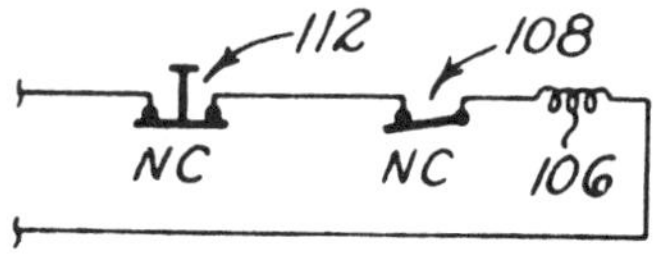

INVENTOR.
WILLARD Z. KENDIG

BY HIS ATTORNEYS.
HARRIS, KIECH, RUSSELL & KERN

PATENTED AUG 14 1973 3,752,451

SHEET 3 OF 3

FIG. 13.

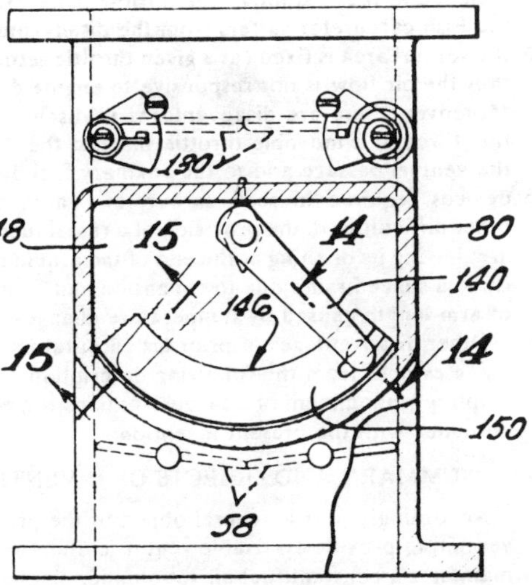

FIG. 14.

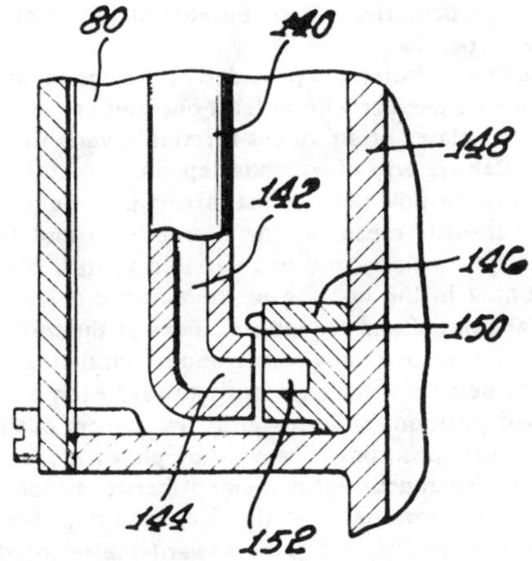

FIG. 17.

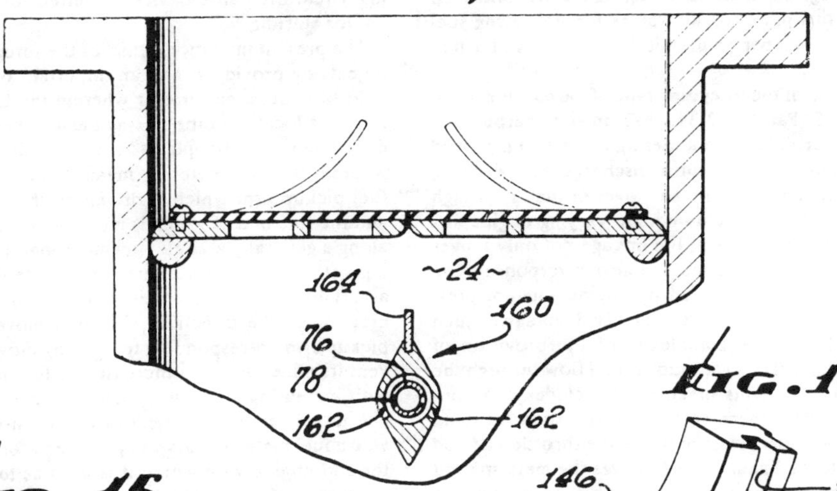

FIG. 16.

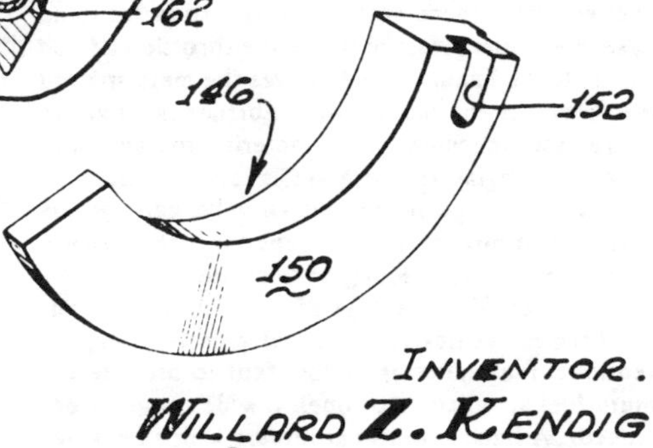

FIG. 15.

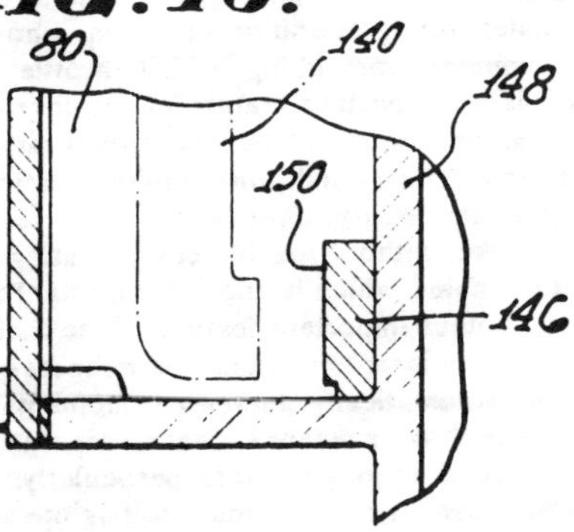

INVENTOR.
WILLARD Z. KENDIG
By
Miketta, Glenny, Poms & Smith
ATTORNEYS.

3,752,451

1

FUEL METERING DEVICE FOR INTERNAL COMBUSTION ENGINE

This is a continuation of application Ser. No. 45,037, filed June 10, 1970 and now abandoned, which was a continuation-in-part of application Ser. No. 865,239, filed Oct. 10,1969 and now abandoned.

BACKGROUND OF INVENTION

The present invention relates in general to a fuel-air metering device for internal combustion engines, and more particularly, to a carburetor of the so-called air valve type.

Such a carburetor is provided at the upstream end of its mixing passage with at least one and preferably two pivoted plates or air valves forming a variable venturi the effective area of which is dependent on the engine-induced air flow through the mixing passage permitted by a throttle means at the downstream end thereof. More particularly, the two variable venturi plates are mounted in the housing of the carburetor at the upstream end of and on opposite sides of the mixing passage for pivotal movement about parallel axes, the plates being pivotable inwardly toward each other into closed positions and outwardly away from each other into open positions. The venturi plates are interconnected for concurrent movement between their closed and open positions and the fuel metering device includes means biasing them toward their closed positions in opposition to the action of an engine-induced vacuum downstream therefrom. A carburetor of this type also includes means for discharging fuel into the mixing passage and includes metering means controlled by the venturi plates or air valves for delivering fuel from a fuel chamber to the discharge means at a metered rate.

An example of the foregoing type of carburetor is disclosed in U.S. Pat. No. 3,342,462. In such carburetor, fuel flow is regulated by a metering rod which is moved in relation to an orifice in a discharge nozzle in response to movement of the air valves or plates through suitable linkage. To accommodate varying engine air-fuel mixture requirements, the linkage not only moves in response to the air plates, but also in response to an accelerator pump controlled by engine vacuum pressure below the throttle plate. One disadvantage of such system is due to the complexity of the provision of mechanism to effect an additional fuel flow through the discharge orifice. This mechanism includes a piston-cylinder assembly (accelerator pump) responsive to an increase in vacuum pressure below the throttle valve so as to actuate the linkage which moves the metering rod so as to increase the orifice area. Further mechanism is required for movement of the metering rod and linkage when the engine temperature is less than normal. Another disadvantage of prior art air valve carburetors is the failure to provide relief means for high pressure gases produced when the engine backfires.

Aside from the disadvantage of the general configuration of the above-described air valve carburetor, the conventional fuel metering system fails to provide the optimum fuel-air mixture through a wide range of engine speeds. A fixed venturi carburetor which provides a more efficient fuel metering system is disclosed in U.S. Pat. No. 2,236,595 and is generally referred to as a Fish carburetor. Such carburetor employs a fixed venturi area and a throttle plate pivotally mounted in the throat having a plurality of fuel discharge openings

2

therein. The throttle plate is connected to a pivotable arm that extends into the fuel chamber. A passage from the throttle plate apertures through the radial arm terminates at the end of the arm in spaced varying relation to the carburetor wall so that as the arm is rotated the orifice area increases to permit a greater fuel flow. In such carburetor the rate of fuel delivery may be matched to the position of the throttle plate. However, the Fish carburetor suffers from the disadvantage that the venturi area is fixed (at a given throttle setting) and thus the air flow is not responsive to engine demands. Moreover, there are disadvantages in discharging the fuel through a movable throttle plate at the throat of the venturi passage and to the auxiliary fuel discharge devices required in such carburetor. There are still other difficulties in the provision of a radial fuel pickup arm having its opening at the end of the arm in proximity to a spaced ramp due to expansion and contraction of arm length caused by temperature changes.

Other disadvantages of prior art carburetors will become evident from the following description of an exemplary embodiment of a carburetor constructed in accordance with the present invention.

SUMMARY AND OBJECTS OF INVENTION

Accordingly, it is a general object of the present invention to provide a variable venturi carburetor which maintains a constant fuel-air ratio under all engine operating conditions from idle to full throttle and which compensates for variation in air density, altitude, temperature and/or humidity without the provision of auxiliary fuel discharge devices required for acceleration and/or starting.

The present invention achieves the foregoing general objects by providing, and an important object thereof is to provide, a carburetor wherein the fuel discharge means is located in the mixing passage between the independent venturi plates and throttle means, and wherein the fuel metering means comprises a pivoted fuel pickup arm which is driven by the venturi plates and the free end of which is movable in a fuel chamber along a generally arcuate fuel metering ramp cooperating with the fuel pickup arm to provide therebetween a variable fuel metering clearance, such clearance increasing in the direction of pivotal movement of the pickup arm corresponding to opening movement of the venturi plates so as to increase the fuel metering rate with increasing air flow through the mixing passage.

Another important object of the invention is to provide a fuel metering ramp so calibrated for the carburetor and engine with which it is used as to provide the same air-fuel ratio for all angular positions of the venturi plates during operation under equilibrium conditions, ranging from idling to full throttle operation, which is an important feature of the invention. This construction also maintains the air-fuel ratio constant under equilibrium conditions despite variations in such things as air density, altitude, temperature, humidity, and the like, without specific compensating means for these variables, which is another important feature.

Still another important feature of the fuel metering device of the invention is that the richness of the mixture is automatically increased momentarily upon opening of throttle means so that no separate acceleration pump is required. More particularly, when the throttle means is opened, the resulting opening movement of the venturi plates is delayed very slightly due to such things as inertia, friction, the resistance of the means biasing the venturi plates toward their closed positions, and the like. This momentary lag results in the development of a momentary vacuum in the mixing passage at the fuel discharge means to draw an increased quantity of fuel into the mixing passage until equilibrium is restored, thereby providing the desired effect of an acceleration pump without any necessity for one.

Another and also important object of the invention is to provide a resilient means for biasing the venturi plates toward their closed positions, which includes temperature responsive means having a biasing force which increases with decreasing temperature. A related object is to provide such a temperature responsive means comprising a bimetallic spring. With this construction, the resistance to opening of the venturi plates is increased when the engine is cold so that the engine induces a greater vacuum at the fuel discharge means between the venturi plates and the throttle means, thus drawing fuel at a greater rate than normal. Consequently, the carburetor of the invention has the effect of a choke at subnormal temperatures with no necessity for specifically providing one, which is an important feature.

A further object of the invention is to provide a carburetor wherein the fuel discharge means comprises a tubular spray bar parallel to the pivot axes of the venturi plates and spanning the mixing passage downstream from the closed positions of the venturi plates and upstream from the throttle means, the spray bar being provided with lateral discharge openings for effective mixing of the fuel with the air.

Additional objects of the invention are to provide a fuel metering device wherein the admission of fuel into the chamber containing the pickup arm is controlled by a float, or a fuel metering device wherein the fuel chamber is pressurized and the admission of fuel thereinto is controlled by an electrically operated valve means, such as a solenoid valve. Another object in connection with the latter is to provide means for preventing pressurization of the fuel chamber until such time as the venturi plates open a predetermined amount, thereby avoiding flooding.

Yet another object of the invention is to provide a throttle means which includes two interconnected throttle plates in side-by-side relation at the downstream end of the mixing passage and mounted in the housing for pivotal movement about axes parallel to the pivot axes of the venturi plates. Such dual throttles or throttle plates reduce the length of the mixing passage, which is a feature of the invention.

Yet one more object of the present invention is to provide an opening to atmosphere from the mixing chamber, for example by venturi plates provided with a plurality of openings, each of the plates being provided with conforming flexible covers so that high pressure gases produced by engine backfire may pass out of the carburetor through the openings by forcing the covers upward.

A still further object of the present invention is to provide a carburetor with a fixed spray bar supported in the mixing chamber having an aerodynamic configuration which promotes atomization of the fuel discharged therethrough.

The foregoing objects, advantages, features and results of the present invention, together with various other objects, advantages, features and results thereof which will be evident to those skilled in the carburetor art in the light of this disclosure, may be achieved with the exemplary embodiments of the invention illustrated in the accompanying drawings and described in detail hereinafter.

DESCRIPTION OF THE DRAWINGS

FIG. 1 is a horizontal sectional view of a fuel metering device or carburetor of the invention and is taken as indicated by the arrowed line 1—1 of FIG. 2;

FIG. 2 is an end elevational view taken as indicated by the arrowed line 2—2 of FIG. 1;

FIG. 3 is a vertical sectional view through a fuel chamber of the carburetor and is taken as indicated by the arrowed line 3—3 of FIG. 1;

FIGS. 4 and 5 are vertical sectional views, taken as indicated by the arrowed line 4—4 of FIG. 1, through the mixing passage of the carburetor and showing various parts in different operating positions;

FIG. 6 is an enlarged, fragmentary sectional view of a fuel metering means of the invention and is taken as indicated by the arrowed line 6—6 of FIG. 3;

FIG. 7 is an enlarged, vertical sectional view of a fuel discharge means of the invention and duplicates a portion of FIGS. 4 and 5 on a larger scale, FIG. 7 being taken along the arrowed line 7—7 of FIG. 1;

FIG. 8 is a view taken as indicated by the arrowed line 8—8 of FIG. 7;

FIG. 9 is a view similar to FIG. 2, but showing another embodiment of the invention;

FIG. 10 is a fragmentary horizontal sectional view taken as indicated by the arrowed line 10—10 of FIG. 9;

FIG. 11 is a schematic wiring diagram of the embodiment shown in FIGS. 9 and 10 of the drawings; and

FIG. 12 is a fragmentary sectional view showing still another embodiment;

FIG. 13 is a vertical sectional view of a modified fuel metering device according to the invention;

FIG. 14 is a sectional view and is taken as indicated by the arrowed line 14—14 of FIG. 13;

FIG. 15 is a sectional view and is taken as indicated by the arrowed line 15—15 of FIG. 13;

FIG. 16 is a perspective view of fixed ramp means; and

FIG. 17 is a vertical sectional view through the mixing passage of the device shown in FIG. 13.

DESCRIPTION OF EXEMPLARY EMBODIMENT OF INVENTION FIGS. 1 to 8

Designated generally by the numeral 20 in FIGS. 1 to 8 of the drawings is a downdraft fuel metering device or carburetor of the invention having a housing 22 provided therethrough with an upright mixing passage 24. The housing 22, which is illustrated as being of simple, rectangular construction, is shown as provided at its lower, downstream end with a mounting flange 26 for attachment to an intake manifold, not shown, and as provided at its upper, upstream end with a mounting flange 28 for attachment of an air filter, or like, to the carburetor.

Two rectangular plates or air valves 30 forming a variable venturi are pivotally mounted in the housing 22 at the upstream end of and on opposite sides of the mixing passage 24, adjacent the corresponding sidewalls of the mixing passage. The variable venturi plates

30 are pivotable about parallel axes inwardly toward each other into closed positions, shown in FIG. 4, and outwardly away from each other into open positions. Partially open positions are shown in solid lines in FIG. 5, and fully open positions in broken lines.

The carburetor 20 includes means 32, FIG. 2, interconnecting the venturi plates 30 for concurrent movement between their closed and open positions. The connecting means 32 includes two sector gears 34 suitably fixed on shafts 36 carrying the venturi plates 30 and respectively meshed with intermeshed gears 38. With this construction, the venturi plates 30 move in unison through identical angular increments.

The carburetor 20 includes resilient means biasing the venturi plates 30 toward their closed positions in opposition to the action of an engine-induced vacuum downstream from the venturi plates tending to pivot them into open positions. Referring to FIGS. 1 and 3, the resilient means comprises two springs 40 and 42 encircling projecting ends of the respective venturi-plate shafts 36. The spring 40 is a rattrap spring having one end suitably connected to its venturi-plate shaft 36 and having its other end in engagement with a stop 44 on a sector 46 which is pivotable about the axis of the corresponding shaft 36 and which may be locked in any desired adjusted position, corresponding to a particular air-fuel ratio, by a screw 48.

The spring 42 is a bimetallic spiral spring having one end suitably connected to its venturi-plate shaft 36, and having its other end in engagement with a stop 50 on a sector 52 which may be locked in any desired adjusted position by a screw 54. The bimetallic spring 42 operates only below a selected temperature and, above that temperature, exerts no biasing force on the venturi plates 30. The bimetallic spring 42 applies a progressively increasing closing bias to the venturi plates 30 with decreasing temperature below the selected cutoff temperature, which bias may be adjusted by loosening the screw 54 and resetting the sector 52. As will be explained hereinafter, the bimetallic spring 42 causes the venturi plates 30 to produce a progressively decreasing choke effect up to the cutoff temperature.

The carburetor 20 includes a throttle means in the housing 22 at the downstream end of the mixing passage 24, such throttle means being shown as comprising two rectangular throttle plates 58 mounted intermediate their edges for pivotal movement, about axes parallel to the pivot axis of the venturi plates 30, between closed positions, FIG. 4, and open positions, the throttle plates being shown partially open in solid lines in FIG. 5 and fully open in broken lines. More particularly, the throttle plates 58 are carried by shafts 60 intermediate their lateral edges. As best shown in FIG. 2, projecting ends of the shafts 60 have fixed thereon intermeshed sector gears 62 which cause the throttle plates 58 to move in unison. One of the shafts 60 also has fixed thereon externally of the housing 22 an arm 64 connected by a suitable linkage 66 to an accelerator pedal, not shown, or other throttle control. One of the sector gears 62 is engageable with an adjustable stop screw 68 which determines the idling speed of the engine on which the carburetor 20 is mounted.

In the mixing passage 24 between the venturi plates 30 and the throttle plates 58 is a fuel discharge means 70 comprising a tubular spray bar 72. The latter parallels the axes of the venturi plates 30 and the throttle plates 58 and is located midway between the sides of

the mixing passage 24, extending between the end walls of the mixing passage and being suitably secured thereto. Lateral discharge openings 74 spray the fuel transversely of the stream of air flowing through the mixing passage 24 for thorough mixing.

Rotatable in the tubular spray bar 72 is a tubular shaft 76 the interior of which is shown as communicating with the interior of the spray bar through radial ports 78, FIG. 8. One end of the tubular shaft 76 projects through an end wall of the mixing passage 24 into a fuel chamber 80 and has fixed thereon a fuel pickup arm 82 the free end of which is movable in the fuel chamber along an arcuate path. A fuel passage 84 extends from the free end of the pickup arm 82 to and communicates with the tubular shaft 76 so that fuel picked up by the arm 82 in a manner to be described is delivered to the interior of the spray bar 72.

The opposite end of the tubular shaft 76 is suitably closed and projects through the opposite end wall of the mixing passage 24. Such other projecting end of the tubular shaft 76 has fixed thereon a gear 86, FIG. 2, meshed with one of the gears 38 interconnecting the venturi plates 30. This construction provides a driving means interconnecting the venturi plates 30 and the fuel pickup arm 82 for pivoting the pickup arm concurrently with the venturi plates. Thus, as the venturi plates 30 pivot from their closed positions, FIG. 4, to their fully open positions, shown in broken lines in FIG. 5, the fuel pickup arm is moved concurrently from the solid line position of FIG. 3 to the broken line position thereof. The fuel chamber 80 contains a generally arcuate fuel metering ramp 88 adjacent and generally parallel to the arcuate path followed by the free end of the fuel pickup arm 82. The fuel metering ramp 88 and the free end of the fuel pickup arm 82 cooperate to provide a variable fuel metering clearance 90 therebetween. The spacing of the metering ramp from the arcuate path of the free end of the pickup arm 82 increases along the arcuate path in the direction of pivotal movement of the pickup arm corresponding to opening movement of the venturi plates 30, thereby increasing the fuel metering clearance 90 with increasing air flow through the mixing passage 24. The spacing of the metering ramp 88 from the path of the free end of the pickup arm 82 is so calibrated as to provide, for the engine on which the carburetor 20 is installed, the same air-fuel ratio for all angular positions of the venturi plates 30 during operation under equilibrium conditions.

In the embodiment under consideration, fuel is admitted to the fuel chamber 80 from a fuel line 92 by a valve means 94, FIG. 1, controlled or operated by a float 96 in a conventional manner.

OPERATION OF EMBODIMENT OF FIGS. 1 TO 8

As it will be apparent, the positions of the venturi plates 30, and thus the position of the fuel pickup arm 82, are determined by the air flow through the mixing passage 24, which is determined by the positions of the throttle plates 58. More particularly, as the throttle plates 58 are opened, an engine-induced vacuum builds up between the throttle plates and the venturi plates 30, so that the resulting pressure differential across the venturi plates causes them to assume corresponding positions. Thus, changes in the positions of the throttle plates 58 produce corresponding changes in the positions of the venturi plates **30**, and in the position of the fuel pickup arm **82**.

More particularly, when the throttle plates **58** are in their closed positions, the venturi plates **30** are also closed, being biased closed by the spring **40** (and by the bimetallic spring **42** at any temperature below normal operating temperatures, as will be discussed hereinafter). If the throttle plates **58** are opened, the result is an increased vacuum in the mixing passage **24** between the throttle plates and the venturi plates **30**, whereupon the venturi plates open correspondingly. This has the effect of moving the fuel pickup arm **82** along the fuel metering ramp **88** in the direction to increase the fuel metering clearance **90** to permit the rate of fuel flow to the spray bar **72** to increase nonlinearly in proportion to the increased air flow. As previously mentioned, the fuel metering ramp **88** is so calibrated that the fuel metering clearance **90** increases with increasing air flow as required to maintain the air-fuel ratio constant under all engine operating conditions, from idling to full throttle, which is an important feature of the invention. It will be understood, of course, that as the throttle plates **58** are closed, the reverse occurs. In other words, the venturi plates **30** close correspondingly as the throttle plates **58** are closed, thereby moving the fuel pickup arm **82** down the metering ramp **88** to progressively decrease the metering clearance **90** in proportion to the decreasing air flow. Thus, for any positions of the throttle plates **58**, the venturi plates **30** assume corresponding positions and, as they do, they correspondingly displace the fuel pickup arm **82** angularly relative to the fuel metering ramp **88** in a manner to maintain the air-fuel ratio of the mixture delivered to the engine constant, under equilibrium conditions.

The carburetor **20** of the invention will maintain the air-fuel ratio constant, under equilibrium conditions, irrespective of variations in altitude, variations in air density due to causes other than altitude variations, variations in temperature, variations in humidity, and the like, which are important features of the invention. It will be understood, of course, that the air-fuel ratio may be adjusted by adjusting the biasing force provided by the spring **40**.

Still considering the operation of the carburetor **20** with the engine in operation at normal operating temperatures, if the throttle plates **58** are opened, or if the degree of opening thereof is increased, the corresponding opening of the venturi plates **30** is delayed very slightly, due to such things as inertia, friction, the opposition offered by the spring **40**, and the like. Consequently, the pressure in the mixing passage **24** at the fuel discharge means **70** decreases so that the flow of fuel to the discharge means is increased before the venturi plates **30** move to cause the fuel metering clearance **90** to increase. This results in a momentary increase in the richness of the mixture produced in the mixing passage **24**, since the venturi plates **30** have not yet moved to increase the air flow, so that the effect of an acceleration pump is provided, without any necessity for a separate device of this nature.

Considering now the operation of the carburetor **20** at temperatures below the cutoff temperature of the bimetallic spring **42**, the venturi plates **30** are biased closed at such temperatures by both the spring **40** and the bimetallic spring **42**. Consequently, a given opening of the throttle plates **58** results in a higher vacuum at the fuel discharge means **70** than under normal temperature operation, since the venturi plates cannot open to maintain the constant air-fuel ratio achieved under normal operating temperature conditions. Thus, under such subnormal operating temperatures, the richness of the mixture produced in the mixing passage **24** is increased to provide the effect of a choke, for starting and warming up, without the provision of a separate choke, this being another important feature. It will be understood that this choke effect also takes place in starting the engine, cranking of the engine resulting in a reduction in pressure at the fuel discharge means **70**, because of the closed venturi plates **30**, to produce the richer mixture necessary for starting.

DESCRIPTION OF EXEMPLARY EMBODIMENT OF INVENTION FIGS. 9 TO 11

These figures of the drawings show a carburetor **100** which, in most respects, is identical to the carburetor **20**, the same reference numerals being applied to identical components. The carburetor **100** differs in that, instead of being provided with the float operated valve means **94** for admitting fuel into the fuel chamber **80**, it is provided with an electrically operated valve means **102** in the fuel line **92** for admitting fuel into the chamber **80**. More particularly, the electrically operated valve means **102** may comprise a valve **104** operated by a solenoid **106**. In this embodiment, the fuel chamber **80** is maintained at a superatmospheric pressure, instead of atmospheric.

In this embodiment, premature energization of the solenoid **106** could result in flooding of the engine, particularly if the latter were equipped with an electrically driven fuel pump, not shown. To prevent this, the solenoid **106** has connected in series therewith a normally closed microswitch **108** which is held open by a shoulder **110** on one of the venturi-plate sector gears **34** until such time as cranking of the engine during starting results in opening of the venturi plates **30** to a predetermined extent. At this point, the microswitch **108** closes to energize the solenoid **106** and thus pressurize the fuel chamber **80** to discharge fuel into the mixing chamber **24**. Preferably, another normally closed, manually openable, pushbutton switch **112** is also connected in series with the solenoid **106** to permit the operator of the engine to prevent the delivery of additional fuel to the carburetor **100** during cranking of the engine to eliminate a flooded condition. Upon releasing the switch **112**, it automatically closes again.

In all respects other than the foregoing, the carburetor **100** operates in the same manner as the carburetor **20** so that a further description is not necessary.

DESCRIPTION OF EXEMPLARY EMBODIMENT OF INVENTION FIG. 12

In this figure is shown a fuel pickup arm **120** similar to the arm **82**, but made of two telescoping parts **122** and **124** adjustably secured together by a set screw **126**. With this construction, the fuel metering clearance **90** may be adjusted to precisely the correct value at idling, for example, by loosening the set screw **126**, inserting a feeler gauge **128** between the arm **120** and the ramp **88**, and tightening the set screw, thereby compensating for manufacturing variations in the fuel arm length.

DESCRIPTION OF EXEMPLARY EMBODIMENT OF INVENTION, FIGS. 13-17

In these figures are shown a modification of the

pickup arm and fuel metering ramp disclosed in the preceding embodiments. A fuel pickup arm **140** has a fuel passage **142** in communication with the tubular shaft **76** so that fuel picked up by arm **140** is delivered to the interior of the spray bar. The arm is driven through gear means to the interconnected venturi plates **30** for concurrent pivotal movement therewith. The pickup arm **140** has a free end movable in the fuel chamber along an arcuate path. The fuel passage **142** terminates in a side opening **144**. The fuel chamber **80** includes a fuel metering ramp **146**, which may be integral with the wall **148**, adjacent the side opening **144** of pickup arm **140**. The face **150** of the ramp cooperates with the free end of the pickup arm to provide a variable fuel metering clearance **152** therebetween. The spacing of the ramp from the path of the free end of the pickup arm increases along the arcuate path in the direction of pivotal movement of the pickup arm (compare FIGS. **14** and **15**) corresponding to opening movement of the venturi plates **30** so as to increase the fuel metering clearance **152** with increasing air flow through mixing passage **24**.

The metering clearance is calibrated to provide the same fuel-air ratio for all angular positions of the plates **30** during operation of the engine on which the carburetor is installed. It will be appreciated that elongation of pickup arm **140**, due to temperature expansion, will not affect the metering clearance **152**, as in the previous embodiment. To provide a fuel rich mixture during starting, the face **150** of ramp **146** may be provided with a slot or groove **154** over a selected arcuate portion of the ramp (corresponding to low air flow through venturi plates **30**), to increase the clearance and there fore to permit increased fuel flow.

There is also shown in FIGS. **13** and **17**, a modified spray bar **160** fixedly mounted in mixing passage **24** as in the previous embodiment, but having a different external configuration. To improve the atomization of fuel discharged through orifices **162** (which communicate through ports **78** with the interior of tubular shaft **76**), the outer surface of the spray bar has a symmetrical, elongated, diamond-shaped cross section as seen best in FIG. **17**. Smoother aerodynamic effect is also accomplished by the provision of vane elements **164** at the upper end of the spray bar.

Further shown in FIGS. **13** and **17** is relief means for exhausting high pressure gases produced by engine backfire through the carburetor. In the exemplary embodiment, such means comprises a plurality of openings **170** in the venturi plates **172** and resilient cover elements **174, 176** secured by screws or the like **178** to each of the plates. During normal operation, the atmospheric pressure is greater than the mixing passage pressure so that the elements **174, 176** conform closely to the upper surfaces of plates **172**. However, should the engine backfire, the mixing passage pressure will exceed atmospheric pressure, forcing the elements **174, 176** to pivot upwardly, allowing the high pressure gas to escape.

Although exemplary embodiments of the invention have been disclosed for illustrative purposes, it will be understood that various changes, modifications, and substitutions may be incorporated in such embodiments.

I claim:

1. In a carburetor for an internal combustion engine, the combination of:

a. a housing providing a mixing passage adapted to have air drawn therethrough by the engine;
b. movable venturi plates pivotally mounted on parallel axes spaced apart and adjacent the walls upstream of said mixing passage;
c. throttle means in said housing movably mounted downstream of said mixing passage;
d. a fuel chamber;
e. a fuel discharge spray bar extending laterally across said mixing passage having an axis parallel to the pivot axes of said venturi plates and fixedly mounted immediately below the venturi plates when in a closed position, said bar having a plurality of spaced apart fuel discharge openings;
f. a fuel pickup arm pivotally mounted in said fuel chamber and having an opening therein in fluid communication with said fuel discharge spray bar, said opening moving in an arcuate path within said fuel chamber;
g. means interconnecting said movable venturi plates and said pickup arm for concurrent pivotal movement thereof; and
h. ramp means in said fuel chamber variably spaced from said pickup arm opening so as to cooperate therewith to provide a non-linear variable fuel metering clearance at different arcuate positions of said pickup arm, said clearance increasing with arcuate movement of said pickup arm to a position corresponding to an increased opening of said venturi to maintain the air-fuel mixture ratio substantially constant.

2. The carburetor of claim 1 wherein said means interconnecting said movable means defining a variable venturi and said pickup arm comprises mechanical linkage for rotating said pickup arm in direct proportion to movement of said movable means.

3. The carburetor of claim 2 wherein said linkage comprises a gear train.

4. The carburetor of claim 1 including electrically operated valve means for admitting fuel under pressure into said fuel chamber and switch means operable by said movable means defining a variable venturi for preventing energization of said electrically operated valve means when said variable venturi is closed.

5. The carburetor of claim 4 including manually operable, normally closed switch means for preventing energization of said electrically operated valve means.

6. A carburetor for an internal combustion engine comprising:

a. a housing including a mixing passage through which air is drawn;
b. variable venturi plates pivotally mounted on parallel axes in the upstream end of said mixing passage for movement from a position at which the free ends of said plates are substantially in contact so as to entirely close said mixing passage to an open position at which said plates are substantially removed from the air flow passage;
c. pivotally mounted throttle means mounted at the downstream end of said mixing passage;
d. a fuel supply chamber;
e. a hollow fuel discharge spray bar extending entirely across said mixing passage having an axis parallel to the pivotal axes of said venturi plates and fixedly mounted immediately below said venturi plates so as to be in close proximity to the free edge of said plates when said plates are partially open,

and including a plurality of fuel discharge openings;

f. a tubular shaft rotatably mounted within said hollow spray bar and having openings in communication with the interior of said spray bar;

g. a fuel pickup arm pivotally mounted in said fuel supply chamber and having a passage therein in fluid communication at one end with said tubular shaft and having an inlet opening at its other end;

h. means interconnecting said variable venturi plates and said fuel pickup arm for concurrent pivotal movement; and

i. ramp means in said fuel chamber variably, nonlinearly spaced from said pickup arm opening so as to cooperate therewith to provide a varying fuel metering clearance upon movement of said arm, said clearance increasing in proportion to increased opening of said variable venturi plates.

7. A carburetor for an internal combustion engine comprising:

a. a housing adapted for mounting to an intake manifold having an elongated mixing passage through which air is drawn and fuel is discharged;

b. a pair of venturi plates each mounted on a shaft, said plates being pivotable toward each other, said shafts being spaced apart, parallel and positioned in the upstream end of said mixing passage;

c. throttle means movably mounted in the downstream end of said mixing passage;

d. a fuel discharge spray bar fixedly mounted in said mixing passage below said venturi plates and above said throttle means and having an axis parallel to said venturi plate shafts and having a plurality of lateral fuel discharge openings;

e. a fuel pickup arm pivotally mounted in a fuel supply chamber, said arm having a fuel passage in communication with said fuel discharge spray bar at one end and an inlet opening at the other end moving in an arcuate path within said fuel chamber adjacent a ramp in said chamber variably and nonlinearly spaced from the pickup arm inlet opening so as to cooperate therewith to provide a variable fuel metering clearance at different arcuate positions of said pickup arm; and

f. means interconnecting said venturi plates and said pickup arm for coordinated concurrent pivotal movement of said arm whereby the clearance between said arm inlet opening and said ramp increases with increasing opening of said venturi so as to control the air-fuel mixture ratio.

8. The carburetor of claim 7 additionally including temperature responsive resilient means biasing said venturi plates towards their closed position with variable force in opposition to the force due to the pressure differential across said plates tending to move said plates to the open position.

* * * * *

Nov. 26, 1974 J. W. MOULDS 3,851,021

CARBURETOR

Filed July 26, 1973

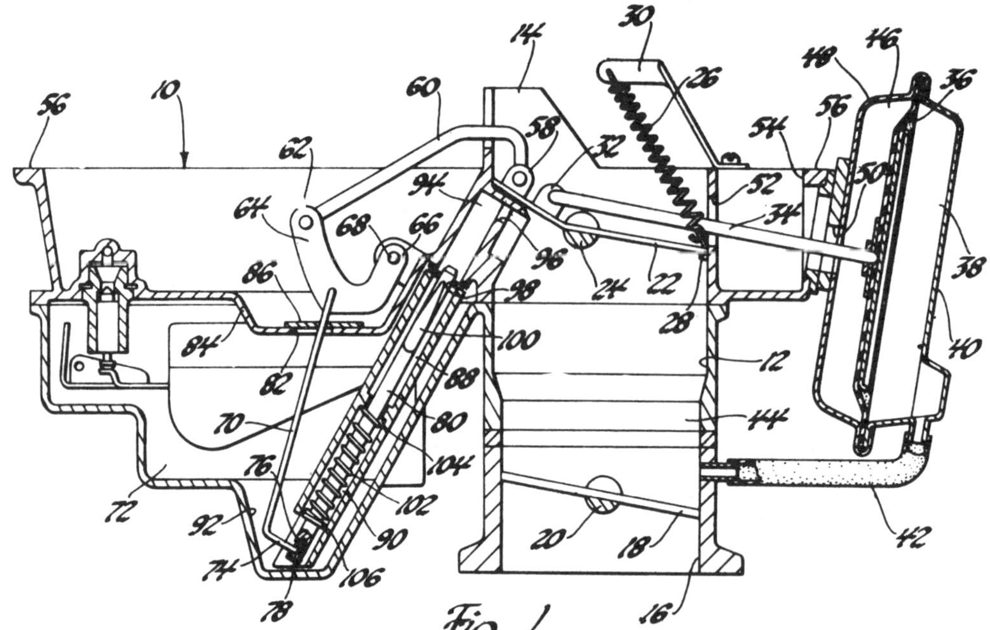

Fig. 1

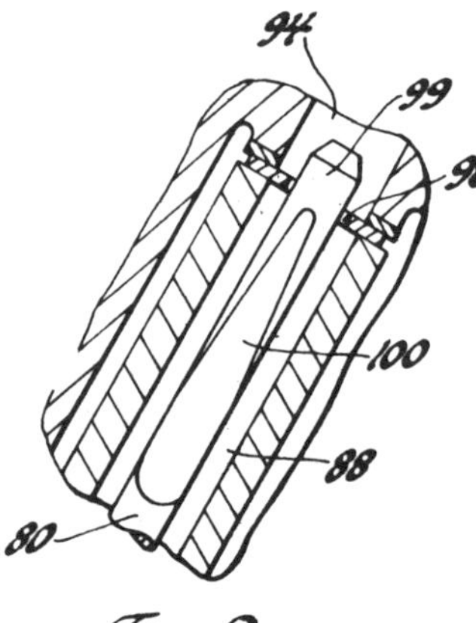

Fig. 2

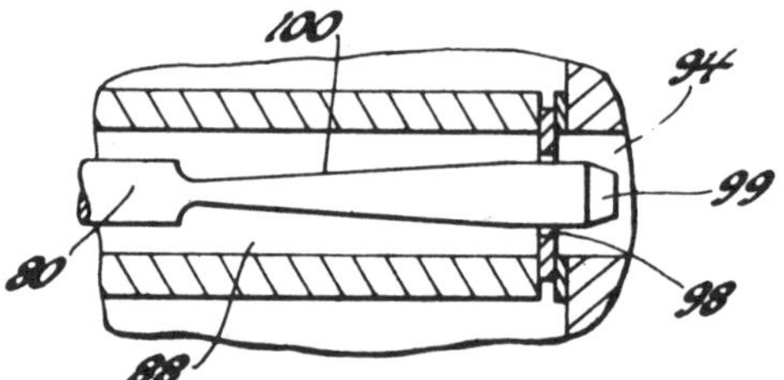

Fig. 3

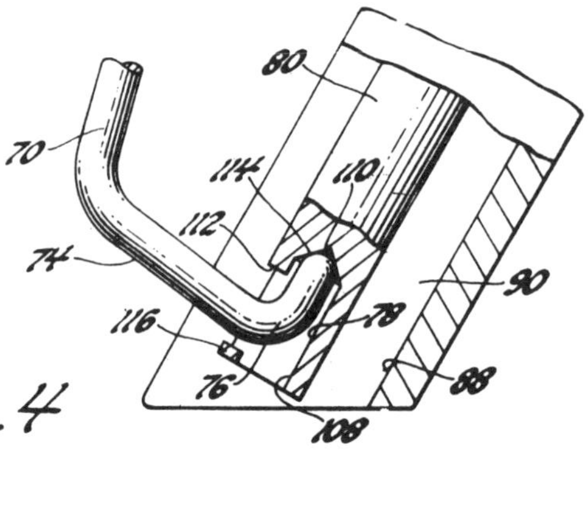

Fig. 4

United States Patent Office

3,851,021

Patented Nov. 26, 1974

1

2

3,851,021
CARBURETOR

John W. Moulds, Penfield, N.Y., assignor to General Motors Corporation, Detroit, Mich.

Filed July 26, 1973, Ser. No. 382,685
Int. Cl. F02m 7/22
U.S. Cl. 261—50 A 4 Claims

ABSTRACT OF THE DISCLOSURE

In an air valve carburetor, the metering rod hanger is received and retained in a socket formed in the metering rod.

This invention relates to an internal combustion engine carburetor and more particularly to a carburetor in which the metering rod is operated by a hanger received and retained in a socket formed in the base of the metering rod. The construction provided by this invention may be more easily assembled during production and repair operations than prior constructions and yet prevents accidental uncoupling during engine operation.

The details as well as other objects and advantages of this invention are set forth in the remainder of the specification and are shown in the drawings in which:

FIG. 1 is a sectional elevational view of the carburetor showing the basic metering linkage;

FIGS. 2 and 3 are enlarged views of the metering rod showing the configuration of the tapered metering portion; and

FIG. 4 is an enlarged view of the hanger-metering rod interconnection.

Referring first to FIG. 1, the carburetor **10** has a mixture conduit **12** including an air inlet **14** and a mixture outlet **16** which discharges to the engine. A throttle **18** is disposed in mixture outlet **16** in the usual manner on a throttle shaft **20**.

An air valve **22** is disposed in air inlet **14** on an air valve shaft **24**. A spring **26** is secured to the downstream edge **28** of air valve **22** and extends to a bracket **30** to bias air valve **22** to the position shown.

A tang **32** reaches upwardly from air valve **22** and is connected by a link **34** to a diaphragm **36**. A chamber **38**, formed between the right side of diaphragm **36** and a cover member **40**, is connected by a tube **42** to a region **44** of mixture conduit **12** defined between air valve **22** and throttle **18**.

A chamber **46**, defined between the left side of diaphragm **36** and a cover member **48**, is subjected to substantially atmospheric pressure, present in air inlet **14** and in the air cleaner (not shown), through openings such as **50**, **52** and **54**. (The air cleaner seats on a rim **56** disposed about the upper portion of carburetor **10**.)

In operation, chamber **38** is subjected to the subatmospheric pressure created in region **44** as throttle **18** is opened, and diaphragm **36** acts through link **34** to pull air valve **22** clockwise to an open position. Spring **26** is effective to balance the opening force of diaphragm **36**, thereby creating a substantially constant subatmospheric pressure in region **44**. By thus establishing a generally constant pressure drop across air valve **22**, the area about air valve **22** and thus the rotative position of air valve **22** is determined by and is a measure of the rate of air flow through mixture conduit **12**.

A tab **58** extends upwardly from air valve **22** and is connected through a link **60** to one end **62** of a lever **64**. The opposite end **66** of lever **64** is pivoted about a pin **68**. Intermediate ends **62** and **66**, a hanger **70** extends from lever **64** into the carburetor fuel bowl **72**. The lower end **74** of hanger **70** has a hook **76** which is received in a socket **78** formed in the lower end of a metering rod **80**.

It may be noted that hanger **70** extends through an opening **82** in the cover **84** for fuel bowl **72**. Opening **82** is closed by a slider **86** which shifts horizontally during movement of hanger **70**.

Metering rod **80** is disposed in a fuel passage **88** having its lower end **90** disposed to receive fuel from a well **92** formed in the bottom of fuel bowl **72**. The upper end **94** of fuel passage **88** has an opening **96** through which fuel is discharged into region **44** of mixture conduit **12**. It will be appreciated, therefore, that the fuel in fuel bowl **72** is subjected to a substantially constant metering head—from the substantially atmospheric pressure in the upper portion of the fuel bowl to the generally constant pressure in region **44**.

A metering jet or orifice **98** is disposed in fuel passage **88** around the tip **99** of metering rod **80**. As best shown in FIGS. 2 and 3, metering rod **80** has flat tapered surfaces **100** on opposite sides which, upon reciprocation of metering rod **80** in jet **98**, varies the area available for fuel flow through jet **98**.

In operation, as air valve **22** opens by clockwise rotation, link **60** rotates lever **64** in a clockwise direction. Lever **64** then lifts hanger **70** to move metering rod **80** generally upwardly and rightwardly in fuel passage **88**. Thus as air valve **22** is opened to increase the area available for air flow through air inlet **14**, metering rod **80** is shifted to increase the area available for fuel flow through metering orifice **98**. By this means, a substantially constant air-fuel ratio may be maintained—the precise proportion being controlled by the geometry of tapered surfaces **100** and of the linkage between air valve **22** and metering rod **80**.

A spring **102** extends from an annular ledge **104** formed in fuel passage **88** to the lower end **106** of metering rod **80** to take up any slack in the linkage and to load metering rod **80** against jet **98**.

It may be noted from FIG. 3 that the thickness of metering rod **80** increases from the end of surfaces **100** most closely adjacent passage inlet **90** to tip **99**. Tip **99** is therefore enlarged and assists in discharging fuel from fuel passage **88** as air valve **22** and metering rod **80** are moved to increase air and fuel flow. This offsets the greater inertia of the fuel which otherwise could create a mixture temporarily leaner than desired.

Referring now to FIG. 4, it may be observed that socket **78** comprises an elongated cylindrical recess or cavity **108** having a conical bearing surface **110** at its upper end and that a window **112** opens laterally into the lower end of cavity **108**. The hooked portion **76** at the lower end **74** of hanger **70** extends through window **112** and has a spherical tip **114** received by conical surface **110** of socket **78**. Spherical tip **114** engages conical surface **110** along a circular line which may translate slightly over tip **114** and surface **110** as the orientation of hanger **70** relative to metering rod **80** varies during operation.

While metering rod **80** could be secured to hanger **70** by the conventional method of wrapping a wire about both metering rod **80** and hanger **70**, the extra material and labor required by such a method is avoided by the construction shown in FIG. 4. Here the lower edge of window **112** is defined by a bar **116** extending below lower end **74** of hanger **70**. Assembly and intentional disassembly of metering rod **80** and hanger **70** is permitted by pivoting hooked portion **76** through window **112**, and rotation of hanger **70** relative to meterin g rod **80** (about tip **114**) is permitted by the length of window **112**. However, accidental uncoupling of metering rod **80** from hanger **70** during engine operation is prevented by bar **116** which precludes axial motion of metering rod **80** relative to hanger **70** sufficient to dislodge hooked portion **76** from socket **78**.

3,851,021

3

It is apparent from the foregoing that this invention provides a metering rod and hanger construction which is easily assembled and disassembled for manufacture and repair and yet which is prevented from accidentally uncoupling during engine operation.

The embodiments of the invention in which an exclusive property or privilege is claimed are defined as follows:

1. A carburetor comprising a mixture conduit, a fuel bowl, a fuel passage having an inlet receiving fuel from said fuel bowl and an outlet discharging fuel into said mixture conduit and extending upwardly from said inlet to said outlet, a metering orifice disposed in said fuel passage, a metering rod reciprocably disposed in said passage inlet, said metering rod having a portion disposed in said orifice for controlling fuel flow from said fuel bowl through said passage and said orifice to said mixture conduit, said metering rod having an axially extending cavity formed in its lower end and a window opening laterally into the lower end of said cavity, and a hanger disposed generally vertically in said fuel bowl and having a hooked portion at its lower end, said hooked portion being received through said window and extending into the upper portion of said cavity and supporting said metering rod for reciprocation within said fuel passage, said metering rod further having means defining a lower edge of said window to prevent accidental uncoupling of said metering rod from said hanger.

2. A carburetor comprising a mixture conduit, a fuel bowl, a fuel passage having an inlet receiving fuel from said fuel bowl and an outlet discharging fuel into said mixture conduit and extending upwardly from said inlet to said outlet, a metering orifice disposed in said fuel passage, a metering rod reciprocably disposed in said passage inlet, said metering rod having a portion disposed in said orifice for controlling fuel flow from said fuel bowl through said passage and said orifice to said mixture conduit, said metering rod having an elongated cylinrical recess extending into its lower end and a window opening laterally into the lower end of said recess, and a hanger disposed generally vertically in said fuel bowl and having a hooked portion at its lower end, said hooked portion being received through said window and extending into the upper portion of said recess and supporting said metering rod for reciprocation within said fuel passage, said metering rod further having a bar extending across and defining a lower edge of said window to prevent accidental uncoupling of said metering rod from said hanger.

3. A carburetor comprising a mixture conduit, a fuel bowl, a fuel passage having an inlet receiving fuel from said fuel bowl and an outlet discharging fuel into said mixture conduit and extending upwardly from said inlet to said outlet, a metering orifice disposed in said fuel passage, a metering rod reciprocably disposed in said passage inlet, said metering rod having a portion disposed in said orifice for controlling fuel flow from said fuel bowl

4

through said passage and said orifice to said mixture conduit, said metering rod having an elongated cylindrical recess extending into its lower end and a window opening laterally into the lower end of said recess, and a hanger disposed generally vertically in said fuel bowl and having a hooked portion at its lower end, said hooked portion being received through said window and extending into the upper portion of said recess and supporting said metering rod for reciprocation within said fuel passage, said upper portion of said recess being formed to define a conical surface and said hooked portion of said hanger terminating in a spherical surface received by said conical surface whereby the line of contact between said surfaces may translate during relative movement of said hanger and said metering rod.

4. A carburetor comprising a mixture conduit, a fuel bowl, a fuel passage having an inlet receiving fuel from said fuel bowl and an outlet discharging fuel into said mixture conduit and extending upwardly from said inlet to said outlet, a metering orifice disposed in said fuel passage, a metering rod reciprocably disposed in said passage inlet, said metering rod having a portion disposed in said orifice for controlling fuel flow from said fuel bowl through said passage and said orifice to said mixture conduit, said metering rod having an elongated cylindrical recess extending into its lower end and a window opening laterally into the lower end of said recess, and a hanger disposed generally vertically in said fuel bowl and having a hooked portion at its lower end, said hooked portion being received through said window and extending into the upper portion of said recess and supporting said metering rod for reciprocation within said fuel passage, said upper portion of said recess being formed to define a conical surface and said hooked portion of said hanger terminating in a spherical surface received by said conical surface whereby the line of contact between said surfaces may translate during relative movement of said hanger and said metering rod, said metering rod further having a bar extending across and defining a lower edge of said window to prevent accidental uncoupling of said metering rod from said hanger.

References Cited

UNITED STATES PATENTS

1,369,419	2/1921	Gibford	261—50 A
3,485,482	12/1969	Fuchs	261—50 A
3,745,983	7/1973	Sweeney	261—50 A
3,362,694	1/1968	Gould	261—50 A

TIM R. MILES, Primary Examiner

U.S. Cl. X.R.

261—Digest 38

Section II.
Vaporizer Type Carburetors

F. JOHNSTON & O. E. KELLUM.
CARBURETER.
APPLICATION FILED DEC. 9, 1912.

1,065,370. Patented June 24, 1913.

Fig. 1.

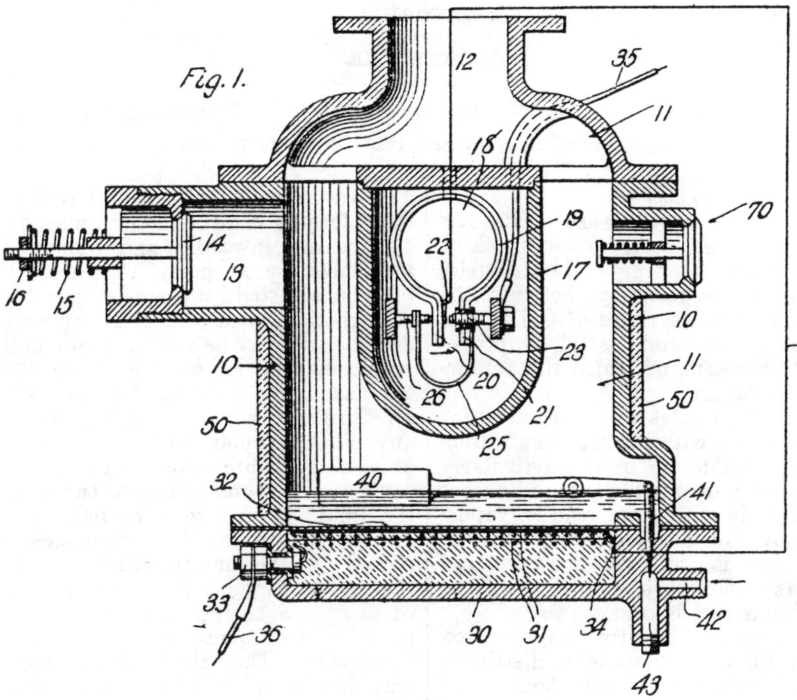

Fig. 2.

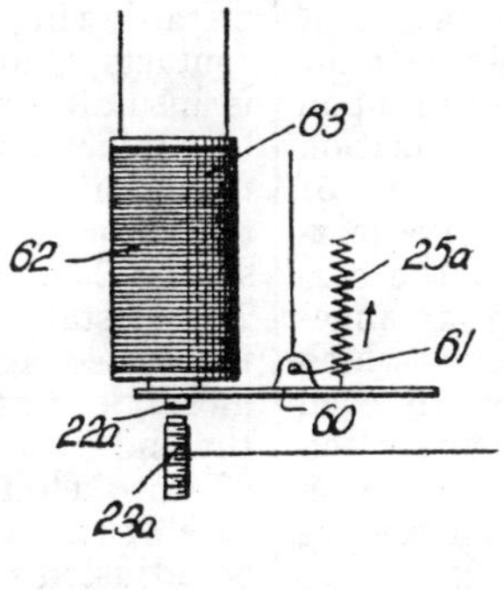

Witnesses
Edward H. Cankelew
D. M. Robarge

Inventors
Orlando E. Kellum, and
Fred Johnston.
by James T. Cankelew
their Attorney.

UNITED STATES PATENT OFFICE.

FRED JOHNSTON, OF LOS ANGELES, AND ORLANDO E. KELLUM, OF LONGBEACH, CALIFORNIA.

CARBURETER.

1,065,370. Specification of Letters Patent. Patented June 24, 1913.

Application filed December 9, 1912. Serial No. 735,832.

To all whom it may concern:

Be it known that we, Fred Johnston and Orlando E. Kellum, both citizens of the United States, residing at Los Angeles and Longbeach, respectively, both in the county of Los Angeles, State of California, have invented new and useful Improvements in Carbureters, of which the following is a specification.

This invention relates to carbureters for use in combination with internal combustion engines; and has to do more particularly with a carbureter in which the liquid fuel is heated so that it may more readily vaporize, thus providing for the use of heavier liquid fuel than is ordinarily the case, and providing for efficient and quick starting when the liquid fuel is cold.

Heretofore great difficulty has been experienced in the use of petroleum distillates heavier than gasolene; largely because of difficulty in obtaining free vaporization of the liquid. We propose to suitably heat the liquid fuel to a temperature depending upon the temperature at which free vaporization takes place, so that a sufficient amount of vapor is obtained to form an explosive mixture with the air passing through the carbureter. Preferably, the heating arrangement is controlled by a thermostat which will maintain the temperature at an even degree; but in some aspects our invention will consist in the combination of an electrical heating device with a carbureter as will be hereinafter set forth.

We have shown in the accompanying drawing the preferred form or forms of our device, typifying thereby other equivalent arrangements which may be used without departing from the scope of our invention.

In these drawings Figure 1 is a vertical section of our complete improved carbureter. Fig. 2 is a diagrammatic view illustrating a modified form of thermostatic arrangement.

In the drawings 10 designates the body of a carbureter containing the liquid fuel chamber. This liquid fuel chamber may be of any desired size and configuration, suitable for the apparatus to which it is designed to feed explosive mixture. Above the body 10 there is fitted the cap piece 11 with outlet 12, to which outlet the manifold or intake pipe of the engine will connect. The air inlet 13 to the liquid fuel chamber is preferably controlled by a suitable valve 14 opening inwardly and opposed in its movement by a spring 15 whose pressure may be adjusted by means of nut 16. This construction is of the ordinary and usual design; and may be varied to suit individual requirements. In fact, we have shown in our carbureter a means for merely passing air through the liquid fuel chamber; but any other common and well known means of passing the air through or over the liquid surface or around a nozzle through which the liquid sprays, may be had. Centrally located within the fuel chamber we have placed a thermostat chamber 17. This thermostat chamber is covered by a portion of cap piece 11 and is arranged to contain the thermostatic mechanism 18, somewhat as illustrated. The thermostatic mechanism may consist of a differentially expanding metallic ring 19 whose ends 20 and 21 carry contacts 22 and 23 respectively. Contact 23 is insulated and is adjustable with relation to contact 22 so that the movement of the thermostat required to close the circuit between the contacts may be varied. A suitable spring 25 is provided and a pressure screw 26 adjustably fixed on the end of the spring is pressed against the end 20 of the thermostat spring, to press the end 20 toward the end 21 and to press the contacts 22 and 23 into electrical engagement. The pressure of the spring 25 on the thermostat may be adjusted by adjustment of screw 26; and the gap between the two contacts 22 and 23 may also be adjusted; and, in this manner, the temperature at which the thermostat acts to close or open the contact switch may be adjusted to a nicety. Normally, when the carbureter is cold, the thermostat will be contracted and the contacts 22 and 23 in engagement with each other. They will remain in engagement with each other until the thermostat sufficiently expands to carry them apart. This will break the electrical circuit; and the arrangement is so made that the circuit will be thus broken when the temperature of the thermostat reaches the point at which it is desired to keep the whole carbureter and the liquid fuel therein.

Immediately below the liquid fuel chamber there is arranged the casing 30 which contains the heating element or elements 31.

These heating elements are preferably arranged flatly and as close as possible to the heat conducting sheet 32 (preferably copper) which separates the liquid fuel chamber from the chamber in casing 30 in which the heating elements are contained. The heating elements may be laid in asbestos cement or other heat insulating material, and be separated by mica or other insulating material from each other and from the copper plate 32. An insulating binding post 33 connects with one end of the heating element while the other end is grounded at 34 to the body of the carbureter. This grounded end at 34 will connect with the grounded contact 22 of the thermostat through the body of the carbureter, or by suitable wire connection shown diagrammatically at 34ª in Fig. 1. A suitable wire connection 35 connects with contact 23 of the thermostat, while a suitable wire connection 36 connects with binding post 33. Any sufficient source of electrical current may be utilized, to which wires 35 and 36 are connected.

The liquid fuel chamber is provided with a pivoted float which controls liquid fuel inlet valve 41. This valve may be arranged in any of the manners common to the art, the liquid fuel inlet being at 42 and there being provided a well 43 below the valve opening top for catching any sediment which might otherwise pass into the carbureter.

The carbureter may be supplied with fuel, and being cold, as soon as the electrical current is supplied to wires 35 and 36, the heating elements start to increase the temperature of the fuel and of the carbureter generally. As soon as the fuel has been heated to a sufficient temperature to vaporize and for the heat being thus distributed throughout the chamber, the thermostatic arrangement 18 becomes heated and will open the circuit between contacts 22 and 23. This immediately stops the action of the heating element until the temperature falls sufficiently to cause the closing of the circuit between the contacts 22 and 23. In this manner the temperature of the carbureter is kept fairly constant at a point where the liquid fuel will freely vaporize. The vapor arising from the liquid fuel is carried and mixed with the air passing through inlet 13 and through outlet 12.

We may use fuels of extremely heavy natures with this our improved device for the reason that it is only necessary to heat such fuels to a sufficient temperature to produce their vaporization. Heretofore, gasolene and similar like products have only been used in carbureters for the reason they alone would freely vaporize at ordinary temperatures. But even such light fuels have been found to vaporize with difficulty when first starting an engine when the temperature is low; and our device is not only useful for utilization of heavy fuel oils; but it is useful in initially heating a carbureter for starting with the lighter fuels, the current being turned off from the carbureter as soon as it is warmed up and started.

In Fig. 2 we have shown a modified form of the thermostatic device illustrating this form as typical of the departures we may make from the specific form shown in Fig. 1. In the form shown in Fig. 2 there is an armature 60 pivoted at 61 and pulled in the direction indicated by the arrow by a spring 25ª. Contact 22ª is carried by the armature and the contact 23ª is adjustable to engage with the contact 22ª. A suitable magnet 62 attracts the armature to open the circuit between contacts 22ª and 23ª. And this magnet is wound with wire 63 of such material as changes its resistance greatly upon the change in temperature. As the temperature rises, the resistance of the wire 63 increases and the magnetic traction due to the current flowing in the wire also decreases. The spring 25ª can be arranged in its amount of tension to pull the armature away from the magnet at any suitable temperature, and to close the circuit between contacts 22ª and 23ª.

In Fig. 1 we have shown a small safety valve 70 which may be used if desired. This valve will allow escape of vapors which may be generated from highly volatile fuels, such as gasolene, when the engine has been suddenly stopped and the carbureter is still warm. In heavier fuels, such as kerosene, for which this device is particularly designed, there may be no use for this valve. Our carbureter is of particularly great utility for kerosene.

Having described our invention, we claim:

1. A carbureter, comprising a chamber for containing liquid fuel, a heat conducting bottom for said chamber, an electrical heating element immediately beneath the heat conducting bottom, a float controlled valve for maintaining a constant amount of liquid fuel in the lower part of the chamber, a thermostat chamber situated entirely within the liquid fuel chamber above the liquid level therein, a thermostat in the chamber, and a switch operated by the thermostat and electrically connected with the electrical heating element to control the supply of current thereto.

2. A carbureter, comprising a liquid fuel chamber, an electric heating element in proximity thereto, means for introducing liquid fuel to said chamber, outlet means for vapors from the liquid fuel, a thermostat within said chamber, and a switch operated by the thermostat and electrically connected to the heating element to control supply of current thereto.

3. A carbureter, comprising a liquid fuel chamber, means for passing air through the chamber over the fuel therein, a heat con-

ducting bottom for the chamber, an electric heating element below the heat conducting bottom, a thermostat chamber within the fuel chamber, a thermostat therein, an electric switch operated by the thermostat, and an electrical circuit including the heating element and the switch in series.

In witness that we claim the foregoing we have hereunto subscribed our names this 2nd day of December 1912.

FRED JOHNSTON.
ORLANDO E. KELLUM.

Witnesses:
ELWOOD H. BARKELEW,
JAMES T. BARKELEW.

F. A. OSOWSKI.

PROCESS OF MAKING GAS.

APPLICATION FILED SEPT. 28, 1917.

1,259,898. Patented Mar. 19, 1918.

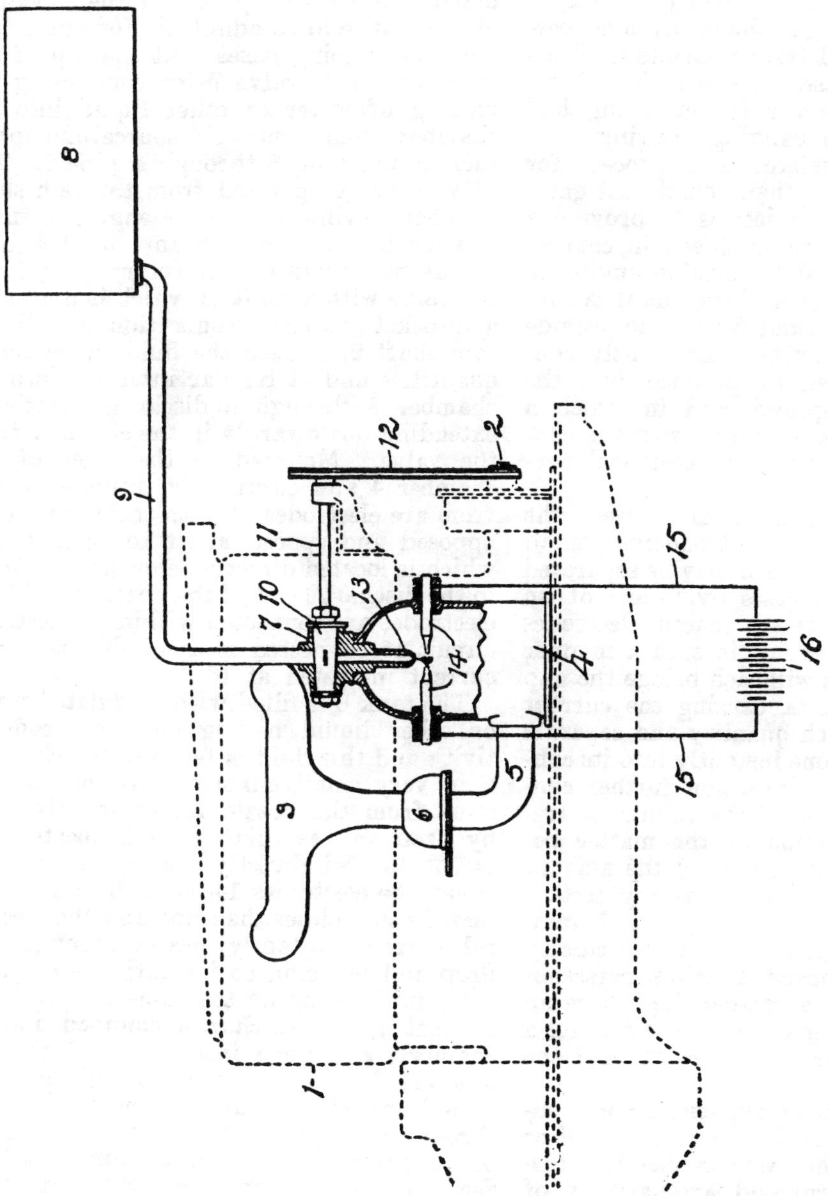

Inventor

Witness

Charles Balg

Chas. W. Stauffiger

Felix A. Osowski

By [signature]

UNITED STATES PATENT OFFICE.

FELIX A. OSOWSKI, OF DETROIT, MICHIGAN.

PROCESS OF MAKING GAS.

1,259,898. Specification of Letters Patent. Patented Mar. 19, 1918.

Application filed September 28, 1917. Serial No. 193,644.

To all whom it may concern:

Be it known that I, Felix A. Osowski, a citizen of the United States of America, residing at Detroit, in the county of Wayne and State of Michigan, have invented certain new and useful Improvements in Processes of Making Gases, of which the following is a specification, reference being had therein to the accompanying drawing.

This invention relates to a process for separating fluids into their constituent gases electrically, and its object is to provide a process which may be employed in connection with an internal combustion engine to supply the gaseous fuel thereto as it is consumed in the engine, and further, to provide a process which may be conveniently controlled or regulated to produce just the amount of gas required, and in which a maximum of gas is produced with the consumption of a minimum of electrical current.

With these and other ends in view, the invention consists in introducing small quantities of matter which may be separated into its constituent gases by means of an electric current, between spaced electrodes within an electric circuit in such a manner that said quantities will each bridge the gap and close the circuits, causing the current to pass through each quantity and separate or transform the same instantly into its constituent gases. The invention further consists in the regulation of the volume or frequency of introduction of the matter between the electrodes to control the amount of gas produced, and in providing certain other new and useful features, all as hereinafter more fully described and particularly set forth in the appended claims, reference being had to the accompanying drawing which is a diagrammatic view illustrating a method of and means for carrying out the invention.

While a specific arrangement and construction is shown in the drawing, it will be understood that the same is merely illustrative of one form and arrangement of apparatus for carrying out my process and that I do not limit myself to any particular apparatus or arrangement thereof for performing my invention.

In the drawing 1 indicates an internal combustion engine, 2 the usual cam shaft thereof and 3 the intake manifold. A gas chamber 4 of any suitable form or construction is located adjacent the manifold and is connected thereto by a pipe 5, an air inlet valve 6 of any suitable form being interposed between this pipe and the inlet end of the manifold to admit air for admixture with the ingoing gases. At the top of the chamber 4 is a valve 7 for controlling the feeding of water or other liquid into the chamber from a suitable source of supply, such as the tank 8 through a pipe 9. The valve may be operated from the cam shaft or other moving part of the engine in timed relation to the engine in any suitable manner as by providing the rotary plug 10 of the valve with a shaft 11 which is driven by a sprocket and chain connection 12 with the cam shaft 2, to feed the fluid in measured quantities and at regular intervals into the chamber 4 through a discharge nozzle 13 extending downwardly in the chamber from the valve. Mounted in the sides of the chamber 4 and electrically insulated therefrom are electrodes 14 with their inner ends opposed and spaced apart to form a gap which is located directly below and adjacent to the discharge end of the nozzle 13. These electrodes are connected within an electrical circuit 15 connected to a suitable source of current indicated at 16.

The tank 8 is filled with acidulated water or other liquid having electrical conductivity and this fluid is fed into the chamber 4 in very small quantities so that it will issue from the nozzle 13, preferably drop by drop and as this nozzle is located adjacent to and directly above the gap between the electrodes 14, each drop as it enters the gap closes the same and the electrical current instantly passes through the drop and in doing so instantly decomposes the liquid, changing the same into its constituent gases which are confined in the chamber, and upon inspiration of the engine are drawn therefrom through the pipe 5 and manifold 3 and in passing therethrough are mixed with the proper amount of air drawn in through the air inlet 6 to form combustible charges. By regulating the size and frequency of the water charges by means of the valve 7, the necessary volume of gas to supply the engine will be made just as it is used, each drop of liquid being converted into gas instantly so that none will remain in the chamber and the gas so made being immediately drawn from the chamber and consumed in the engine.

The electrodes are spaced just far enough apart to prevent the current from jumping across and so that each charge of liquid when it drops between them will complete the circuit by closing the gap. The circuit is therefore normally open and current is consumed only during those intervals in which the gap is closed by the drops or small quantities of liquid. A minimum of current is therefore required and as the volume of liquid is so very small and the current is concentrated therein, each charge or drop is instantly decomposed or converted into gas by a current of comparatively low voltage and amperage.

Obviously any desired number of pairs of electrodes and water feeds may be employed if it be found that a sufficient volume of gas to supply the particular engine is not produced by one pair of electrodes.

This process of separating liquids or other matter into constituent gases, comprises the introduction of small quantities of matter between spaced terminals of an electric circuit, to close the circuit and cause decomposition and liberation of the gases by the momentary passage of the current therethrough, and it is not limited to any particular means for carrying out the process or any particular construction or arrangement of apparatus.

What I claim is:—

1. The process of making gases which consists in introducing limited quantities of matter between the spaced terminals of an electric circuit to cause current to pass through and instantly liberate the gases of each quantity introduced.

2. The process of converting matter into its constituent gases which consists in causing a limited quantity of the matter to form the electrical connection between the terminals of a normally open electric circuit.

3. The process of converting liquids into their constituent gases which consists in causing the liquid in separate small amounts to intermittently pass into the gap between spaced terminals of an electric circuit and form the conductor for the current therebetween.

4. The process of making gas electrolytically which consists in intermittently feeding limited quantities of matter between spaced terminals of a normally open circuit to cause each quantity of matter to be converted into its constituent gases by the momentary passage of the current.

5. The process of making gas by electrolytically changing liquids into their constituent gases which consists in providing an electric circuit with terminals spaced to provide a gap in the circuit and feeding small bodies of the liquid into the gap.

6. The process of making gas by electrolytically changing liquids into their constituent gases which consists in providing an electric circuit with spaced terminals and feeding liquid drop by drop into the space between the terminals, each drop forming a conductor for the current between the terminals.

7. The process of making gas by electrolytically changing liquids into their constituent gases which consists in feeding the liquid in small bodies intermittently, into a gap between terminals of an electric circuit to momentarily close the circuit by each body of liquid and cause the body to be instantly decomposed by the passage of current therethrough.

8. The process of making gas by electrolytically decomposing liquid, which consists in providing an electric circuit with terminals spaced to form a gap in the circuit and provide a normally open circuit, and intermittently feeding liquid into the gap in bodies of sufficient size to bridge the gap.

9. The process of electrolytically producing gas from water which consists in providing a normally open circuit having a gap between terminals thereof and feeding acidulated water intermittently, drop by drop into the gap within a chamber.

10. The process of electrically producing gas from water which consists in providing an electric circuit having spaced terminals forming a gap, acidulating water, and measuring and intermittently feeding the water, drop by drop, between the terminals to bridge the gap and cause the current to flow through and decompose each drop separately.

In testimony whereof I affix my signature in the presence of two witnesses.

FELIX A. OSOWSKI.

Witnesses:
ANNA M. DORR,
KARL H. BUTLER.

G. R. HOLLAND.
FUEL MIXER.
APPLICATION FILED NOV. 8, 1917.

1,281,962. Patented Oct. 15, 1918.

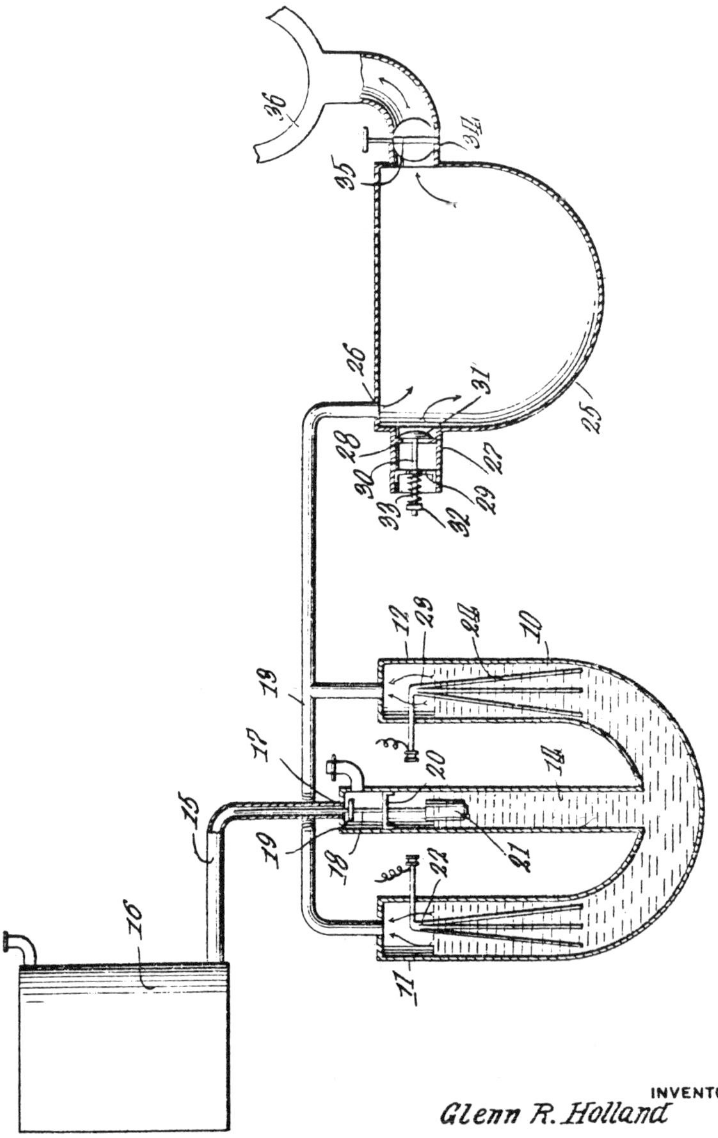

INVENTOR
Glenn R. Holland

WITNESSES
Guy M. Spring
Wm Zeaman

BY Richard B. Owen.
ATTORNEY

UNITED STATES PATENT OFFICE.

GLENN R. HOLLAND, OF GENESEE, ILLINOIS.

FUEL-MIXER.

1,281,962. Specification of Letters Patent. Patented Oct. 15, 1918.

Application filed November 8, 1917. Serial No. 201,001.

To all whom it may concern:

Be it known that I, Glenn R. Holland, a citizen of the United States, residing at Genesee, in the county of Henry and State of Illinois, have invented certain new and useful Improvements in Fuel-Mixers, of which the following is a specification.

This invention has relation to gas generating apparatus, with particular reference to the use of such apparatus in connection with automobiles and internal combustion engines, and has for an object to provide, an apparatus for generating gas by the electrical decomposition of water so as to form a source of fuel supply for the engine.

Another object of the invention is to provide a gas generating apparatus of the nature above set forth including a generating chamber with a valve controlled fluid supply therefor, and a mixing chamber in communication with the gas generating chamber having a valve to permit the introduction of predetermined quantities of fresh air with the combustible gases, to dilute the same whereby they may be fed in a diluted condition to the internal combustion engine for explosion in its chambers.

In addition to the foregoing my invention comprehends improvements in the details of construction and arrangement of the parts to be hereinafter more fully described and particularly set forth in the appended claim.

With reference to the drawings, forming a part of this application, in which I have illustrated in section a gas generating apparatus, constructed in accordance with my invention, 10 indicates the gas generating apparatus generally in the form of a U-shaped tube, the columns 11 and 12 of which are directed upwardly and formed with openings at their upper ends which are connected to a supply pipe 13. At the intermediate or bight portion of the gas generating apparatus proper, a tube 14 is erected vertically in communication with the interior thereof, and in communication at its upper end by means of a pipe 15 with a container 16, mounted at an elevation above said tube 14. A valve seat 17 is formed at the upper end of the tube 14, and a valve stem 18 is movable within the pipe 15 with a valve 19 secured thereto for engagement upon the seat 17, the valve stem being guided in a bearing 20, and having a float 21 at its lower end.

Electrodes 22 and 23 are provided in the gas generating apparatus, one in each vertical column 11 and 12, each electrode being in the form of downwardly diverging fingers 24, with the upper end of the electrode extending laterally through an opening in the column for connection to a source of current supply, (not shown).

A mixing chamber 25 is provided with an opening 26 to which the pipe 13 is connected. An intake branch 27 is also provided in said mixing chamber and formed with a valve seat 28 and a guide 29 in said branch. The guide supports for movement a valve stem 30 having a valve 31 thereon, for engagement upon the valve seat 28, the outer end of the stem being threaded to receive a nut 32. A coil spring 33 is then interposed between said nut and the guide 29 whereby to lightly retain the valve 31 upon its seat.

An outlet branch 34 is provided for the mixing chamber at a point opposite the intake branch 27, said outlet branch being controlled by means of a throttle valve 35 and adapted for connection to the intake manifold 36 of an internal combustion engine.

In operation, the container 16 is filled with water which flows therefrom, through the pipe 15 into the tube 14 of the gas generating apparatus, filling said tube and the columns 11 and 12 to a point adjacent the upper end thereof. When the level of the water reaches the float 21 the valve 19 is held upon its seat and the further supply of water discontinued. The water is then rendered conductive to electricity by dissolving potassium hydroxid, or other chemical for a similar purpose, and the terminals of the source of current supply of high amperage are connected to the electrode 22 and 23. Oxygen and hydrogen gases are then evolved by decomposition of water and conducted through the pipe 13 to the mixing chamber 25. At each intake stroke of the engine a certain portion of the mixed hydrogen and oxygen gases are drawn, together with a quantity of fresh air into the intake manifold of the internal combustion engine and thence into the cylinders for explosion.

The tension of the spring 33 is adjusted by means of a nut 32 so as to regulate the proportion of fresh air to mixed gases. The throttle valve 35 may also be adjusted so as to regulate the quantity of gases admitted to the intake manifold. As the level of water in the gas generating apparatus falls through the electrical action thereon, it is replenished from the container 16 the float 21 falling with the level of the water so as to slightly unseat the valve 19.

Thus it will be seen that I have provided an apparatus which is inherently capable of generating gas from easily obtained material, the apparatus being especially desirable in automobiles using internal combustion engines, the electrical generator being utilizable to generate the current for the decomposition of water. This apparatus may be used in connection with the regulation carbureter thereby permitting the use of gaseous fuel formed partly of hydrocarbon saturated air and properly mixed hydrogen and oxygen gases with a requisite portion of fresh air to dilute the hydrogen and oxygen gases, and to form an explosive mixture for the carbureted air. As the combustion of hydrogen and oxygen results in the production of a high temperature it will readily be seen that the efficiency of an internal combustion engine will be greatly increased due to an increased expansive force within the cylinder, not to mention the desirable effect in ridding the cylinders of deposits of carbon and to retain them in a clean condition.

While I have illustrated and described my invention with some degree of particularity, I realize that in practice various alterations thereover may be made, and I therefore desire to reserve the right and privilege of changing the form of the details of construction, or otherwise altering the arrangement of the correlative parts without departing from the spirit of the invention or the scope of the appended claim.

Having thus described my invention what I claim as new and desire to secure by Letters Patent is:—

In a gas generating apparatus, the combination with a U-shaped water-container having a tube upstanding from its bend between its columns, the upper ends of the latter being closed, a pipe leading from said upper ends, and pendant electrodes within the columns; of a source of water supply, a pipe leading from the same into the upper end of the intermediate tube, a control valve for said pipe, and a float within said tube connected with the valve, for the purpose set forth.

In testimony whereof I affix my signature in presence of two witnesses.

GLENN R. HOLLAND.

Witnesses:
FRANK HOLLAND,
THOMAS H. TAYLOR.

Copies of this patent may be obtained for five cents each, by addressing the "Commissioner of Patents, Washington, D. C."

Dec. 9, 1924. | 1,518,572

H. V. CROON

VAPORIZER AND CARBURETOR

Filed Jan. 21, 1922

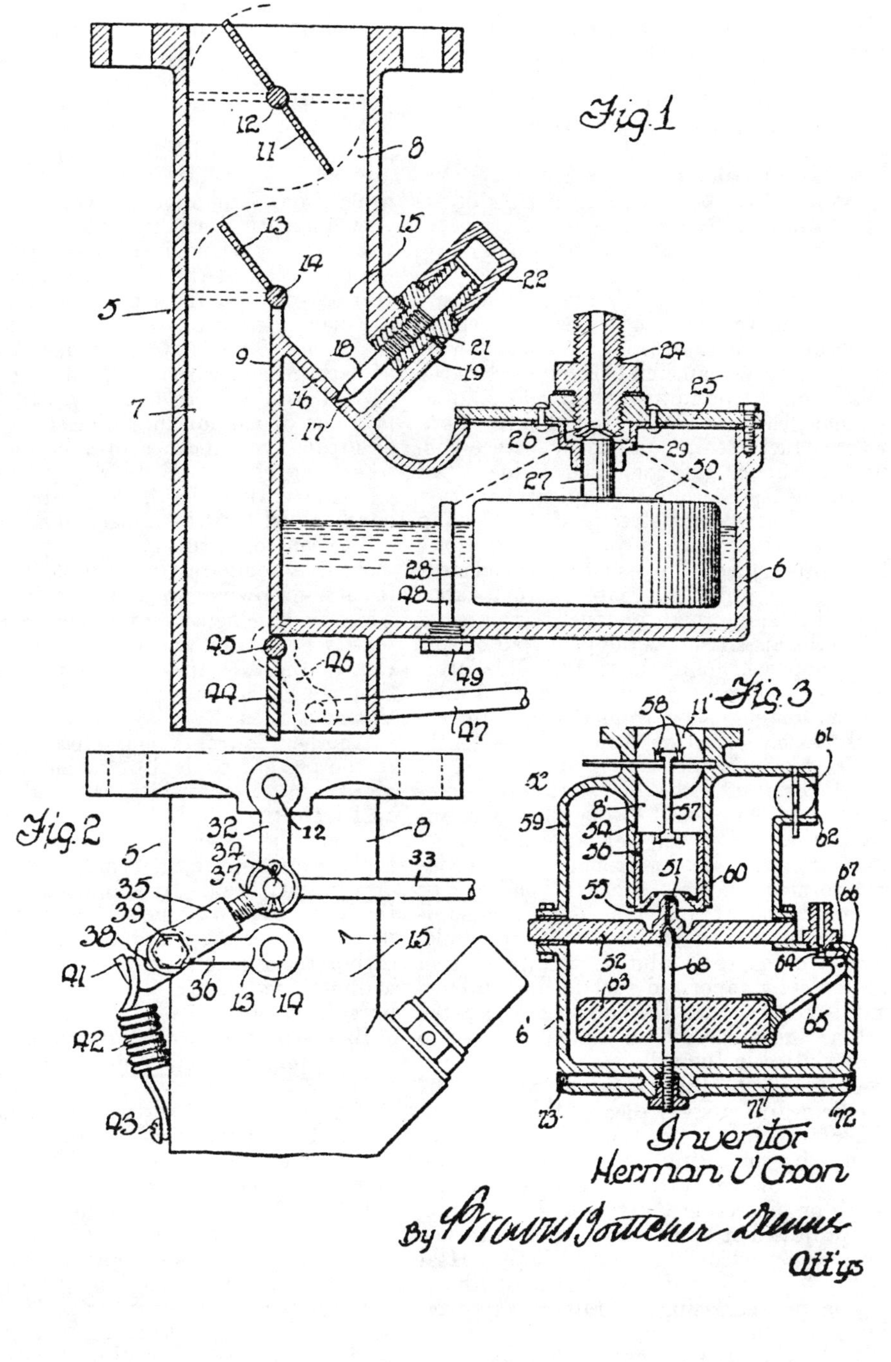

Patented Dec. 9, 1924. **1,518,572**

UNITED STATES PATENT OFFICE.

HERMAN V. CROON, OF CHICAGO, ILLINOIS.

VAPORIZER AND CARBURETOR.

Application filed January 21, 1922. Serial No. 530,761.

To all whom it may concern:

Be it known that I, HERMAN V. CROON, a citizen of the United States, residing at Chicago, in the county of Cook and State of Illinois, have invented a certain new and useful Improvement in Vaporizers and Carburetors, of which the following is a full, clear, concise, and exact description, reference being had to the accompanying drawings, forming a part of this specification.

My invention pertains to vaporizing carburetors.

The theory upon which the present carburetor and vaporizer operates is an entire departure from the theory of operation of practically all commercial carburetors of the present day. Heretofore, the feed of liquid fuel has generally been in liquid spray form from a fuel feeding nozzle or orifice discharging under the combined influence of the engine suction and the aspirating action of the high velocity stream of air flowing past the fuel nozzle. In the present instance the fuel is subjected to a suction of sufficient intensity to produce a gaseous fuel vapor from the liquid fuel, this fuel vapor being fed to the air stream as the principal or sole fuel component of the mixture. This latter method produces a completely gasified mixture as distinguished from the wet, imperfectly formed mixture which so frequently results from the projection of the fuel in liquid form into the air stream. Consequently, by the present theory there is avoided a large part of the well known carburetion difficulties arising from these improperly formed mixtures, such as "loading" or accumulations of condensation on the manifold walls; lack of responsiveness to acceleration; crank case dilution resulting from incompletely vaporized fuel failing to burn and descending into the crank case past the piston; and other difficulties well known to those skilled in the art.

According to the preferred manner of practicing the present theory, the vaporization of the fuel by suction may be accomplished by subjecting a substantially quiescent body of fuel to the suction; or by agitating or projecting the fuel and subjecting this projected or moving body of fuel to suction; or by a combination of both. Heat may also be employed in connection with the suction for increasing the rate of vaporization of the fuel.

In the embodiments hereinafter shown, it will be noted that the proper production of vapor according to this theory is not dependent in any sense upon the maintenance of a strictly accurate fuel level. Where the vaporization is from a substantially quiescent body of fuel it is only necessary to maintain a body of fuel with a sufficiently extensive area or free vaporizing surface to insure the generation of the requisite volume of vapor. Any fluctuation or tilting of the level of this vaporizing fuel surface does not influence the operation of the carburetor, as the generation of vapor is not materially affected by such fluctuations or inclinations of the fuel level. This is of advantage in all fields of carburetion, where inclination or change of fuel level will frequently modify the proportions of the mixture undesirably, but is of greatest significance in aviation practice where the motion and the various maneuvers of the aircraft make the maintenance of any definite fuel level extremely difficult.

The control of this type of vaporizing carburetor is comparatively simple and certain in action, owing to the definite relation which will generally exist between the intensity of suction and the rate of vaporization of the fuel, this rate of fuel vaporization being primarily responsive to the intensity of suction.

Another desirable characteristic accruing to the present method of carburetion is the ability to automatically raise or draw the fuel from the supply tank to the carburetor without the intervention of vacuum feed tanks, a pressure feed from the supply tank, or a gravity head. By subjecting the fuel in the vaporizer to a suction of a vaporizing degree sufficient suction is available at the carburetor during all engine speeds to afford a direct suction lift of the fuel to the carburetor.

In the accompanying drawings illustrating a preferred embodiment of my invention:

Figure 1 is a vertical sectional view through the present carburetor;

Fig. 2 is a fragmentary side elevational view of the same; and

Fig. 3 is a vertical sectional view of a modified construction.

The vaporizer and carburetor is embodied in a casting 5 comprising a fuel vaporizing chamber 6 and an air passageway 7. The passageway 7 is illustrated as being vertical, but it will be obvious that the carburetor may be of the horizontal tube type as well. The air passageway 7 opens into a carbureting chamber 8 of greater diameter, the air passageway 7 being preferably formed by dividing the tubular chamber 8 by a vertical wall or partition 9, extending down to the bottom of the vaporizing chamber 6. The discharge of combustible mixture from the carbureting chamber 8 is controlled by a conventional throttle valve 11 mounted on a transverse shaft 12. The discharge of air into the carbureting chamber 8 is controlled by a throttle or swinging flap valve 13 which is also mounted on a control shaft 14 extending through the side walls of the casting. The two valves 11 and 13 are connected together for joint operation in such a manner as to maintain a vaporizing suction in the carbureting chamber 8, as I shall presently describe.

On the other side of the wall or partition 9 a vapor passageway 15 extends down into proximity to the top of the vaporizing chamber 6. This vapor passageway is separated from the interior of the vaporizing chamber 6 by an inclined wall 16 having a vapor feeding port 17. The discharge of vapor through this port is controlled by a needle valve 18 which projects diagonally out of the carburetor through a diagonal enlargement or boss 19. The valve 18 may have direct screw threaded engagement in the boss 19, but is preferably threaded in a bushing 21 which is in turn threaded into the boss 19. A cap 22 threads down over the bushing 21 and seals the needle adjustment against leakage of suction. In lieu of the arrangement shown, it will be obvious to those skilled in the art that this needle valve may be arranged for manual control from the dash or steering column through the intervention of a suitable control rod extending thereto, or this needle valve may be arranged to operate simultaneously with the throttle valve through any well known system of linkage for controlling the size of the vapor port 17 in accordance with the position of the throttle valve.

The fuel is supplied to the vaporizing chamber 6 through an inlet plug 24 which screws into the removable cover 25 closing the top of the chamber 6. The supply pipe is adapted to have threaded connection in an obvious manner with the upper end of the plug, 24, the lower end of the bore of this plug constituting an inlet port 26 adapted to be controlled by a needle valve 27 and float 28. The stem of the needle valve 27 is guided in the hub of the guide bracket 29 secured to the under side of the cover 25. The inlet port 26 and the end of the needle v 've 27 may be formed on a relatively obtuse taper. In the operation of the carburetor a relatively high suction prevails above the fuel level in the vaporizing chamber 6 and this causes any fuel entering through the port 26 to be projected into the chamber at a relatively high velocity owing to the considerable difference of pressure between the chamber 6 and the supply tank. This fuel is projected by the obtuse taper on the end of the needle valve 27 into a conical curtain of liquid fuel broken up into a finely subdivided state. This is an alternative feature which may or may not be employed to augment the volume of fuel vapor obtained from the free vaporizing surface in the chamber 6.

The requisite suction for obtaining the aforesaid vaporizing action is obtained by linking the valves 11 and 13 so that the throttle valve 11 will have a certain lead of opening over the air valve 13. The purpose of this is to maintain in the carbureting chamber 8 and passageway 15 a vacuum or suction of sufficient degree to vaporize the fuel in the chamber 6 throughout substantially the entire range of engine speeds. As shown in Fig. 2, the outer end of the throttle shaft 12 carries an operating arm 32 which has pivotal connection with an operating rod 33 extending to the control mechanism, such as the accelerator pedal and steering column control lever. The pivot 34 has connection through a link 35 with an operating arm 36 on the projecting end of the air valve shaft 14. An eye 37 engages over the pivot 34 and is screw threaded for reception in the threaded socket in the end of the link 35, whereby the link may be conveniently extended or shortened. The end of the operating arm 36 is extended down diagonally to receive a slot 38 through which extends a bolt 39 passing through the end of the link 35. The engagement of the bolt 39 in the slot 38 affords a lost motion connection between the link 35 and operating arm 36, whereby the throttle valve 11 is opened to a limited extent prior to opening movement of the air valve 13. The end of the operating arm 36 is provided with a hook or lug 41 over which engages one end of a tension spring 42, the other end of this spring being secured to the body of the carburetor at 43.

The admission of air to the air passageway 7 is controlled by a choke valve 44 which swings on a transverse valve shaft 45. The outer end of this shaft has an operating arm 46 connecting through a rod 47 with any suitable dash or steering column control. The needle valve 27 has no pivotal or fixed connection with the float 28 but merely rides on a plate 50 on the top of the float. This permits the float to incline or shift under change of level without binding the needle valve. The float is confined against shifting entirely out from under the needle valve by a guide post 48 which projects upwardly from a removable nut or plug 49 screwing into the bottom of the vaporizing chamber. The removal of the plug 49 affords a convenient drain outlet in the bottom of the vaporizing chamber 6 for draining the liquid contents thereof.

In the operation of the device, the cooperative relation between the valves 11 and 13 always maintains a relatively high suction in the carbureting chamber 8 and the vapor passageway 15. When the throttle 11 is initially opened, the air valve 13 lags behind the opening movement of the throttle to such an extent that the volume of air admitted from the passageway 7 is insufficient to satisfy the suction in the carbureting chamber 8, with the result that a relatively high suction prevails in this chamber and the vapor passageway 15 at this time. Similarly, when the throttle 11 has been moved to its full open position, the flow of air from the passageway 7 is still insufficient to completely satisfy the suction in the carbureting chamber 8, with the result that a relatively high suction is also maintained in the carbureting chamber at this time. This high suction is transmitted through the vapor port 17 into the vaporizing chamber 6. Here it lowers the boiling point of the liquid fuel and results in the relatively broad surface of fuel giving up a considerable volume of fuel vapor. This vapor is drawn out through the port 17 under the control of the needle valve 18 and, passing upwardly through the passageway 15, mingles with the air stream entering the carburetion chamber 8 from the air passageway 7. The fuel which is thus supplied to the air stream is an ideal vapor completely vaporized and devoid of liquid particles which form such a large part of the discharge from the conventional aspirating fuel jet, and which cause the loading, deposition of carbon and crank case dilution hereinbefore mentioned. If desired, the emission of vapor may be augmented by arranging the entering fuel to discharge in a finely subdivided state whereby the suction in the chamber 6 will have intimate contact with the fuel for vaporizing the same. It will be obvious that the high suction in the upper part of the chamber 6 will insure a reliable suction feed of the fuel from the supply tank to the carbureter at all engine speeds.

In Figure 3 I have illustrated a modified construction wherein the fuel vapor is discharged from a jet or nozzle over which sweeps the air stream in its flow through the carbureting chamber. The carbureting chamber 8′ has a vapor nozzle 51 projecting into the lower end of the chamber, this nozzle extending upwardly from an intermediate plate 52 which closes the bottom of the carbureting chamber 8′ and the top of the fuel vaporizing chamber 6′. The complete carburetor assembly in this form shown in Figure 3 consists of the upper casting 53, the plate 52 and the bowl 6′ forming the fuel vaporizing chamber. The carbureting chamber 8′ is defined in a sleeve 54 constituting part of the casting 53, and this sleeve terminates short of the plate 52 so as to provide an annular air admission port 55 to the bottom of the carbureting chamber 8′. This annular port is below the tip of the vapor nozzle 51 so that the high velocity stream of air entering through this port will sweep upwardly and laterally across the tip of the nozzle 51, a venturi baffle 60 aiding in directing the flow of air, thereby adding the velocity suction of the air stream to the static suction of the carbureting chamber in producing a high aggregate suction available at the tip of the nozzle. An adequate static suction is maintained in the carbureting chamber 8′ by a sleeve valve 56 which reciprocates either inside or outside of the sleeve 54 and controls the size of the air admission port 55. A link 57 has pivotal connection with this sleeve valve 56 and also with lugs 58 which extend from the throttle 11′. As a result of this arrangement, operation of the throttle also operates the air admission valve 56 to control the suction in the carbureting chamber, analogously to the operation of the valves 11 and 13 of the preceding embodiment. A choker control of the air supply may be obtained by providing a jacket 59 to surround the air admission port 55, and arranging a choker valve 61 in the intake 62 to this jacket.

As previously remarked, the high suction transmitted into the vaporizing chamber 6′ through the nozzle 51 results in the fuel feeding into the chamber 6′ under a considerable head. To insure complete closure of the fuel intake valve when the level reaches a relatively high point, I have arranged a multiplied leverage between the float 63 and the inlet valve 64. An arm 65 extends from the float and has pivotal connection at 66 in the vaporizing chamber, this arm having a short extension 67 which operates to control the seating or unseating of the valve 64. This valve and its cooperating valve seat may be tapered to secure the fuel dispersing action previously described, if desired. A needle valve 68 extends upwardly from the bottom of the vaporizing chamber, through a hole in the float 63, and controls the size of the port in the nozzle 51.

As remarked in the general statement of the invention, I also contemplate employing heat for increasing the rate of vaporization of the fuel. This may be provided for by subjecting all or a restricted part of the vaporizing chamber 6 or 6′ to exhaust heat or heat from the cooling water circulating through the cylinder jackets. To this end the vaporizing chamber 6 or 6′ may be provided with a jacket 71 for heating a large or small area of the vaporizing bowl, and having an inlet 72 and an outlet 73 adapted for connection with the exhaust manifold or with the cooling circuit, in a manner obvious to those skilled in the art. However, in the actual operation of a device such as shown in Figure 1 I have found that the suction alone is capable of vaporizing a relatively large quantity of fuel.

I claim:

1. In a vaporizer and carburetor for internal combustion engines, a carbureting chamber having an air inlet and a mixture outlet adapted to be connected to the intake of an internal combustion engine whereby said carbureting chamber is under suction when the engine is operating, a vaporizing chamber having liquid fuel inlet means and vapor outlet means, said vaporizing chamber being otherwise closed and said vapor outlet means being in open communication with said carbureting chamber to effect suction in said vaporizing chamber by the suction created in said carbureting chamber, said vaporizing chamber having non-absorbent walls defining a body of liquid except for its level, a float in said vaporizing chamber, and a valve operated by said float and controlling said inlet means, whereby the liquid fuel in said vaporizing chamber is prevented from rising at any point above the level determined by said float and valve.

2. In a vaporizer and carburetor for internal combustion engines, a carbureting chamber having an air inlet and a mixture outlet adapted to be connected to the intake of an internal combustion engine whereby said carbureting chamber is under suction when the engine is operating, a vaporizing chamber having liquid fuel inlet means and vapor outlet means, said vaporizing chamber being otherwise closed and said vapor outlet means being in open communication with said corbureting chamber to effect suction in said vaporizing chamber by the suction created in said carbureting chamber, said vaporizing chamber having non-absorbent walls defining a body of liquid except for its level, a float in said vaporizing chamber, a valve operated by said float and controlling said inlet means, whereby the liquid fuel in said vaporizing chamber is prevented from rising at any point above the level determined by said float and valve, a throttle in said mixture outlet, and a valve in said air inlet operatively connected with said throttle.

3. In a vaporizer and carburetor for internal combustion engines, a carbureting chamber having an air inlet and a mixture outlet adapted to be connected to the intake of an internal combustion engine, whereby said carbureting chamber is under suction when the engine is operating, a vaporizing chamber having liquid fuel inlet means and vapor outlet means, said vaporizing chamber being otherwise closed and said vapor outlet means being in open communication with said carbureting chamber to effect suction in said vaporizing chamber by the suction created in said carbureting chamber, said vaporizing chamber having non-absorbent walls defining a body of liquid except for its level, a float in said vaporizing chamber, a valve operated by said float and controlling said inlet means, whereby the liquid fuel in said vaporizing chamber is prevented from rising at any point above the level determined by said float and valve, and means for applying heat to said vaporizing chamber.

In witness whereof, I hereunto subscribe my name this 14th day of January, 1922.

HERMAN V. CROON.

Aug. 20, 1929. A. C. SWEENY **1,725,111**

OIL VAPORIZER

Filed Aug. 9, 1926 2 Sheets-Sheet 1

Fig. 1.

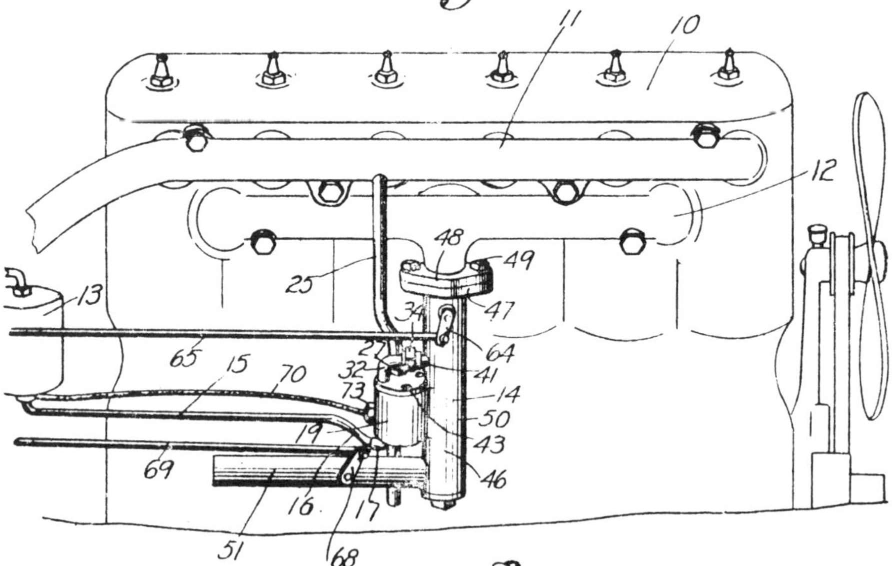

Fig. 2.

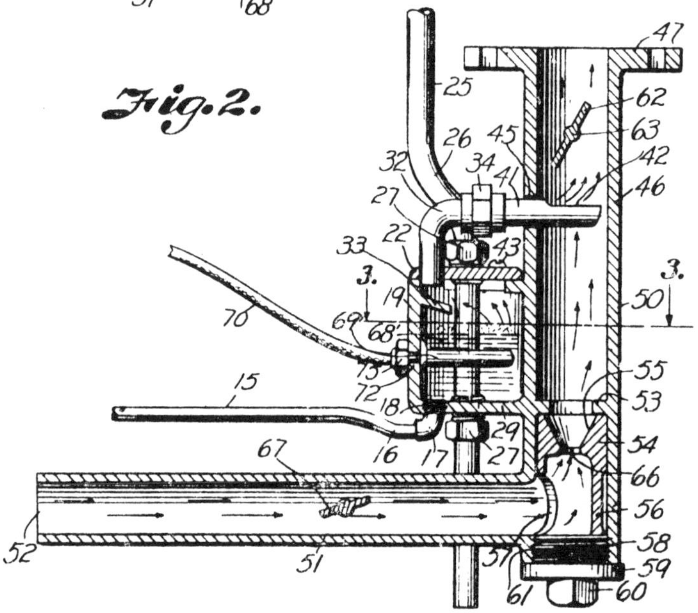

INVENTOR
Arthur C. Sweeny.
BY Albert C. Brown
ATTORNEY

Aug. 20, 1929. A. C. SWEENY **1,725,111**

OIL VAPORIZER

Filed Aug. 9, 1926 2 Sheets-Sheet 2

Fig. 3.

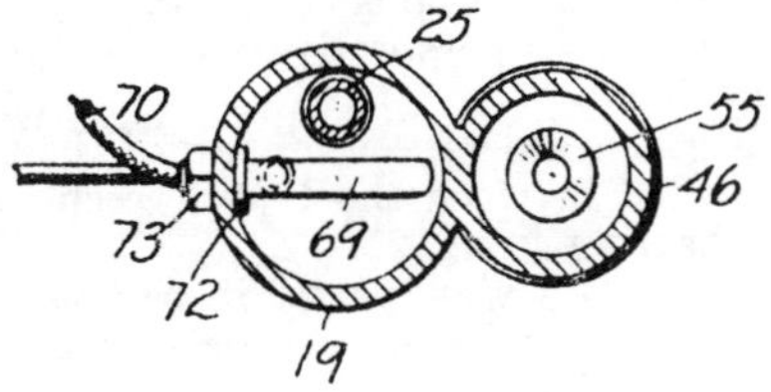

Fig. 4.

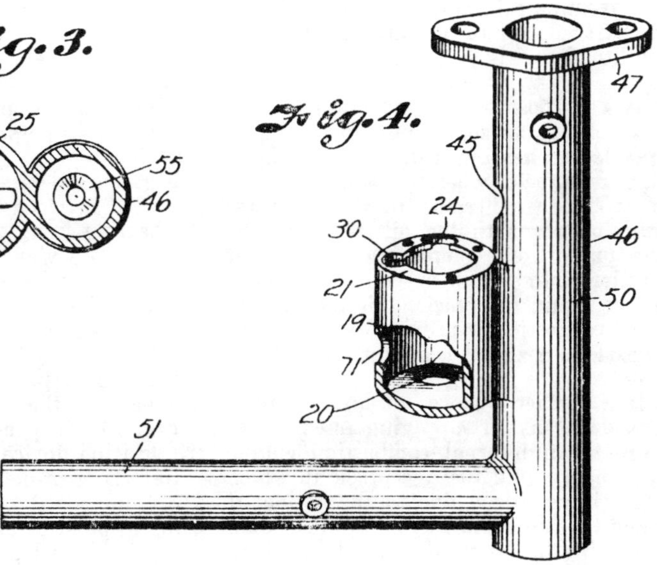

Fig. 5.

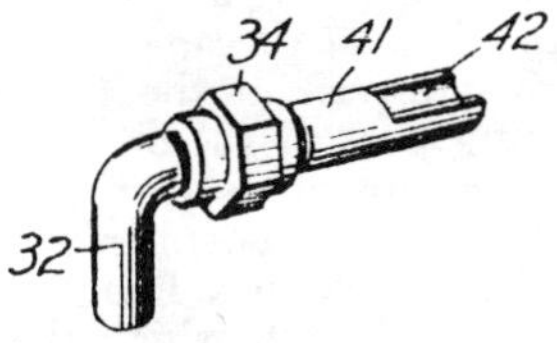

Fig. 6.

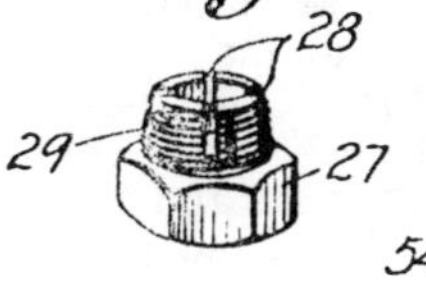

Fig. 7.

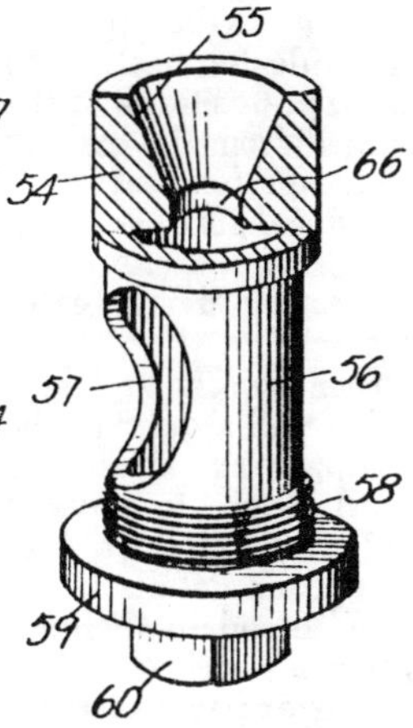

Fig. 9.

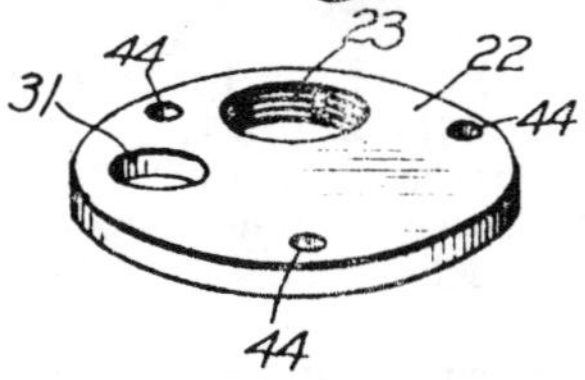

Fig. 8.

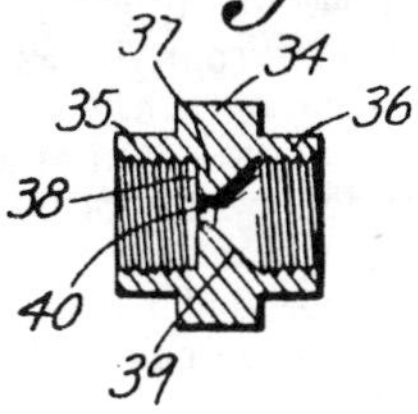

Fig. 10.

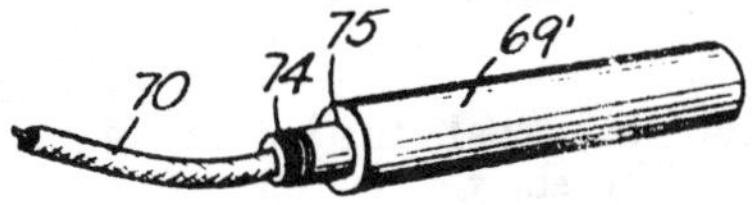

INVENTOR
Arthur C. Sweeny.
BY Arthur C. Brown
ATTORNEY

Patented Aug. 20, 1929. **1,725,111**

UNITED STATES PATENT OFFICE.

ARTHUR C. SWEENY, OF TULSA, OKLAHOMA.

OIL VAPORIZER.

Application filed August 9, 1926. Serial No. 128,097.

My invention relates to oil vaporizers and more particularly to an oil vaporizer in which liquid fuel is vaporized by the action of heat.

It is a purpose of my invention to provide an oil vaporizer comprising a liquid receptacle and heating means for the liquid in the receptacle whereby said liquid is vaporized, and furthermore to provide means for mixing air with the oil vapor in the proper proportions comprising an air inlet connection into which a conduit extends from the liquid receptacle. Preferably the air inlet connection is provided with a Venturi nozzle as is also the conduit leading into the same.

It is a further purpose of my invention to provide means for supplying an explosive mixture to an internal combustion engine comprising a liquid fuel receptacle in which the fuel is vaporized due to the action of heat and the vaporized fuel is supplied to an inlet leading to the intake manifold which is provided with an intake open to the outer air and which is provided with a valve regulating the effective opening in the inlet passage which is so located that the same does not vary the proportions of the air and vaporized fuel in the explosive mixture.

It is a further purpose of my invention to provide for heating the oil that is to be vaporized by means of the heat of the engine obtained from the exhaust gases and to provide means for heating said oil independent of said exhaust gases prior to and during starting of the engine, comprising an electric heating device extending into the oil receptacle.

Other objects and advantages of the invention will appear as the description of the accompanying drawings proceeds. However, I desire to have it distinctly understood that I do not intend to limit myself to the exact details shown or described, but that I intend to include as part of my invention all such obvious changes and modifications of parts as would occur to a person skilled in this art and as would fall within the scope of the claims.

In the drawings:

Fig. 1 is a fragmentary side elevational view of an internal combustion engine showing my improved oil vaporizer applied thereto.

Fig. 2 is a fragmentary vertical sectional view thereof, taken at right angles to Fig. 3.

Fig. 3 is a section taken on the line 3—3 of Fig. 2.

Fig. 4 is a perspective view partly broken away of the oil vaporizer, showing the cover and pipe connections removed.

Fig. 5 is a perspective view of the vapor inlet tube.

Fig. 6 is a perspective view of the bushing provided between the liquid receptacle and the heating tube.

Fig. 7 is a view partly in perspective and partly in section of a nozzle member provided in the air inlet.

Fig. 8 is a transverse sectional view of the nozzle provided in the vapor conduit.

Fig. 9 is a perspective view of the cover member for the liquid receptacle, and

Fig. 10 is a perspective view of the electric heating device detached.

Referring in detail to the drawings:

My improved oil vaporizer is shown as being applied to an internal combustion engine 10 which has the exhaust manifold 11 and the intake manifold 12, and which is provided with the usual vacuum tank 13 for supplying liquid fuel to the engine. Leading from the vacuum tank 13 to the oil vaporizer 14 is a liquid fuel supply tube 15, said tube being provided with an offset end portion 16 having a connecting member such as an elbow 17 threaded into a screw-threaded opening 18 in the liquid receptacle 19 forming part of the oil vaporizer 14. Said liquid receptacle is further provided with a screw-threaded opening 20 in the bottom thereof and with a flange 21 on the upper end thereof with which the cover plate 22 engages. Said cover plate 22 is provided with a threaded opening 23 that aligns with the cut away portion 24 in the flange 21 and with the threaded opening 20. A tubular member 25 extends from the exhaust manifold 11 through the openings 23 and 20, said tube being offset as indicated at 26. Mounted in the threaded openings 20 and 23 are the bushings 27 which are slotted as at 28 and externally threaded as indicated at 29. When the bushings engage with the threads in the openings 20 and 23, the same will provide a tight joint between the pipe 26 and the bottom 29 and cover plate 22 of the receptacle 19.

The receptacle 19 is further provided with an opening 30 in the flange 21 which aligns with the opening 31 in the cover plate 2 and through which the vapor feeding conduit 32 extends. An inclined partition 33 extends outwardly from the side wall portion of the receptacle 19 below the opening in the end of the conduit 32, thus acting as an inclined baffle member to prevent the drawing of the liquid fuel up through the tube 32 by suction. Screw-threadedly engaging the tube 32 is the nozzle member 34 which has the internally threaded end portion 35 engaging with the tubular member 32 and an internally threaded end portion 36. The central portion of the nozzle 36 is provided with an inwardly directed annular rib portion 37 which has a substantially perpendicular wall portion 38 and a conical wall portion 39 leading from the reduced opening 40 therein toward the threaded end 36 thereof. Screw-threadedly engaging with the internally threaded end 36 of the member 34 is the tubular member 41 which is cut away as at 42 adjacent the end thereof to thus provide a lateral opening therein as well as an opening in the end thereof. The cover plate 22 is screwed in position in any suitable manner as by means of the screws 43 passing through the openings 44 in the member 22. The tubular member 41 extends through an opening 45 in the side wall portion of the inlet tube 46, said inlet tube having a flange 47 which is adapted to be connected with a flange 48 on the intake manifold by means of the bolts 49, so as to provide a gas-tight connection between the intake manifold 12 and the intake tube 46.

The intake tube 46 is provided with a vertically extending portion 50 and a horizontally extending portion 51, said portions being of substantially uniform diameter and the portion 51 being open at the end 52 thereof remote from the vertically extending portion 50. An abutment 53 in the form of a flange or lip extending inwardly from the vertically extending portion 50 is provided in the inlet connection, against which the upper end portion of the plug 54 is adapted to seat, said plug having a thickened upper end portion providing the flaring passage 55 therein and being provided with a tubular central portion 56 having an opening 57 in the side thereof and being further provided with an externally threaded portion 58 between said tubular portion 56 and the flange 59 provided at the lower end thereof. The lower end of the plug 54 is provided with a downwardly extending projection 60 which is preferably polygonal in form so that the same may be engaged by a wrench for turning the same. The intake tube is provided with an internally threaded opening 61 at the junction of the vertically extending portion 50 and the horizontally extending portion 51 thereof, the threaded portion 58 engaging with the threaded portion 61 when the plug 54 is in position. By the provision of the plug 54 a Venturi nozzle is provided in the air inlet connection, and by the provision of the nozzle 34 a Venturi nozzle is provided in the vapor conduit.

A valve member 62 is mounted in the vertically extending portion 50 of the member 46 to pivot around the central portion 63 thereof and is provided with a lever 64 actuated by means of the rod 65 for controlling the position of the valve 62 to thus control the effective opening through the passage 50, thus regulating the supply of explosive mixture supplied to the engine 10 through the intake manifold 12. The valve 62, it will be noted, is located between the point where the tubular member 41 enters the tubular member 50 and the intake manifold. Thus the position thereof has no effect on the relative amount of vaporized fuel and air in the explosive mixture, this being determined by the relative size of the opening 40 and the opening 66 in the plug 54, and as these openings do not vary, the relative proportion of vapor and air in the mixture is substantially constant. The opening provided between the cut away portion 42 in the member 41 faces toward the intake manifold end of the member 46. Thus the stream of oil vapor passing out through the opening 42 is directed toward the intake manifold. The tubular member 51 is preferably provided with a valve 67 serving to choke off the supply of air when this is desired, movement of the valve 67 being accomplished by the lever 68 actuated by the rod 69.

The heated exhaust gases passing through the tube 25 will ordinarily vaporize the liquid fuel 68′ contained in the receptacle 19 when the engine is running, but in starting the engine and in providing vaporized fuel prior to starting the engine, it is also necessary to heat the liquid fuel 68′, and in order to do this an electric heating device 69′ is mounted in the receptacle 19 below the level of the liquid fuel 68′ therein, electric current being supplied thereto to heat the same by means of the conductor 70. The member 69′ is secured in liquid-tight engagement with the wall of the receptacle 19 where the same passes through the openings 17 by means of the washer or gasket 72 on the member 69′ which is drawn tight against the wall of the receptacle 19 by means of the nut 73 screw-threadedly engaging with the reduced screw-threaded end 74 on the member 69′, the reduced end portion 74 providing a shoulder 75 against which the gasket 72 will seat.

What I claim and desire to secure by Letters Patent is:—

1. In an oil vaporizer, an oil container having a baffle plate at its upper end, a tube extending through the container for conducting heat therethrough to vaporize the oil, a mixing chamber exterior to said container for connection to a manifold and hav-

ing one end open to atmospheric air, a removable Venturi fitting controlling admission of air into the mixing chamber, a conduit for oil vapor produced in said container leading from above the baffle plate in the container and into said mixing chamber, and a coupling in said conduit having a Venturi port controlling flow of vaporized oil.

2. In an oil vaporizer, an oil container having a baffle plate at its upper end, a tube extending through the container for conducting heat therethrough to vaporize the oil, a mixing chamber exterior to said container for connection to a manifold and having an inner annular flange forming a seat, an angularly disposed tubular member communicating with said chamber, a Venturi fitting in the chamber and engaging said seat, a conduit for oil vapor produced in said container leading from above said baffle plate in the container into said mixing chamber, and a coupling in said conduit having a Venturi opening controlling flow of vaporized oil from the container to the mixing chamber.

In testimony whereof I affix my signature.

ARTHUR C. SWEENY.

OVER

April 9, 1935. C. N. POGUE **1,997,497**

CARBURETOR

Filed Nov. 3, 1934 2 Sheets-Sheet 1

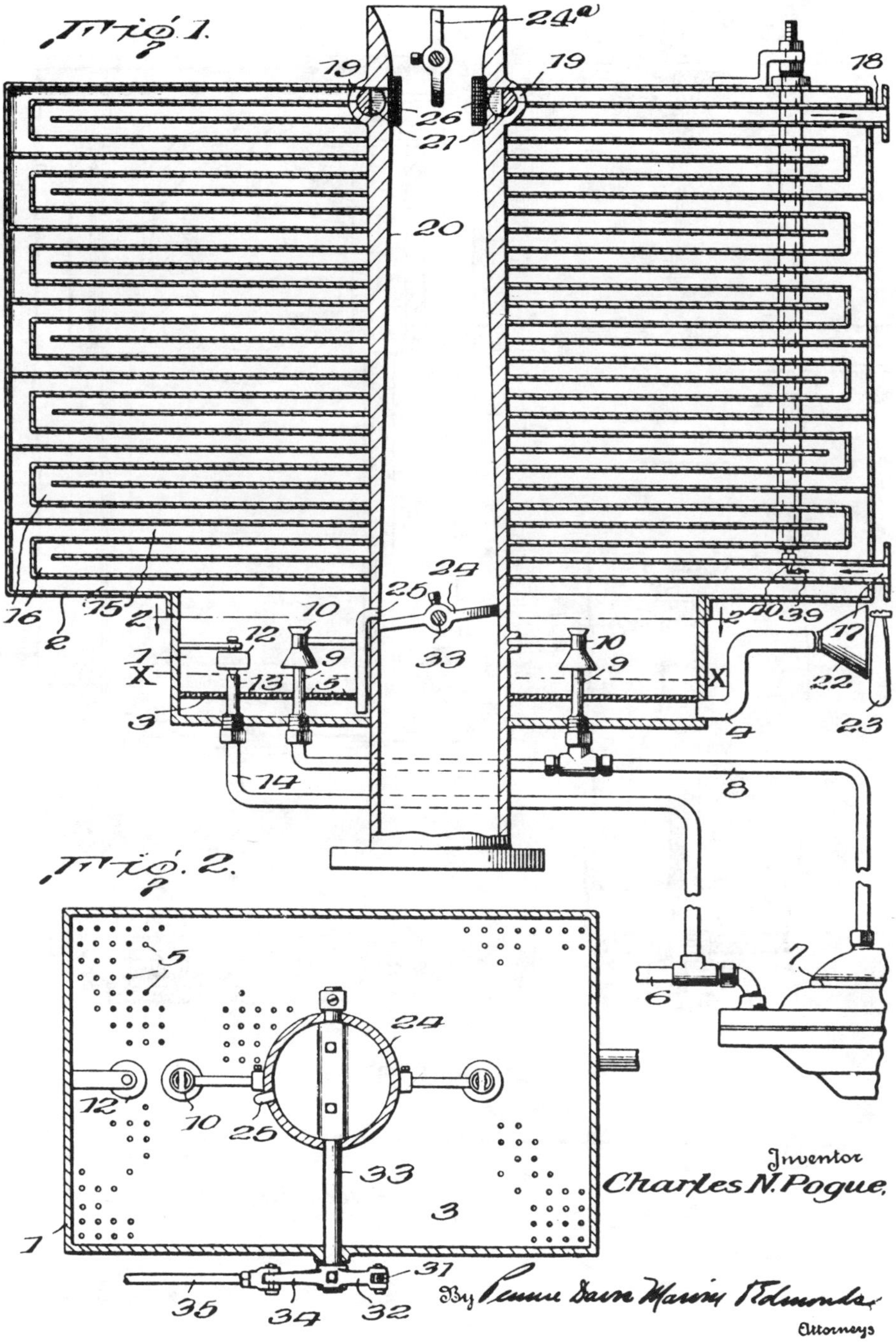

April 9, 1935. C. N. POGUE 1,997,497

CARBURETOR

Filed Nov. 3, 1934 2 Sheets-Sheet 2

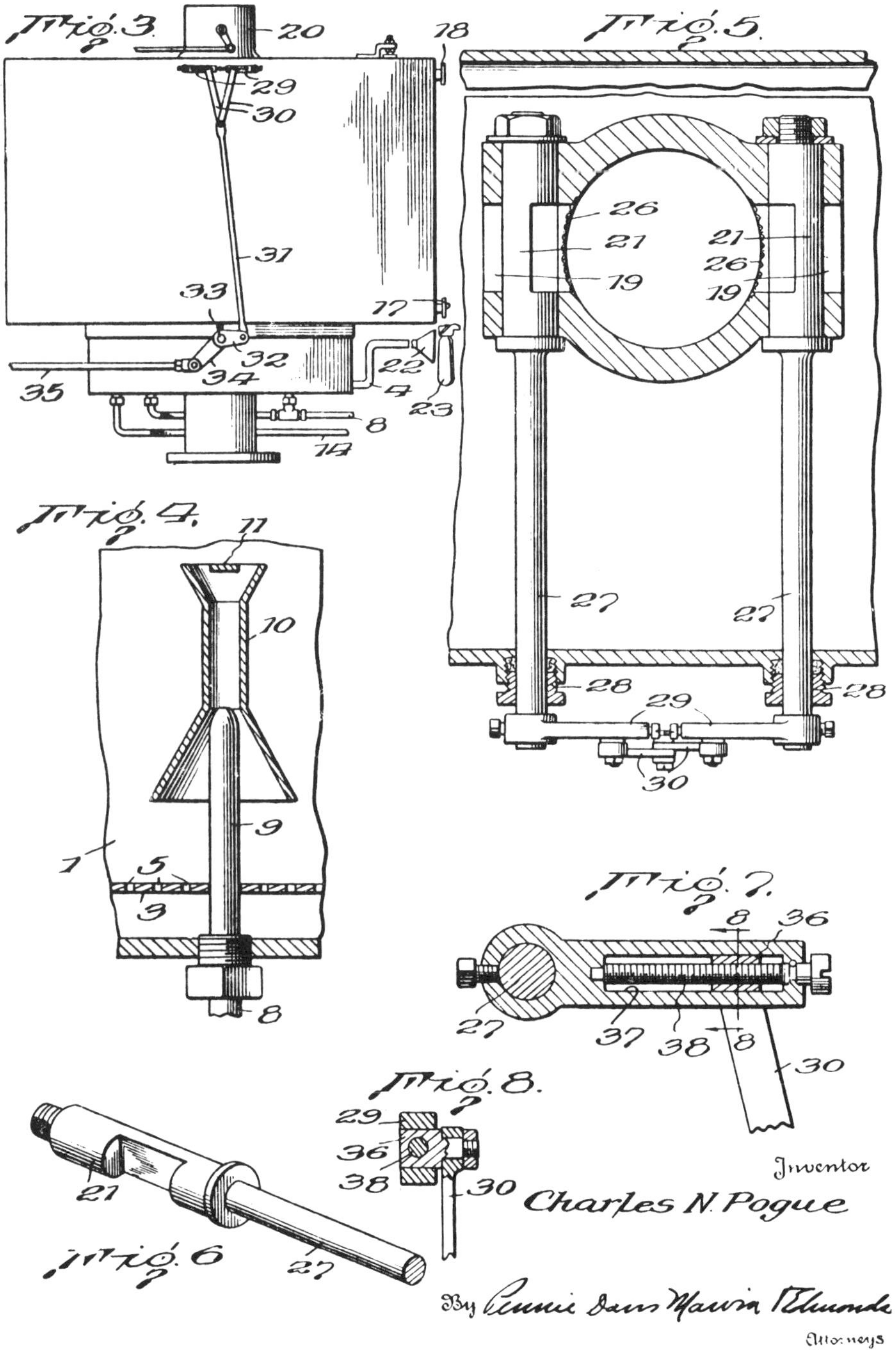

Patented Apr. 9, 1935 1,997,497

UNITED STATES PATENT OFFICE

1,997,497

CARBURETOR

Charles Nelson Pogue, Norwood, Manitoba, Canada

Application November 3, 1934, Serial No. 751,394

8 Claims. (Cl. 261—15)

This invention relates to a device for obtaining intimate contact between a liquid in a truly vaporous state and a gas, and particularly to such a device which may serve as a carburetor for internal combustion engines, and is an improvement on the form of device shown in my Patent No. 1,938,497, granted December 5, 1933.

In carburetors as commonly used for supplying a combustible mixture of air and liquid fuel to internal combustion engines a relatively large amount of the atomized liquid fuel is not vaporized and enters the engine cylinder more or less in the form of microscopic droplets. When such a charge is "fired" in the engine cylinder only that portion of the liquid fuel which has been converted into the vaporous and consequently the molecular state, combines with the air to give an explosive mixture. The remaining portion of the liquid fuel which is drawn into the engine cylinders and remains in the form of small droplets does not explode and thereby impart power to the engine, but burns with a flame and raises the temperature of the engine above that at which the engine operates most efficiently, i. e., from 160° to 180° F.

In my aforesaid patent there is shown and described a form of carburetor in which the liquid fuel is substantially completely vaporized prior to its introduction into the engine cylinders, and in which means are provided for maintaining a reverse supply of "dry" vapors available for introduction into the engine cylinder. Such a carburetor has been found superior to the standard type of carburetor referred to above and to give better engine performance with far less consumption of fuel.

It is an object of the present invention to provide a carburetor in which the liquid fuel is broken up and prepared in advance of and independent of the suction of the engine and in which a reserve supply of dry vapors will be maintained under pressure ready for introduction into the engine cylinder at all times. It is also an object of the invention to provide a carburetor in which the dry vapors are heated to a sufficient extent prior to being mixed with the main supply of air which carries them into the engine cylinder to cause them to expand so that they will be relatively lighter and will become more intimately mixed with the air prior to their explosion in the engine cylinders.

I have found that when the reserve supply of dry vapors is heated and expanded prior to being admixed with the atmospheric air a greater proportion of the potential energy of the fuel is obtained and the mixture of air and fuel vapors will explode in the engine cylinders without any apparent burning of the fuel which would result in unduly raising the operating temperature of the engine.

More particularly, the present invention comprises a carburetor in which liquid fuel vapors are passed from a main vaporizing chamber under at least a slight pressure into and through a heated chamber where they are caused to expand and in which droplets of liquid fuel are either vaporized or separated from the vapors, so that the fuel finally introduced into the engine cylinders is in true vapor phase. The chamber in which the liquid fuel vapors are heated and caused to expand preferably comprises a series of passages through which the vapors and the exhaust gases from the engine pass in tortuous paths and in such manner that the exhaust gases are brought into heat interchange relation with the vapors and give up a part of their heat to the vapors to cause their heating and expansion.

The invention will be further described in connection with the accompanying drawings, but this further disclosure and description is to be taken merely as an exemplification of the invention, and the same is not limited thereby, except as is pointed out in the appended claims.

In the drawings, Fig. 1 is a vertical cross-sectional view through a carburetor embodying my invention, Fig. 2 is a horizontal sectional view through the main vaporizing or atomizing chamber, the same being taken on line 2—2 of Fig. 1, Fig. 3 is a side elevation of the carburetor, Fig. 4 is a detail sectional view of one of the atomizing nozzles and its associated parts, Fig. 5 is a detail cross-sectional view showing the means for controlling the passage of gases from the vapor-expanding chamber into the intake manifold of the engine, Fig. 6 is a perspective view of one of the valves shown in Fig. 5, Fig. 7 is a cross-sectional view showing means for adjusting the valves shown in Fig. 5 and Fig. 8 is a cross-sectional view on line 8—8 of Fig. 7.

Referring now to the drawings, the numeral 1 indicates a main vaporizing and atomizing chamber for the liquid fuel located at the bottom of and communicating with a vapor heating and expanding chamber 2.

The vaporizing chamber is provided with a perforated false bottom 3 and is normally filled with liquid fuel to the level x. Atmospheric air from a conduit 4 enters the space below the false bottom 3 and passes upwardly through perforations 5 in said bottom and then bubbles up through the liquid fuel vaporizing a portion of it.

Liquid fuel for maintaining the level x in the chamber 1 passes from the usual fuel tank (not shown) through a pipe 6, and is forced by a pump 7 through a pipe 8 into and through a pair of nozzles 9 having their outlets located in the chamber 1, just above the level of the liquid fuel therein. The pump 7 may be of any approved form but is preferably of the diaphragm type, as such fuel pumps are now standard equipment on most automobiles.

The nozzles 9 are externally threaded at their lower ends to facilitate their assembly in the chamber 1 and to permit them to be removed readily, should cleaning be necessary.

The upper ends of the nozzles 9 are surrounded by Venturi tubes 10 having a baffle 11 located at their upper ends opposite the outlets of the nozzles. The liquid fuel being forced from the ends of the nozzles 9 into the restricted portions of the Venturi tubes causes a rapid circulation of the air and vapors in the chamber through the tubes 10 and brings the air and vapors into intimate contact with the liquid fuel, with the result that a portion thereof is vaporized. Unvaporized portions of the liquid fuel strike the baffles 11 and are thereby further broken up and deflected downwardly into the upwardly flowing current of air and vapors.

The pump 7 is regulated to supply a greater amount of liquid fuel to the nozzles 9 than will be vaporized. The excess over that vaporized will drop into the chamber 1 and cause the liquid to be maintained at the indicated level. When the liquid fuel rises above that level, a float valve 12 will be lifted and the excess will flow through an overflow pipe 13 into a pipe 14 leading back to the pipe 6 on the intake side of the pump 7. Such an arrangement permits a large amount of liquid fuel to be circulated by the pump 7 without more fuel being withdrawn from the fuel tank than is actually vaporized and consumed in the engine. As the float valve 12 will set upon the end of the outlet pipe 13 as soon as the liquid level drops below the indicated level, there is no danger of vapors passing into the pipe 14 and hence into the pump 7 to interfere with its normal operation.

The upper end of the vaporizing and atomizing chamber 1 is open and vapors formed by the atmospheric air bubbling through the liquid fuel in the bottom of the chamber and those formed as the result of the atomization at the nozzles 9 will pass into the heating and expanding chamber 2. As is clearly shown in Fig. 1, the chamber 2 comprises a series of tortuous passages 15 and 16 leading from the bottom to the top. The vapors pass through the passages 15 and the hot exhaust gases of the engine pass through the passages 16, a suitable entrance 17 and exit 18 being provided for that purpose.

The vapors passing upwardly in a zigzag path through the passages 15 will be brought into heat interchange relation with the hot walls of the passages 16 for the exhaust gases. The total length of the passages 15 and 16 is such that a relatively large reserve supply of the liquid fuel is always maintained in the chamber 2, and by maintaining the vapors in heat interchange relation with the hot exhaust gases for a substantial period, the vapors will absorb sufficient heat from those gases to cause the vapors to expand, with the result that when they are withdrawn from the top of the chamber 2, they will be in the true vapor phase and, due to their expansion, relatively light.

Any minute droplets of liquid fuel entrained by the vapors in the chamber 1 will precipitate out in the lower passages 15 and flow back into the chamber 1, or else be vaporized by heat which the vapors absorb from the hot exhaust gases in their passage through the chamber 2.

The upper end of the vapor passage 15 communicates with openings 19 adjacent the upper end of a down-draft air tube 20 leading to the intake manifold of the engine. Valves 21 are interposed in the openings 19, so that the passage of the vapors therethrough into the air tube may be controlled. The valves 21 preferably are of the rotary plug type and are controlled as hereinafter described.

Suitable means are provided for causing the vapors to be maintained in the chamber 2 under a pressure greater than atmospheric so that when the valves 21 are opened, the vapors will be forced into the air tube 20 independently of the suction of the engine. Such means may comprise an air pump (not shown) for forcing the atmospheric air through the pipe 4 into the chamber 1, beneath the false bottom 3, but I prefer merely to provide the pipe 4 with a funnel-shaped inlet end 22 and located just back of the usual fan 23 of the engine. That will cause the air to pass through the pipe 4 with sufficient force to maintain the desired pressure in the chamber 2, and the air being drawn through the radiator by the fan will be preheated prior to its introduction into the chamber 1 and hence will vaporize greater amounts of the liquid fuel. If desired, the pipe 4 may be surrounded by an electric or other heater, or exhaust gases from the engine may be passed around it to further preheat the air passing therethrough prior to its introduction into the liquid fuel in the bottom of the chamber 1. The air tube 20 is provided with a butterfly throttle valve 24 and a choke valve 24*a*, as is customary with carburetors, used for internal combustion engines. The upper end of the air tube 20 extends above the chamber 2 a distance sufficient to receive an air filter and/or silencer, if desired.

A low speed or idling jet 25 has its upper end communicating with the passage through the air tube 20 adjacent the throttling valve 24 and its lower end extending into the liquid fuel in the bottom of the chamber 1 for supplying fuel to the engine when the valves are in a position such as to close the passages 19. However, the passage through the idling jet 25 is so small that under normal operations the suction thereon is not sufficient to lift liquid fuel from the bottom of the chamber 1.

To prevent the engine from backfiring into the vapor chamber 2, the ends of the passages 19 are covered with a fine mesh screen 26 which, operating on the principle of a miner's lamp, will prevent the vapors in the chamber 2 from exploding in case of a backfire, but will not interfere substantially with the passage of the vapors from the chamber 2 into the air tube 20 when the valves 21 are in open position. The air tube 20 preferably is in the form of a venturi with the greatest restriction being at that point where the openings 19 are located, so that when the valves 21 are opened there will be a pulling force on the vapors because of the increased velocity of the air at the restricted portion of the air tube 20 opposite the openings 19, as well as an expelling force on them due to the pressure in the chamber 2.

As shown in Fig. 3, the operating mechanism for the valves 21 is so connected to the operating mechanism for the throttle valve 24 that they are opened and closed simultaneously with the opening and closing of the throttle valve, so that the amount of vapor supplied to the engine will at all times be in proportion to the demands placed upon the engine. To that end, each valve 21 has an extension or operating stem 27 protruding through one of the side walls of the vapor-heating and expanding chamber 2. Packing glands 28 of the ordinary construction surround the stems 27 where they pass through the chamber wall to prevent leakage of vapors at those points.

Operating arms 29 are rigidly secured to the outer ends of the stems 27 and extend towards each other. The arms are pivotally and adjustably connected to a pair of links 30 which at their lower ends are pivotally connected to an operating link 31, which in turn is pivotally connected to an arm 32 rigidly secured on an outer extension 33 of the stem of the throttle valve 24. The extension 33 also has rigidly secured thereto an arm 34, to which is connected an operating link 35 leading from the means for accelerating the engine.

The means for adjustably connecting the upper ends of the links 30 to the valve stems 27 of the valves 21, so that the amount of vapors delivered from the chamber 2 may be regulated to cause the most efficient operation of the particular engine to which the carburetor is attached, comprises angular slides 36 to which the upper ends of the links 30 are fastened, and which are slidably, but non-rotatably mounted in guideways 37 in the arms 29. The slides 36 have threaded bores through which screws 38 pass. The screws 38 are rotatably mounted in the arms 29, but are held against longitudinal movement so that when they are rotated the slides 36 will be caused to move along the guideways 37 and change the relative position of the links 30 to the valve stems 27, so that a greater or less movement, and consequently a greater or less opening of the ports 19 will take place when the throttle valve 24 is operated.

For safety, and for most efficient operation of the engine, the vapors in the chamber 2 should not be heated or expanded beyond a predetermined amount, and in order to control the extent to which the vapors are heated, and consequently the extent to which they are expanded, a valve 39 is located in the exhaust passage 16 adjacent the inlet 17. The valve 39 is preferably thermostatically controlled, as, for example, by an expanding rod thermostat 40 which extends through the chamber 2. However, any other means may be provided, for reducing the amount of hot exhaust gases entering the passages 16 when the temperature of the vapors in the chamber reaches or exceeds the optimum.

The carburetor has been described in detail in connection with a down-draft type of carburetor, but it is to be understood that its usefulness is not restricted to that particular type of carburetor, and that the manner in which the mixture of atmospheric air and vapors is introduced into the engine cylinders is immaterial as far as the advantages of the carburetor are concerned.

The term "dry vapor" is used herein to define the physical condition of the liquid fuel vapor after removal of liquid droplets, or the mist which is frequently entrained in what is ordinarily termed a vapor.

From the foregoing description it will be seen that the present invention provides a carburetor in which the breaking up of the liquid fuel for subsequent use is independent of the suction created by the engine, and that after the liquid fuel is broken up it is maintained under pressure in a heated space for a length of time sufficient to permit all entrained liquid particles to be separated or vaporized and to permit the dry vapors to expand prior to their introduction into, and admixture with the main volume of atmospheric air passing into the engine cylinders.

I claim:

1. A carburetor for internal combustion engines comprising a vaporizing chamber, means for vaporizing liquid fuel therein, a vapor-heating and expanding chamber communicating with the vaporizing chamber, a passage for atmospheric air extending entirely through said heating and vaporizing chamber and adapted to communicate with the intake manifold of the engine, and means for conducting vapors from the heating and expanding chamber into said air passage.

2. A carburetor for internal combustion engines comprising means for vaporizing liquid fuel, a vapor-heating and expanding chamber into which the vaporized liquid fuel passes, means for creating a pressure in said heating and expanding chamber greater than atmospheric during normal operation of the carburetor, and means for conducting heated and expanded gases from the heating and expanding chamber to the intake manifold of the engine upon a suction being created in said manifold.

3. A carburetor for internal combustion engines comprising means for vaporizing liquid fuel, a vapor-heating and expanding chamber into which the vaporized liquid fuel passes, a passage for atmospheric air adapted to be connected with the intake manifold of the engine, a connecting passage between the heating and expanding chamber and said atmospheric air passage, a valve controlling the passage of vapors through said connecting passage, and means for creating a pressure in said heating and vaporizing chamber greater than atmospheric, so that when said valve is opened the pressure under which the vapors are maintained in the heating and expanding chamber will cause at least a part of them to be expelled from said chamber and into the atmospheric air passage.

4. A carburetor for internal combustion engines comprising a vaporizing chamber, means for vaporizing liquid fuel therein, a vapor-heating and expanding chamber communicating with the vaporizing chamber and adapted to receive vapors therefrom; said heating and vaporizing chamber having tortuous passages therethrough for said vapors and for a heating fluid which is maintained in heat interchange relation with said vapor passages, a passage for atmospheric air adapted to be connected with the intake manifold of the engine and extending through said heating and vaporizing chamber, a pair of connecting passages extending through the walls of said atmospheric air passage through which vapors from the vapor passage in the heating and expanding chamber may pass into said air passage, a valve in each of said connecting passages, a valve in said atmospheric air passage controlling the flow of air and vapors to the engine, and means for operating the valves in said connecting passages in unison with said air and vapor control valve.

5. A carburetor for internal combustion engines comprising a vaporizing chamber, a liquid

fuel nozzle in said vaporizing chamber, a pump for forcing liquid fuel through said nozzle, an overflow passage for liquid fuel located above the bottom of said vaporizing chamber, whereby the liquid fuel which is projected from said nozzles and remains unvaporized forms a body in the bottom of the vaporizing chamber, means for introducing atmospheric air into said body of liquid fuel whereby the air bubbles upwardly through said body and vaporizes a portion thereof, means for conducting liquid fuel from said overflow to the intake of said pump, so that said pump may be regulated to circulate an amount of liquid fuel in excess of that vaporized by the nozzles, a vapor heating and expanding chamber communicating with the vaporizing chamber and receiving vapors therefrom, and means for conducting heated and expanded vapors from the heating and expanding chamber into the intake manifold of the engine.

6. A carburetor for internal combustion engines comprising a vaporizing chamber, means for maintaining a body of liquid fuel in the bottom of said chamber, a heating and expanding chamber communicating with the vaporizing chamber, means for introducing atmospheric air into the body of liquid in the vaporizing chamber under a pressure greater than atmospheric, whereby said air will bubble upwardly through the liquid fuel, vaporize a portion thereof and build up a pressure greater than atmospheric in the heating and expanding chamber, and means for conducting heated and expanded vapors from the heating and expanding chamber into the intake manifold of the engine.

7. A carburetor for internal combustion engines comprising a vaporizing chamber, a liquid fuel atomizing nozzle in said chamber, a pump for forcing liquid fuel through said nozzle, a Venturi tube surrounding said nozzle and having a baffle opposite the outlet of said nozzle for further breaking up liquid fuel projected from said nozzle, a heating and expanding chamber communicating with the vaporizing chamber and receiving liquid fuel vapors therefrom, and means for conducting heated and expanded vapors from the heating and expanding chamber into the intake manifold of the engine.

8. A carburetor for internal combustion engines comprising a vaporizing chamber, means for vaporizing liquid fuel therein, a vapor heating and expanding chamber communicating with the vaporizing chamber and adapted to receive vapors therefrom, said heating and vaporizing chamber comprising a relatively long zig-zag passage for the vapors, a relatively long chamber for a heating fluid interleaved with the zig-zag passages for the vapors and in heat interchange relation with said vapor passage, and means for conducting heated and expanded vapors from the heating and expanding chamber into the intake manifold of the engine.

CHARLES NELSON POGUE.

July 2, 1935. C. H. GARRETT 2,006,676

ELECTROLYTIC CARBURETOR

Original Filed July 1, 1932 2 Sheets-Sheet 1

FIG. 1.

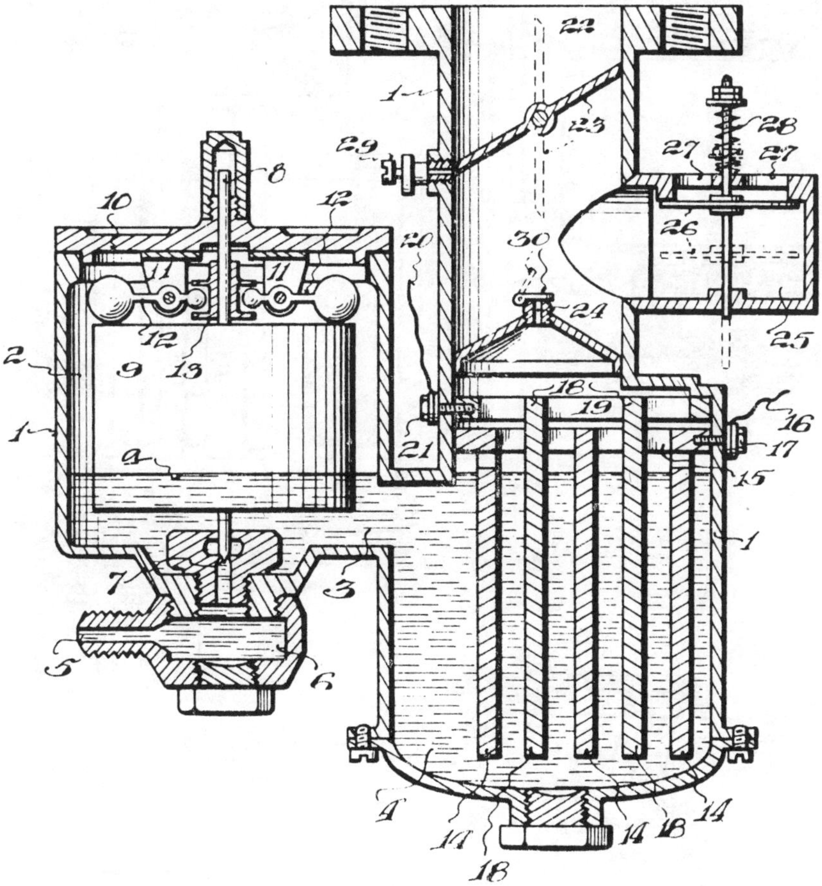

Charles H. Garrett
INVENTOR

BY Ernest G. Wood
ATTORNEY

July 2, 1935. C. H. GARRETT 2,006,676

ELECTROLYTIC CARBURETOR

Original Filed July 1, 1932 2 Sheets-Sheet 2

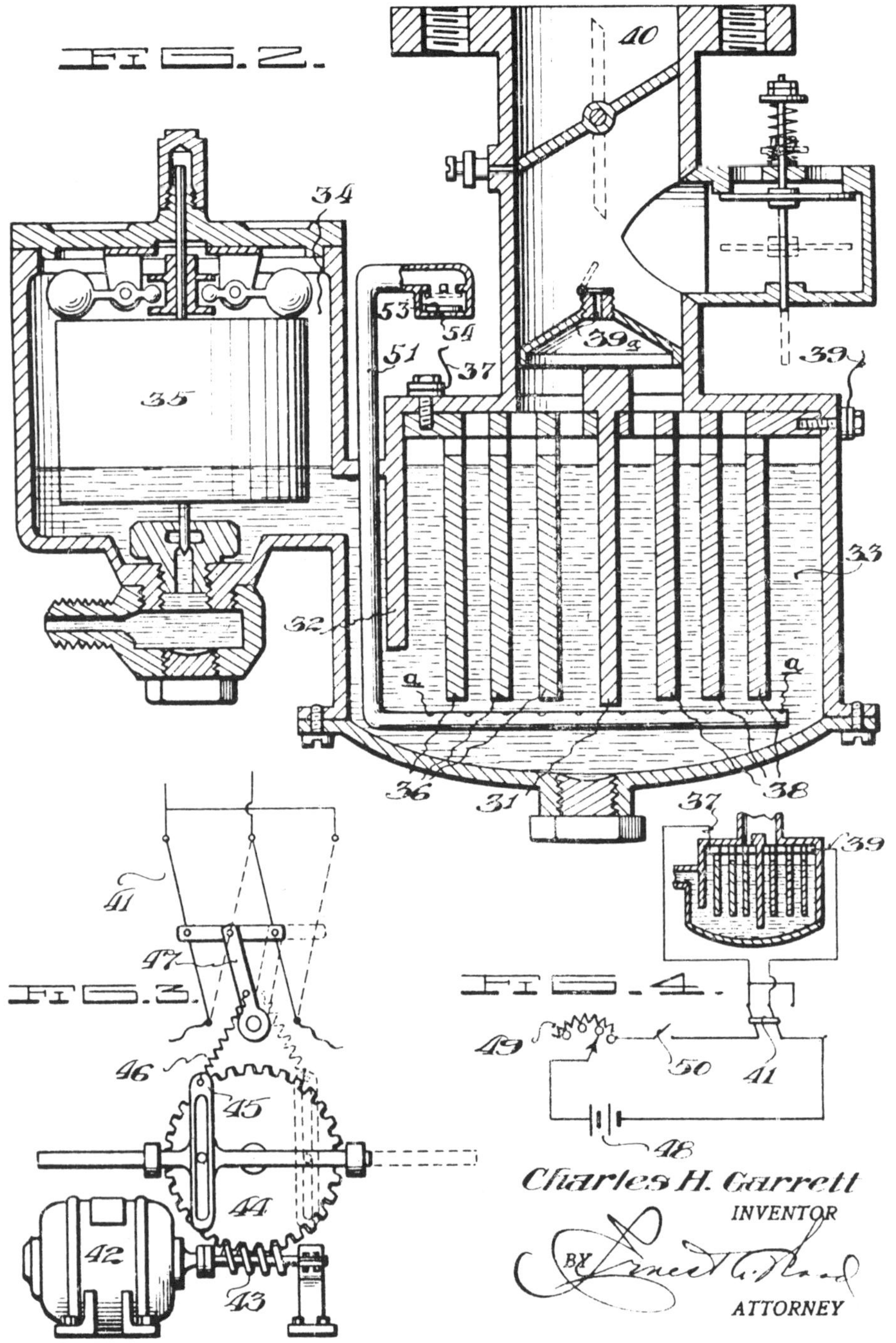

Charles H. Garrett
INVENTOR

BY [signature]
ATTORNEY

Patented July 2, 1935 **2,006,676**

UNITED STATES PATENT OFFICE

2,006,676

ELECTROLYTIC CARBURETOR

Charles H. Garrett, Dallas, Tex.

Application July 1, 1932, Serial No. 620,364
Renewed November 30, 1934

5 Claims. (Cl. 204—5)

This invention relates to carburetors and it has particular reference to an electrolytic carburetor by means of which water may be broken up into its hydrogen and oxygen constituents and the gases so formed suitably mixed with air.

The principal object of the invention is to provide in a device of the character described, a mechanism by means of which water may be readily decomposed into its constituents, and the constituents intimately mixed with each other and with air.

Another object of the invention is to provide means whereby the electrolyte level in the carburetor may be maintained at a more or less constant level regardless of fluctuations in fluid pressure at the fluid inlet of the carburetor.

Another object of the invention is to provide means whereby the relative amount of air mixed with the hydrogen and oxygen may be regulated as desired.

Still another object of the invention is the provision of means to prevent loss of hydrogen and oxygen gases during periods in which these gases are not being drawn from the carburetor.

Still another object of the invention is the provision of means whereby the hydrogen and oxygen resulting from electrolysis may be formed in separate compartments, and a further object of the invention is the provision of means to periodically reverse the direction of current flow and thereby alternate the evolution of the gases in the separate compartments, to be later intermingled.

With the foregoing objects as paramount, the invention has particular reference to its salient features of construction and arrangement of parts, taken in connection with the accompanying drawings, wherein:—

Figure 1 is a view in vertical section of one form of carburetor.

Figure 2 is a modified form.

Figure 3 is a diagrammatic view of a pole changer, showing its actuating mechanism, and

Figure 4 is a wiring diagram for the modified form of carburetor shown in Figure 2.

Continuing more in detail with the drawings, reference is primarily directed to Figure 1 in which the reference numeral 1 designates the carburetor housing, which is preferably constructed of bakelite or other suitable insulating material. The housing 1 is so designed as to divide the carburetor into a float chamber 2 and gas generating chamber 4, connected by a fluid passage 3.

Water under pressure is forced into the carburetor through an opening 5 which communicates with the float chamber 2 through the medium of the sediment chamber 6 and the needle valve orifice 7, which is closed by a needle valve 8 when the device is not in operation. A float 9 surrounds the needle valve 8 and is free to move vertically relative thereto. Depending from the cover 10 to the float chamber 2 are two ears 11, located at spaced intervals on opposite sides of the needle valve 8. The members 12 are pivoted to the ears 11, as shown. The weighted outer ends of the members 12 rest on top of the float 9, and their inner ends are received in an annular groove in the collar 13 which is rigidly attached to the needle valve 8.

Within the gas generating chamber 4, a series of spaced, depending plates 14 are suspended from a horizontal member 15 to which a wire 16 has electrical contact through the medium of the bolt 17, which extends inwardly through the housing 1 and is threaded into the horizontal member 15.

A second series of plates 18 is located intermediate the plates 14 and attached to the horizontal member 19, and has electrical contact with the wire 20 through the bolt 21.

A gas passageway 22, in which a butterfly valve 23 is located, communicates with the gas generating chamber 4 through an orifice 24. An air inlet chamber 25 has communication with the gas passageway 22 above the orifice 24. A downwardly opening check valve 26 is in control of the openings 27, and is held inoperatively closed by means of light spring 28.

An adjustable auxiliary air valve 29 is provided in the wall of the gas passageway 22, which air valve is closed by the butterfly valve 23 when the butterfly valve is closed, but communicates with the outside air when the butterfly valve is open.

The operation of the device is as follows:

The chambers 2 and 4 are first filled to the level *a* with a solution of weak sulphuric acid or other electrolyte not changed by the passage of current therethrough, and the opening 5 is connected to a tank of water, not shown.

The wire 16 is next connected to the positive pole of a storage battery or other source of direct current and the wire 20 to the negative pole. Since the solution within the carburetor is a conductor of electricity, current will flow therethrough and hydrogen will be given off from the negative or cathode plates 18 and oxygen from the positive or anode plates 14.

The butterfly valve 23 is opened and the gas passageway 22 brought into communication with a partial vacuum. Atmospheric pressure acting on the top of the check valve 26 causes it to be forced downwardly as shown in dotted lines. The hydrogen and oxygen liberated from the water at the plates 18 and 14 are drawn upwardly through the orifice 24 covered by the check valve 30 where they are subsequently mixed with air entering through the openings 27 and through the auxiliary air valve 29.

When it is desired to reduce the flow of hydrogen and oxygen from the plates 18 and 14, the current flowing through the device is reduced, and when the current is interrupted the flow ceases. When the butterfly valve 23 is moved to closed position, the check valve 26 is automatically closed by the spring 28. Any excess gas given off during these operations is stored in the space above the fluid where it is ready for subsequent use.

Water is converted into its gaseous constituents by the device herein described, but the dilute sulphuric acid or other suitable electrolyte in the carburetor remains unchanged, since it is not destroyed by electrolysis, and the parts in contact therewith are made of bakelite and lead or other material not attacked by the electrolyte.

The structure shown in Figure 2 is substantially the same as that shown in Figure 1 with the exception that the modified structure embraces a larger gas generating chamber which is divided by means of an insulating plate 31 and is further provided with a depending baffle plate 32 which separates the gas generating chamber 33 from the float chamber 34 in which the float 35 operates in the same manner as in Figure 1. Moreover, the structure shown in Figure 2 provides a series of spaced depending plates 36 which are electrically connected to the wire 37, and a second series of similar plates 38 which are electrically connected to the wire 39 and are spaced apart from the plates 36 by the insulating plate 31. Gases generated on the surfaces of the plates 36 and 38 pass upward through the orifice 39*a* into the gas passageway 40 where they are mixed with air as explained in the description of Figure 1.

A pipe 51 bent as shown in Figure 2 passes downwardly through the housing of the carburetor and has a series of spaced apertures *a* in its horizontal portion beneath the plates 36 and 38. An upwardly opening check valve 53 is in control of the air inlet 54. When a partial vacuum exists in the chamber 33, air is drawn in through the opening 54 and subsequently passes upwardly through the apertures *a*. This air tends to remove any bubbles of gas collecting on the plates 36 and 38 and also tends to cool the electrolyte. The check valve 53 automatically closes when a gas pressure exists within the carburetor and thereby prevents the electrolyte from being forced out of the opening 54.

In order to provide for alternate evolution of the gases from the plates 36 and 38, a pole changer 41 shown in Figure 3 is provided, which is actuated periodically by the motor 42 which drives the worm 43 and the gear 44 and causes oscillations of the member 45 which is connected by a spring 46 to the arm 47, thereby causing the pole changer to snap from one position to the other.

In operation, the carburetor shown in Figure 2 is connected as shown in the wiring diagram of Figure 4. A storage battery 48 or other suitable source of direct current is connected to a variable rheostat 49, switch 50, pole changer 41 and to the carburetor as shown. Thus the rate of evolution of the gases can be controlled by the setting of the rheostat 49 and the desired alternate evolution of the gases in the compartments of the carburetor is accomplished by means of the periodically operated pole changer 41.

Manifestly, the construction shown is capable of considerable modification and such modification as is considered within the scope and meaning of the appended claims is also considered within the spirit and intent of the invention.

What is claimed is:

1. An electrolytic carburetor including an anode and a cathode, float means to control the level of the electrolyte within said carburetor, means to mix the gases resulting from electrolysis with air, and a check valve independent of said float means to control ingress of air to said carburetor.

2. An electrolytic carburetor including anode and cathode plates, a float actuated valve in control of the electrolyte level within said carburetor, means to mix the gases resulting from electrolysis with air, a check valve in control of said means, a second check valve independent of said float actuated valve to prevent loss of gases from said carburetor.

3. An electrolytic carburetor for producing mixtures of hydrogen, oxygen and air, including a series of spaced and electrically connected anode plates partially immersed in the electrolyte within said carburetor, a series of electrically connected cathode plates spaced between said anode plates, a float operated valve in control of the electrolyte level within said carburetor, an air inlet to said carburetor and a check valve in control of said air inlet.

4. An electrolytic carburetor for generating hydrogen and oxygen gases from water and for mixing the said gases with air, including an anode and a cathode partially immersed in an electrolyte within said carburetor, float actuated means to replace the water consumed whereby to maintain a constant fluid level within said device, means for mixing the said gases with air, and a check valve in control of said latter means.

5. An electrolytic carburetor including an anode and a cathode, float actuated means to control the level of the electrolyte within said carburetor, means to mix the gases resulting from electrolysis with air, a check valve in control of said latter means and means to periodically reverse the direction of current through said carburetor.

CHARLES H. GARRETT.

Jan. 7, 1936. C. N. POGUE 2,026,798

CARBURETOR

Filed Sept. 27, 1935 2 Sheets-Sheet 1

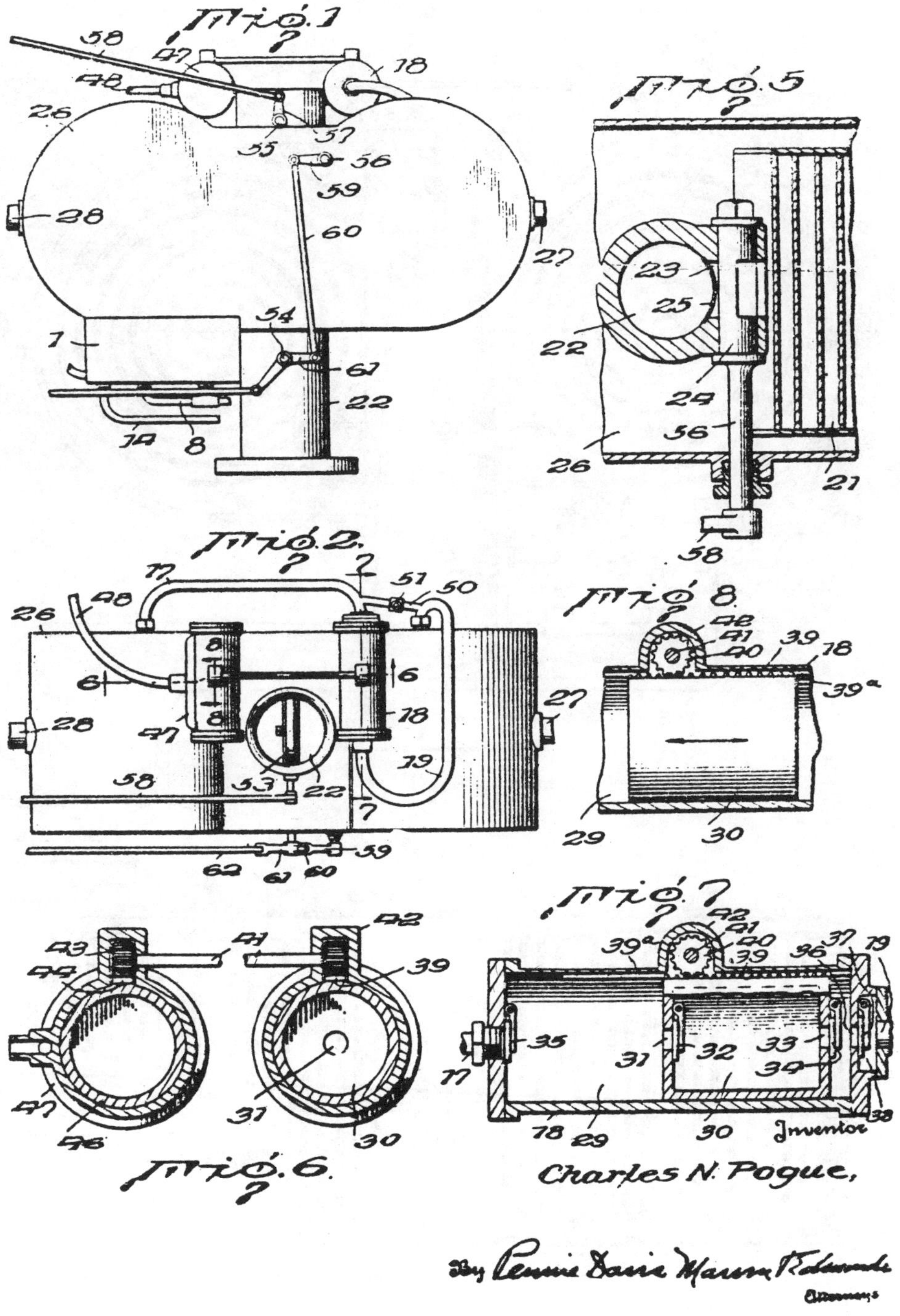

Jan. 7, 1936. C. N. POGUE 2,026,798

CARBURETOR

Filed Sept. 27, 1935 2 Sheets-Sheet 2

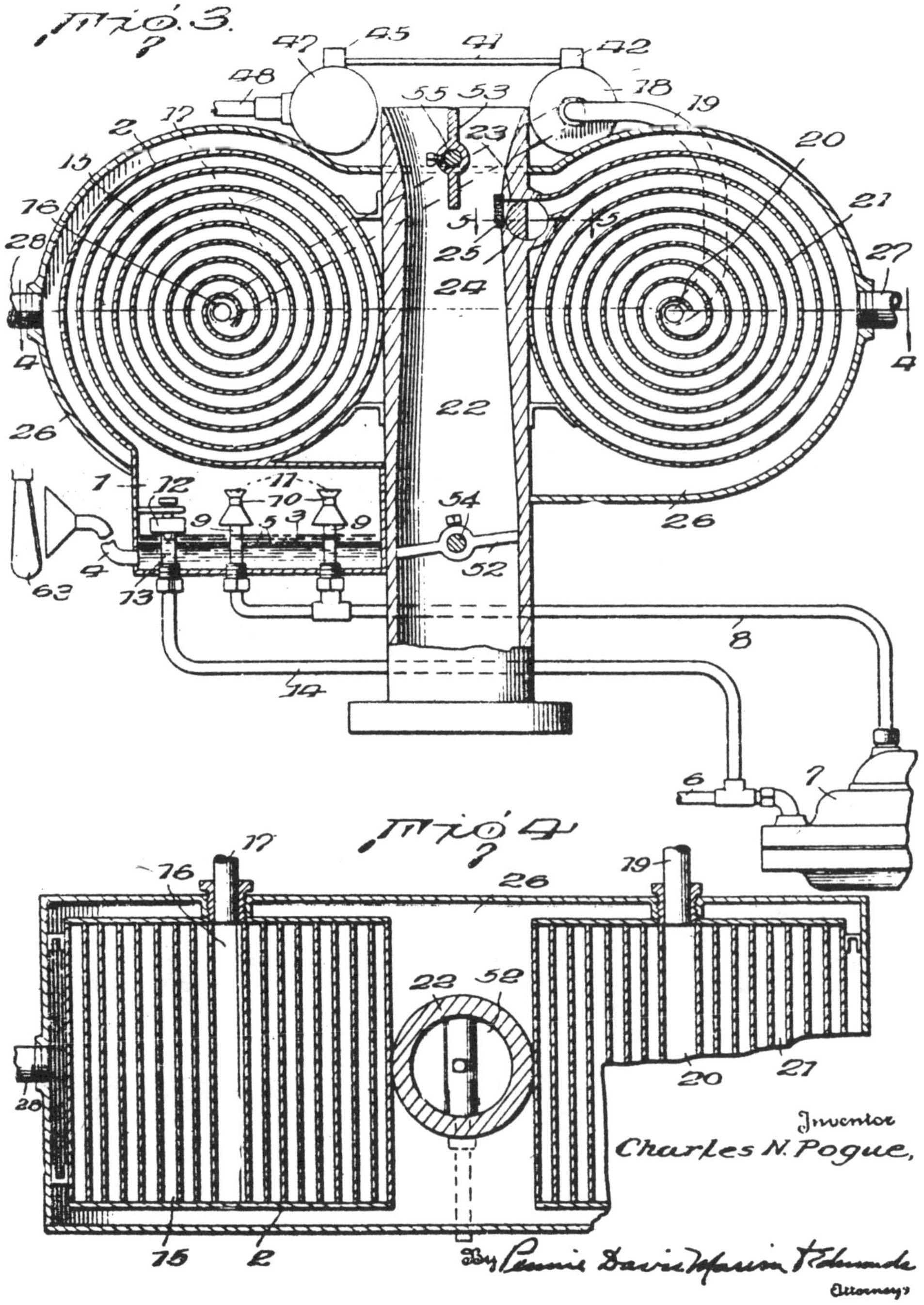

Patented Jan. 7, 1936 2,026,798

UNITED STATES PATENT OFFICE

2,026,798

CARBURETOR

Charles N. Pogue, Norwood, Winnipeg, Manitoba, Canada

Application September 27, 1935, Serial No. 42,540

8 Claims. (Cl. 261—13)

This invention relates to carburetors suitable for use with internal combustion engines and is an improvement on the carburetors shown in my Patents Nos. 1,938,497, granted December 5, 1933, and 1,997,497, granted April 9, 1935.

In my aforesaid patents an intimate contact between a liquid such as the fuel used for internal combustion engines, and a gas, such as air, is obtained by causing the gas to bubble up through a body of the liquid. The vaporized liquid passes into a vapor chamber which preferably is heated, and any liquid droplets are returned to the body of liquid, with the result that the fuel introduced into the combustion chambers is free of liquid particles, and in the molecular state so that an intimate mixture with the air is obtained to give an explosive mixture from which nearer the maximum energy contained in the liquid fuel is obtained. Moreover, as there are no liquid particles introduced into the combustion chambers there will be no burning of the fuel and consequently the temperature of the engine will not be increased above that at which it operates most efficiently.

In my Patent No. 1,997,497, the air which is to bubble up through the body of liquid fuel is forced into and through the fuel under pressure and the fuel vapors and air pass into a chamber where they are heated and caused to expand. The introduction of the air under pressure and the expansion of the vaporous mixture insures a sufficient pressure being maintained in the vapor heating and expanding chamber to cause at least a portion of it to be expelled therefrom into the intake manifold as soon as the valve controlling the passage thereto is opened.

In accordance with the present invention, improved means are provided for maintaining the vaporous mixture in the vapor heating chamber under a predetermined pressure, and for regulating such pressure so that it will be at the optimum for the particular conditions under which the engine is to operate. Such means preferably comprises a reciprocating pump operated by a vacuum-actuated motor for forcing the vapors into and through said chamber. The pump is provided with a suitable pressure-regulating valve so that when the pressure in the vapor-heating chamber exceeds the predetermined amount a portion of the vaporous mixture will be by-passed from the outlet side to the inlet side of the pump and recirculated.

The invention will be described further in connection with the accompanying drawings, but such further disclosure and description is to be taken merely as an exemplification of the invention, and the same is not limited thereby except as is pointed out in the subjoined claims.

In the drawings:

Fig. 1 is a side elevation of a carburetor embodying the invention.

Fig. 2 is a plan view thereof.

Fig. 3 is an enlarged vertical sectional view.

Fig. 4 is a transverse sectional view on the line 4—4 of Fig. 3.

Fig. 5 is a detail sectional view on line 5—5 of Fig. 3.

Fig. 6 is a transverse sectional view through the pump and actuating motor therefor, taken on line 6—6 of Fig. 2.

Fig. 7 is a longitudinal sectional view through the pump, taken on line 7—7 of Fig. 2, and

Fig. 8 is a longitudinal sectional view through a part of the pump cylinder, showing the piston in elevation.

In the accompanying drawings, a vaporizing and atomizing chamber 1 is located at the bottom of the carburetor and has an outlet at its top for the passage of fuel vapors and air into a primary vapor heating chamber 2.

The vaporizing chamber 1 is provided with a perforated false bottom 3 and is normally filled with liquid fuel to the level indicated in Fig. 1. Atmospheric air from a conduit 4 is introduced into the space below the false bottom 3 and passes upwardly through the body of liquid fuel below the false bottom 3, and then through the perforations 5 in said false bottom, which breaks it up into a myriad of fine bubbles, which pass upwardly through the liquid fuel above the false bottom.

Liquid fuel for maintaining the level indicated in the chamber 1 passes from the usual fuel tank (not shown) through a pipe 6, and is forced by a pump 7 through a pipe 8 into and through a pair of nozzles 9 having their outlets located in the chamber 1, just above the level of the liquid fuel therein. The pump 7 may be of any approved form but is preferably of the diaphragm type, as such fuel pumps are now standard equipment on most automobiles.

The nozzles 9 are externally threaded at their lower ends to facilitate their assembly in the chamber 1 and to permit them to be removed readily, should cleaning be necessary.

The upper ends of the nozzles 9 are surrounded by Venturi tubes 10 having baffles 11 located at their upper ends opposite the outlets of the nozzles, as is shown and described in detail in my aforesaid Patent No. 1,997,497. The liquid fuel being forced from the ends of the nozzles 9 into the restricted portions of the Venturi tubes causes a rapid circulation of the air and vapors in the chamber through the tubes 10 and brings the air and vapors into intimate contact with the liquid fuel, with the result that a portion thereof is vaporized. Unvaporized portions of the liquid fuel strike the baffles 11 and are thereby further broken up and deflected downwardly into the upwardly flowing current of air and vapors.

The pump 7 is regulated to supply a greater amount of liquid fuel to the nozzles 9 than will be vaporized. The excess over that vaporized will drop into the chamber 1 and cause the liquid to be maintained at the indicated level. When the liquid fuel rises above that level, a float valve 12 will be lifted and the excess will flow through an overflow pipe 13 into a pipe 14 leading back to the pipe 8 on the intake side of the pump 7. Such an arrangement permits a large amount of liquid fuel to be circulated by the pump 7 without more fuel being withdrawn from the fuel tank than is actually vaporized and consumed in the engine. As the float valve 12 will set upon the end of the outlet pipe 13 as soon as the liquid level drops below the indicated level, there is no danger of vapors passing into the pipe 14 and hence into the pump 7 to interfere with its normal operation.

The amount of liquid fuel vaporized by the nozzles 9 and by the passage of air through the liquid body thereof is sufficient to provide a suitably enriched vaporous mixture for introducing into the passage leading to the intake manifold of the engine through which the main volume of atmospheric air passes.

Vapors formed by the atmospheric air bubbling through the liquid fuel in the bottom of the chamber 1 and those formed as the result of the atomization at the nozzles 9 pass from the top of that chamber into the primary heating chamber 2. As is clearly shown in Fig. 1, the chamber 2 comprises a relatively long spiral passage 15 through which the vaporous mixture gradually passes inwardly to a central outlet 16 to which is connected a conduit 17 leading to a reciprocating pump 18 which forces the vaporous mixture under pressure into a conduit 19 leading to a central inlet 20 of a secondary heating chamber 21 which like the primary heating chamber comprises a relatively long spiral. The vaporous mixture gradually passes outwardly through the spiral chamber 21 and enters a downdraft air tube 22, leading to the intake manifold of the engine, through an outlet 23 controlled by a rotary plug valve 24.

To prevent the engine from backfiring into the vapor chamber 2, the ends of the passages 19 are covered with a fine mesh screen 25, which, operating on the principle of a miner's lamp, will prevent the vapors in the chamber 2 from exploding in case of a backfire, but will not interfere substantially with the passage of the vapors from the chamber 21 into the air tube 23 when the valve 24 is in open position. The air tube 22 preferably is in the form of a venturi with the greatest restriction being at that point where the outlet 23 is located, so that when the valve 24 is opened there will be a pulling force on the vaporous mixture due to the increased velocity of the air at the restricted portion of the air tube opposite the outlet 23, as well as an expelling force on them due to the pressure maintained in the chamber 21 by the pump 18.

Both the primary and secondary spiral heating chambers 15 and 21 and the central portion of the air tube 22 are enclosed by a casing 26 having an inlet 27 and an outlet 28 for a suitable heating medium such as the gaseous products of combustion from the exhaust manifold.

The pump 18 for forcing the vaporous mixture from the primary heating chamber 2 into and through the secondary chamber 21 includes a working chamber 29 for a hollow piston 30 provided with an inlet 31 controlled by a valve 32, and an outlet 33 controlled by a valve 34. The end of the working chamber 29 to which is connected the conduit 17, which conducts the vaporous mixture from the primary heating chamber 2, has an inlet valve 35, and the opposite end of the working chamber has an outlet 36 controlled by a valve 37 positioned in an auxiliary chamber 38, to which is connected the outlet pipe 19 which conducts the vaporous mixture under pressure to the secondary heating chamber 21. Each of the valves 32, 34, 35 and 37 is of the one-way type. They are shown as being gravity-actuated flap valves, but it will be understood that spring-pressed or other types of one-way valves may be used if desired.

One side of the piston 30 is formed with a gear rack 39 which is received in a groove 39a of the wall forming the cylinder of the pump. The gear rack 39 engages with an actuating spur gear 40 carried on one end of the shaft 41 and operating in a housing 42 formed on the pump cylinder. The other end of the shaft 41 carries a spur gear 43, which engages and is operated by a gear rack 44 carried on a piston 46 of a double-acting motor 47. The particular construction of the double-acting motor 47 is not material, and it may be of a vacuum type commonly used for operating windshield wipers on automobiles, in which case a flexible hose 48 would be connected with the intake manifold of the engine to provide the necessary vacuum for operating the piston 45.

Under the influence of the double-acting motor 47, the piston 30 of the pump has a reciprocatory movement in the working chamber 29. Movement of the piston towards the left in Fig. 7 tends to compress the vaporous mixture in the working chamber between the end of the piston and the inlet from the pipe 17, and causes the valve 35 to be forced tightly against the inlet opening. In a like manner, the valves 32 and 33 are forced open and vaporous mixture in that portion of the working chamber is forced through the inlet 31 in the end of the piston 30, into the interior of the piston, where it displaces the vaporous mixture there and forces it into the space between the right-hand end of the piston and the right-hand end of the working chamber. The passage of the vaporous mixture into the right-hand end of the working chamber is supplemented by the partial vacuum created there when the piston moves towards the left. During such movement of the piston, the valve 37 is maintained closed and prevents any sucking back of the vaporous mixture from the secondary heating chamber 21.

When the motor 47 reverses, the piston 30 moves to the right and the vaporous mixture in the right-hand end of the working chamber is forced past the valve 37 and through the pipe 19 into the secondary heating chamber 21. At the same time, a vacuum is created behind the piston 30 and results in the left-hand end of the working chamber again being filled with the vaporous mixture from the primary heating chamber 2.

As the operation of the pump 47 will vary in accordance with the suction created in the intake manifold, it preferably is regulated to actuate the pump at such a speed that the vaporous mixture will always be pumped into the secondary heating chamber at a rate sufficient to maintain a greater pressure there than is desired. In order that the pressure in the working chamber may at all times be maintained at the optimum, a pipe

50 having an adjustable pressure-regulating valve 51 is connected across the inlet and outlet pipes 17 and 19. The valve 51 will permit a portion of the vaporous mixture discharged from the pump to be bypassed to the inlet 17 so that a pressure, predetermined by the seating of the valve 51, will at all times be maintained in the second heating chamber 21.

The air tube 22 is provided with a butterfly throttle valve 52 and a choke valve 53, as is usual with carburetors adapted for use with internal combustion engines. Operating stems 54, 55, 56 for the valves 52, 53 and 24 respectively, extend through the casing 26. An operating arm 57 is fixedly secured to the outer end of the stem 55 and is connected to a rod 58 which extends to the dashboard of the automobile or some other place conveniently located to the driver of the automobile. The outer end of the stem 56 of the valve 24 which controls the outlet 23 from the secondary heating chamber 21 has one end of an operating arm 59 fixedly secured thereto. The other end of the arm 59 is pivotally connected to a link 60 which extends downwardly and pivotally connects to one end of a bell crank lever 61, fixedly secured to the end of the stem 54 of the throttle valve 52. The other end of the bell crank lever 61 is connected to an operating rod 62 which, like the rod 58, extends to a place conveniently located to the driver. The valves 24 and 52 are connected for simultaneous operation so that when the throttle valve 52 is opened to increase the speed of the engine the valve 24 will be opened to admit a larger amount of the heated vaporous mixture from the secondary heating chamber 21.

While the suction created by the pump 18 ordinarily will create a sufficient vacuum in the primary heating chamber 2, to cause atmospheric air to be drawn into and upwardly through the body of liquid fuel in the bottom of the vaporizing chamber 1, in some instances it may be desirable to provide supplemental means for forcing the atmospheric air into and through said body of liquid and in such cases an auxiliary pump may be provided for that purpose, or the air conduit 4 may be provided with a funnel-shaped intake which is positioned behind the fan 63 which is customarily placed behind the radiator of the engine.

The foregoing description has been given in connection with a downdraft type of carburetor, but it is to be understood that the invention is not limited to use with such type carburetors and that the manner in which the mixture of atmospheric air and vapors is introduced into the engine cylinders is immaterial as far as the advantages of the carburetor are concerned.

Before the carburetor is put into use the pressure-regulating valve 51 in the bypass pipe 50 will be adjusted so that the pressure best suited for conditions under which the engine is to operate will be maintained in the secondary heating chamber 21. When the valve 51 has thus been set and the engine started, the pump 18 will create a partial vacuum in the primary heating chamber 2 and cause atmospheric air to be drawn through the conduit 4 and to bubble upwardly through the liquid in the bottom of the vaporizing and atomizing chamber 1 with resultant vaporization of a part of the liquid fuel therein. At the same time, the pump 7 will be set into operation and liquid fuel will be pumped from the fuel tank through the nozzles 9 which will result in an additional amount of the fuel being vaporized. The vapors resulting from such atomization of the liquid fuel and the passage of the air through the body of the liquid will pass into and through the spiral chamber 1 where they will be heated by the products of combustion in the surrounding chamber formed by the casing 28. The fuel vapors and air will gradually pass inwardly to the outlet 16 and thence through the conduit 17 to the pump 18 which will force them into the secondary heating chamber 21 in which they will be maintained at the predetermined pressure by the pressure-regulating valve 51. The vaporous mixture is further heated in the chamber 21 and passes spirally outwardly to the valve-controlled outlet 23 which opens into the air tube 22 which conducts the main volume of atmospheric air to the intake manifold of the engine.

The heating of the vaporous mixture in the heating chambers 2 and 21 tends to cause them to expand, but expansion in the chamber 21 is prevented due to the pressure maintained in that chamber by the regulating valve 51. However, as soon as the heated vaporous mixture passes the valve 24 and is introduced into the air flowing through the intake tube 22, it is free to expand and thereby become relatively light so that a more intimate mixture with the air is obtained prior to the mixture being exploded in the engine cylinders. Thus it will be seen that the present invention not only provides means wherein the vaporous mixture from the heating chamber 21 is forced into the air passing through the air tube 22 by a positive force, but it is also heated to such an extent that after it leaves the chamber 21 it will expand to such an extent as to have a density less than it would if introduced directly from the vaporizing and atomizing chamber 1 into the air tube 22.

The majority of the liquid particles entrained by the vaporous mixture leaving the chamber 1 will be separated in the first half of the outermost spiral of the primary heating chamber 2 and drained back into the body of liquid in the tank 1. Any liquid particles which are not thus separated will be carried on with the vaporous mixture and due to the circulation of that mixture and the application of heat, will be vaporized before the vaporous mixture is introduced into the air tube 22 from the secondary heating chamber 21. Thus "dry" vapors only are introduced into the engine cylinders and any burning of liquid particles of the fuel in the engine cylinder, which would tend to raise the temperature of the engine above that at which it operates most efficiently is avoided.

While the fullest benefits of the invention are obtained by using both a primary and a secondary heating chamber, the primary heating chamber may, if desired, be eliminated, and the vaporous mixture pumped directly from the vaporizing and atomizing chamber 1 into the spiral heating chamber 21.

From the foregoing description it will be seen that the present invention provides an improvement over the carburetor disclosed in my aforesaid Patent 1,997,497, in that it is possible to maintain the vaporous mixture in the heating chamber 21 under a predetermined pressure, and that as soon as the vaporous mixture is introduced into the main supply of air passing to the intake manifold of the engine, it will expand and reach a density at which it will form a more intimate mixture with the air. Furthermore, the introduction of the vaporous mixture into the air stream in the tube 22 causes a certain amount of turbulence which also tends to give a more intimate mixture of the vapor molecules with the air.

I claim:

1. A carburetor for internal combustion engines comprising a vaporizing chamber, means for vaporizing fuel therein, a vapor-heating chamber, a pump for pumping vapors from the vaporizing chamber into the vapor-heating chamber, a bypass for fuel vapors from the outlet side of said pump to the inlet side thereof, means for maintaining vapors in said vapor-heating chamber during normal operation of the carburetor under a predetermined pressure, a passage for atmospheric air adapted to communicate with the intake manifold of the engine and means for introducing vapors from said vapor-heating chamber into said atmospheric air passage.

2. A carburetor for internal combustion engines comprising a vaporizing chamber, means for vaporizing fuel therein, a vapor-heating chamber, a pump for pumping vapors from the vaporizing chamber into the vapor-heating chamber, a bypass for fuel vapors from the outlet side of the pump to the inlet side thereof, a pressure-regulating valve in said bypass for maintaining vapors in said vapor-heating chamber during normal operation of the carburetor under predetermined pressure, a passage for atmospheric air adapted to communicate with the intake manifold of the engine, and means for introducing vapors from said vapor-heating chamber into said atmospheric air passage.

3. A carburetor for internal combustion engines comprising a vaporizing chamber, means for vaporizing fuel therein, a primary vapor-heating chamber, a secondary vapor-heating chamber, means for conducting vapors from the vaporizing chamber to the primary vapor-heating chamber, means for conducting vapors from the primary vapor-heating chamber to the secondary vapor-heating chamber, means for maintaining vapors in said secondary vapor-heating chamber during normal operation of the carburetor under a predetermined superatmospheric pressure, means for varying the pressure at which vapors are maintained in said secondary vapor-heating chamber independently of the speed of the engine, a passage for atmospheric air adapted to communicate with the intake manifold of the engine, and means for introducing vapors from said secondary vapor-heating chamber into said atmospheric air passage.

4. A carburetor for internal combustion engines comprising a vaporizing chamber, means for vaporizing fuel therein, a vapor-heating chamber, means for conducting vapors from the vaporizing chamber into the vapor-heating chamber, means for maintaining vapors in said vapor-heating chamber during normal operation of the carburetor under superatmospheric pressure, means for varying the pressure at which vapors are maintained in said vapor-heating chamber independently of the speed of the engine, and means for introducing vapors from said vapor-heating chamber into the intake manifold of the engine.

5. A carburetor for internal combustion engines comprising a vaporizing chamber, means for maintaining a body of liquid fuel in the bottom of said chamber, means for introducing atmospheric air into the body of liquid in the vaporizing chamber whereby said air will bubble upwardly through the liquid fuel and vaporize a portion thereof, a vapor-heating chamber, means for conducting vapors from the vaporizing chamber into the vapor-heating chamber, means for maintaining vapors in said vapor-heating chamber during normal operation of the carburetor under a predetermined superatmospheric pressure, means for varying the pressure at which vapors are maintained in said vapor-heating chamber independently of the speed of the engine, and means for introducing vapors from said vapor-heating chamber into the intake manifold of the engine.

6. A carburetor for internal combustion engines comprising a vaporizing chamber, means for maintaining a body of liquid fuel in the bottom of said chamber, means for introducing atmospheric air into the body of liquid in the vaporizing chamber whereby said air will bubble upwardly through the liquid fuel and vaporize a portion thereof, a primary vapor-heating chamber, a secondary vapor-heating chamber, means for passing vapors from the vaporizing chamber to the primary vapor-heating chamber, means for conducting vapors from the primary vapor-heating chamber to the secondary vapor-heating chamber, means for maintaining vapors in said secondary vapor-heating chamber under pressure, and means for introducing vapors from said secondary vapor-heating chamber into the intake manifold of the engine.

7. A carburetor for internal combustion engines comprising a vaporizing chamber, means for maintaining a body of liquid fuel in the bottom of said chamber, means for introducing atmospheric air into the body of liquid in the vaporizing chamber whereby said air will bubble upwardly through the liquid fuel and vaporize a portion thereof, a spiral primary vapor-heating chamber, a spiral secondary-vapor-heating chamber, means for passing vapors from the vaporizing chamber into an outer spiral of the primary vapor-heating chamber, means for passing vapors from the center of the primary vapor-heating chamber to the center of the secondary vapor-heating chamber, and means for passing vapors from an outer spiral of the secondary vapor-heating chamber to the intake manifold of the engine.

8. A carburetor for internal combustion engines comprising a vaporizing chamber, means for maintaining a body of liquid fuel in the bottom of said chamber, means for introducing atmospheric air into the body of liquid in the vaporizing chamber whereby said air will bubble upwardly through the liquid fuel and vaporize a portion thereof, atomizing nozzles positioned in said atomizing chamber, means for forcing liquid fuel through said atomizing nozzles whereby an additional amount of liquid fuel is vaporized, a primary vapor-heating chamber, a secondary vapor-heating chamber, means for passing vapors from the vaporizing chamber to the primary vapor-heating chamber, means for passing vapors from the primary vapor-heating chamber to the secondary vapor-heating chamber, means for maintaining vapors in said secondary vapor-heating chamber under a predetermined pressure, and means for passing vapors from said secondary vapor-heating chamber to the intake manifold of the engine.

CHARLES N. POGUE.

Oct. 22, 1940. I. B. NEWBERY 2,218,922

VAPORIZER FOR COMBUSTION ENGINES

Filed June 29, 1938 2 Sheets—Sheet 1

Fig. 1.

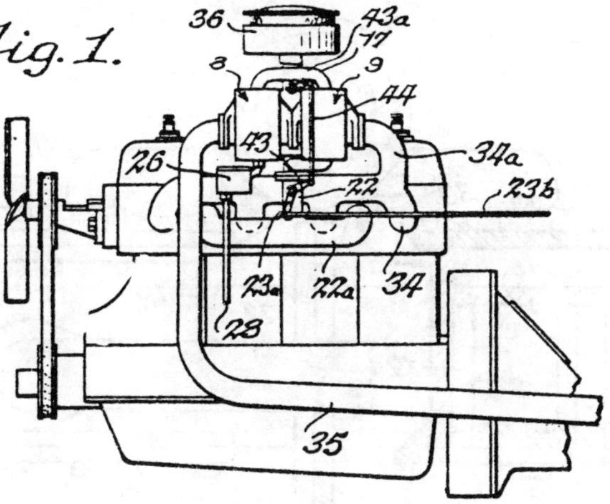

Fig. 2.

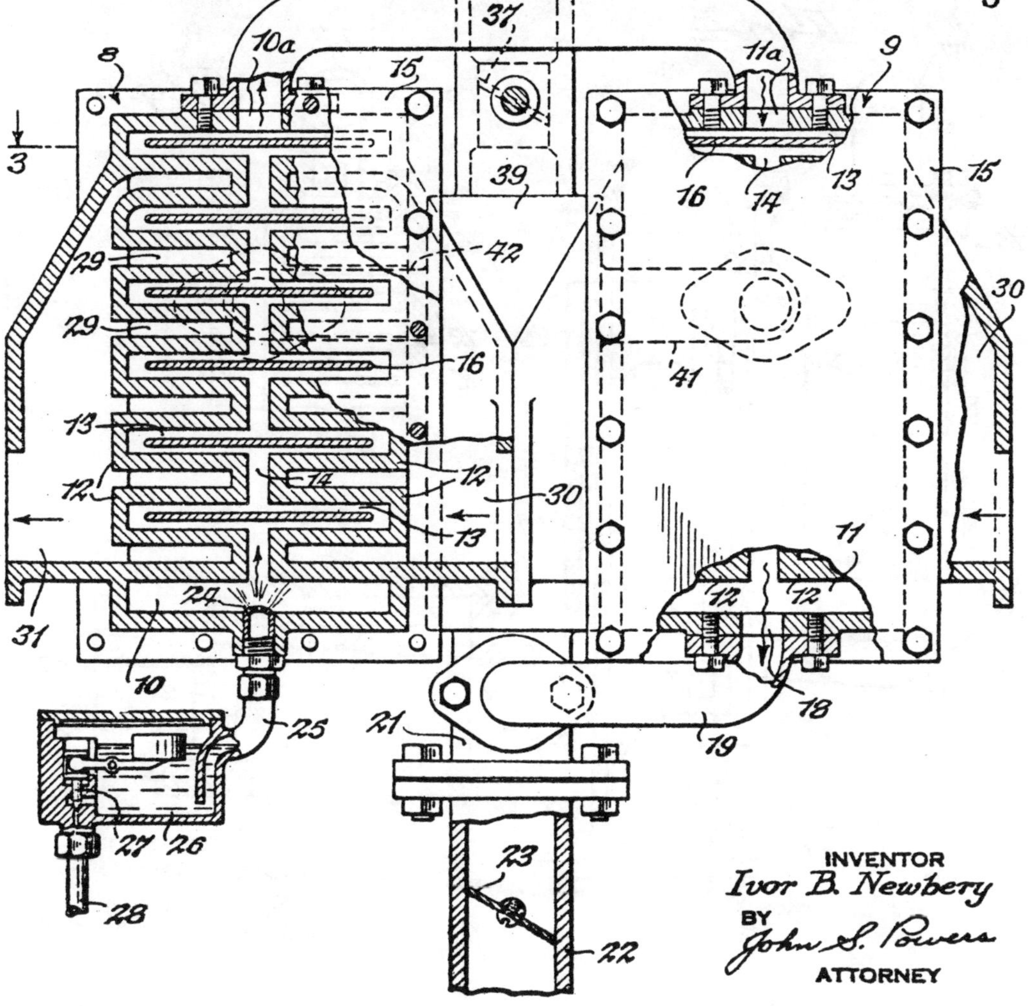

INVENTOR
Ivor B. Newbery
BY
John S. Powers
ATTORNEY

Oct. 22, 1940. I. B. NEWBERY 2,218,922

VAPORIZER FOR COMBUSTION ENGINES

Filed June 29, 1938 2 Sheets-Sheet 2

Fig. 3.

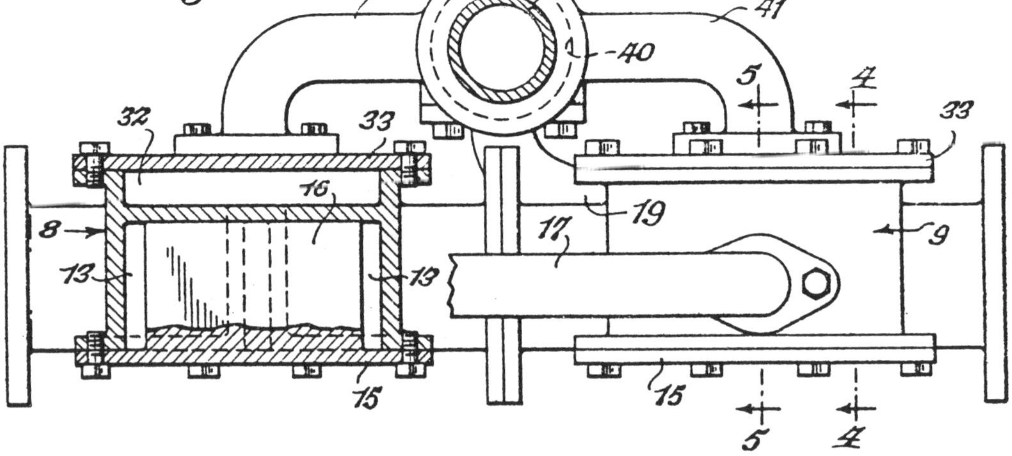

Fig. 4.

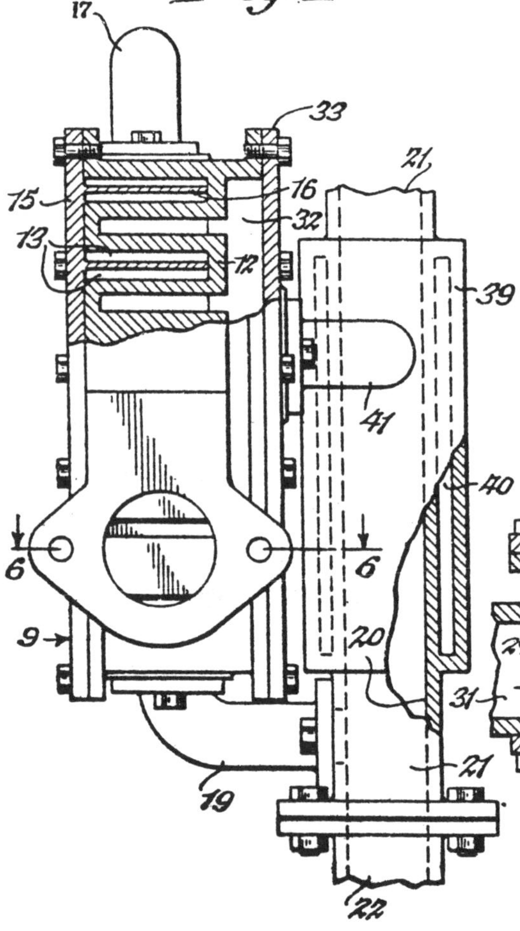

Fig. 5.

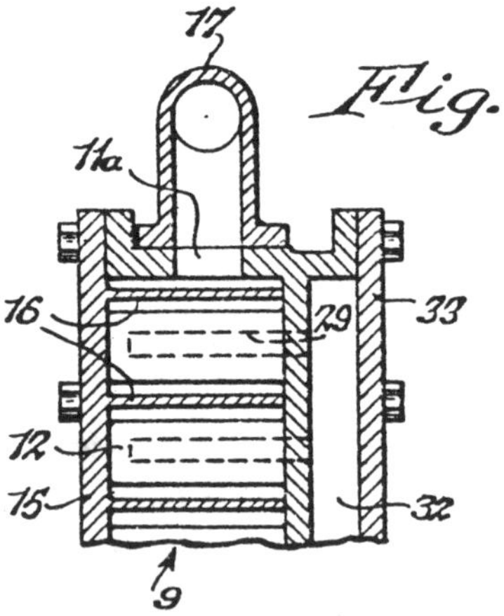

Fig. 6.

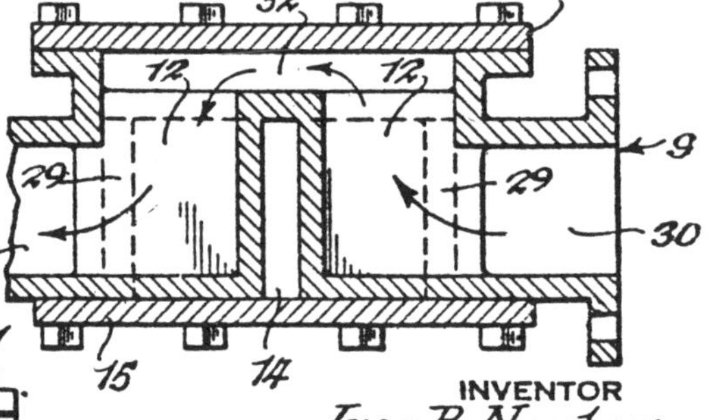

INVENTOR
Ivor B. Newbery
BY
John S. Powers
ATTORNEY

Patented Oct. 22, 1940 **2,218,922**

UNITED STATES PATENT OFFICE

2,218,922

VAPORIZER FOR COMBUSTION ENGINES

Ivor B. Newbery, Buffalo, N. Y., assignor to Universal Research Corporation, Kenmore, N. Y., a corporation of New York

Application June 29, 1938, Serial No. 216,546

5 Claims. (Cl. 48—107)

This invention relates to fuel vaporizing devices for combustion engines and more particularly is concerned with improvements in devices of the kind wherein provision is made for utilizing the exhaust gases of the engines as a heating medium to aid in the vaporization of the fuel.

One object of the invention is to provide a device which will condition the fuel in such a manner that its potential energy may be fully utilized, thereby insuring better engine performance and a saving in fuel consumption and preventing the formation of carbon in the cylinders of the engine and the production of carbon monoxide and other objectionable gases.

A further object is to provide a device which is so designed that the fuel is delivered to the cylinders of the engine in a highly vaporized, dry and expanded state, this object contemplating a device which is available as a muffler and in which the vaporization and expansion of the liquid components are effected at sub-atmospheric pressures and prior to their mixture with the air component.

A still further object is to provide a device which will condition the components of the fuel in such a manner that they may be uniformly and intimately mixed without the use of a carburetor.

A still further object is to provide a device which will enable the use of various inferior and inexpensive grades of fuel.

The invention is illustrated in the accompanying drawings, in which—

Figure 1 is a view in elevation of the device as applied to the engine of a motor vehicle;

Figure 2 is an enlarged view of the device, partially in elevation and partially in section;

Figure 3 is a section taken along line 3—3 of Figure 2;

Figure 4 is a section taken along line 4—4 of Figure 3;

Figure 5 is a fragmentary section taken along line 5—5 of Figure 3; and

Figure 6 is a section taken along line 6—6 of Figure 4.

The device, as illustrated, includes similar casings 8 and 9 which are secured together as a unit and which are formed to provide vaporizing chambers 10 and 11, respectively, it being understood that the number of casings may be varied. Two series of ribs 12 are formed in each of the vaporizing chambers, the ribs of each series being spaced from one another to provide branch passages 13 and being spaced from the ribs of the adjacent series to provide main passages 14 with which the said branch passages communicate. The vaporizing chambers are closed by cover plates 15. The latter carry baffles 16 which are supported in the spaces between the ribs 12, the said baffles extending across the main passages 14 and into, but short of the ends of, the branch passages 13 to provide tortuous paths. The outlet 10a of the chamber 10 is connected by a conduit 17 with the inlet 11a of the chamber 11, the outlet 18 of the latter chamber being connected by a conduit 19 with a mixing chamber 20 which is located at the lower end of a pipe 21. The said pipe is connected to an extension 22 of the intake manifold 22a of the engine. A valve 23 is arranged in the said extension and is connected by a lever 23a (Figure 1) and rod 23b to a conventional throttle (not shown).

The gasoline or other fuel is introduced into the vaporizing chamber 10 through a nozzle 24. The latter is connected by a pipe 25 with a reservoir 26 in which a predetermined amount of the fuel is maintained by a float controlled valve 27, the fuel being supplied to the reservoir through a pipe 28.

In accordance with the invention the ribs 12 are hollow, each being formed to provide a cell 29. The cells in one series of ribs open at one side into an inlet chamber 30 while the cells of the companion series open at one side into an outlet chamber 31, the cells of both series of ribs opening at their backs into a connecting chamber 32 which is located behind the ribs and which is closed by a cover plate 33. The casings 8 and 9 are arranged end-to-end so that the outlet chamber of the latter communicates with the inlet chamber of the former, the gases from the exhaust manifold 34 being introduced into the inlet chamber of the casing 9 through an extension 34a. The exhaust gases, therefore, enter the series of cells at the right side of the said casing. They pass through the said cells into the connecting chamber at the rear and then enter the cells in the companion series of cells at the left side of the casing. The gases then enter the inlet chamber of the casing 8. They pass successively through the two series of cells in the said casing and enter an exhaust pipe 35 as they leave the outlet chamber 31, the path traversed by the exhaust gases being clearly indicated by the arrows in Figure 6. As the said gases pass through the casings 8 and 9 in the manner described their speed is reduced to such a degree that a muffler or other silencing device is rendered unnecessary.

It will be apparent that when the engine is operating at a normal temperature the gasoline or other fuel introduced into the chamber 10 will be immediately vaporized by contact with the hot walls of the ribs 12. The vapors thus produced are divided into two streams one of which is caused to enter each of the branch passages at one side of the casing and the other of which is caused to enter each of the branch passages at the opposite side of the casing. The two vapor streams merge as they pass around the final baffle and enter the conduit 17 but are again divided and heated in a similar manner as they flow through the casing 9. Each of the vapor streams is constantly in contact with the highly heated walls of the ribs 12. In their passage through the casings 8 and 9, therefore, the vapors are heated to such a degree that a dry highly vaporized gas is produced. In this connection it will be noted that the vaporizing chambers are maintained under a vacuum and that vaporization is effected in the absence of air. Conversion of the liquid into highly expanded vapors is thus insured. The flow of the exhaust gases through the casings 8 and 9 is counter to that of the vapors. The latter, therefore, are heated in stages and are introduced into the mixing chamber 20 when at their highest temperature.

The air which is mixed with the fuel vapors enters the pipe 21 after passing through a conventional filter 36, the amount of air entering the said pipe being regulated by a valve 37. The invention contemplates the heating of the air prior to its entry into the mixing chamber 20. To this end a jacket 39 is formed or provided around the pipe 21. The said jacket provides a chamber 40 which communicates with the chamber 32 of the casing 9 through an inlet pipe 41 and with the corresponding chamber of the casing 8 through an outlet pipe 42. A portion of the exhaust gases is thus caused to pass through the chamber 40 to heat the air as it passes through the conduit 21 on its way to the mixing chamber, the valve 37 being connected to the valve 23 by arms 43 and 43a and a link 44 so that the volume of air admitted to the mixing chamber is increased proportionately as the volume of vapors is increased. As the fuel vapors and air are both heated to a high temperature and are in a highly expanded state when they enter the mixing chamber they readily unite to provide a uniform mixture, the use of a carburetor or similar device for this purpose being unnecessary.

From the foregoing it will be apparent that the components of the fuel mixture are separately heated prior to their entry into the mixing chamber 20. As the vapors which are produced are dry and highly expanded complete combustion is insured. The potential energy represented by the said vapors may thus be fully utilized, thereby insuring better engine performance and a saving in fuel consumption. At the same time the formation of carbon in the combustion chambers and the production of carbon monoxide and other objectionable exhaust gases are prevented. The device has the further advantage that, owing to the high temperature to which the fuel is heated prior to its admission into the combustion chambers, various inferior and inexpensive grades may be employed with satisfactory results.

I claim as my invention:

1. A device of the character described including a casing, a vaporizing chamber in said casing which is connected to the intake manifold of a combustion engine and into which a liquid fuel is introduced, an inlet chamber at one side of said vaporizing chamber into which exhaust gases from said engine are introduced, an outlet chamber at the opposite side of said vaporizing chamber into which said exhaust gases are discharged, a plurality of substantially parallel series of ribs which define tortuous paths through said vaporizing chamber and which are formed to provide cells, the cells of one of said series of ribs communicating directly with said inlet chamber while the cells of a second of said series of ribs communicate directly with said outlet chamber, a connecting chamber with which the cells of both of said series of ribs communicate, said inlet chamber, outlet chamber and connecting chamber providing a path for the exhaust gases which extends transversely with respect to said series of ribs, a mixing chamber to which the vapors produced in said vaporizing chamber are conducted and means for introducing air into said last named chamber to provide a fuel mixture having the proper proportions of each component.

2. A device of the character described including a vaporizing chamber which is connected to the intake manifold of a combustion engine and into which a liquid fuel is introduced, a plurality of substantially parallel series of hollow ribs, the ribs of each series being spaced from one another and from the ribs of an adjacent series to provide main and branch passages, baffles which are arranged in said passages and which provide two tortuous paths, means for causing exhaust gases from said engine to simultaneously enter all of the ribs of one of said series and thereafter to simultaneously enter all of the ribs of the adjacent series, whereby to heat the walls of said passages to convert said fuel into vapors, said baffles causing one part of the vapors to traverse one of said paths and a second part to traverse the second of said paths, a mixing chamber to which the vapors thus produced are conducted and means for introducing air into said last named chamber to provide a fuel mixture having the proper proportions of each component.

3. A device of the character described including a casing, a vaporizing chamber in said casing which is connected to the intake manifold of a combustion engine and into which a liquid fuel is introduced, a cover plate for said vaporizing chamber, a plurality of substantially parallel series of hollow ribs, the ribs of each series being spaced from one another and from the ribs of an adjacent series to provide main and branch passages, baffles carried by said cover plate, said baffles being supported in said passages and providing tortuous paths, means for causing exhaust gases from said engine to simultaneously enter all of the ribs of one of said series and thereafter to enter all of the ribs of an adjacent series, whereby to heat the walls of said passages to convert said fuel into vapors, a mixing chamber to which the vapors thus produced are conducted and means for introducing air into said last named chamber to provide a fuel mixture having the proper proportions of each component.

4. A device of the character described including a casing, a vaporizing chamber in said casing which is connected to the intake manifold of a combustion engine and into which a liquid fuel is introduced, a cover plate for said vaporizing chamber, a plurality of substantially parallel series of hollow ribs, the ribs of each series being spaced from one another and from the ribs of an adjacent series to provide main and branch passages, baffles carried by said cover plate, said baffles being supported in said passages and providing two tortuous paths, means for causing ex-

haust gases from said engine to simultaneously enter all of the ribs of one of said series and thereafter to enter all of the ribs of an adjacent series, whereby to heat the walls of said passages to convert said fuel into vapors, said baffles causing one part of the vapors to traverse one of said paths and a second part to traverse the second of said paths, a mixing chamber to which the vapors thus produced are conducted and means for introducing air into said last named chamber to provide a fuel mixture having the proper proportions of each component.

5. A device of the character described including a casing, a vaporizing chamber in said casing which is connected to the intake manifold of a combustion engine and into which a liquid fuel is introduced, a cover plate for said vaporizing chamber, an inlet chamber at one side of said vaporizing chamber into which exhaust gases from said engine are introduced, an outlet chamber at the opposite side of said vaporizing chamber into which said exhaust gases are discharged, a plurality of substantially parallel series of ribs which are formed to provide cells, the ribs of each series being spaced from one another and from the ribs of an adjacent series to provide main and branch passages, the cells of one of said series of ribs communicating directly with said inlet chamber while the cells of the other of said series of ribs communicate directly with said outlet chamber, a connecting chamber with which the cells of both of said series of ribs communicate, whereby said exhaust gases are caused to flow first through the cells of one of said series of ribs and then through the cells of a second series of said said ribs to heat the walls of said passages, said inlet chamber, outlet chamber and connecting chamber providing a path for the exhaust gases which extends transversely with respect to said series of ribs, baffles carried by said cover plate, said baffles being supported in said passages and providing two tortuous paths, said baffles causing one part of the vapors which are produced to traverse one of said paths and a second part to traverse the other of said paths, a mixing chamber to which said vapors are conducted and means for introducing air into said last named chamber to provide a fuel mixture having the proper proportions of each component.

IVOR B. NEWBERY.

April 3, 1945. B. KLEIN **2,373,032**

ELECTROLYTIC DECOMPOSING GAS GENERATOR

Filed June 29, 1942

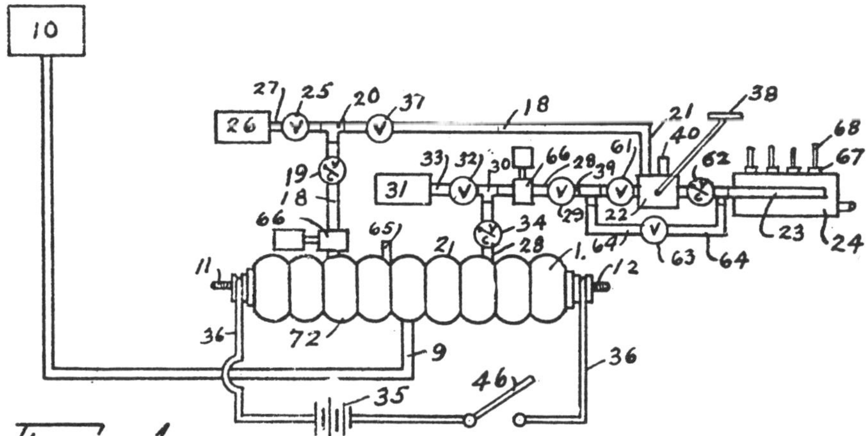

FIG. 1.

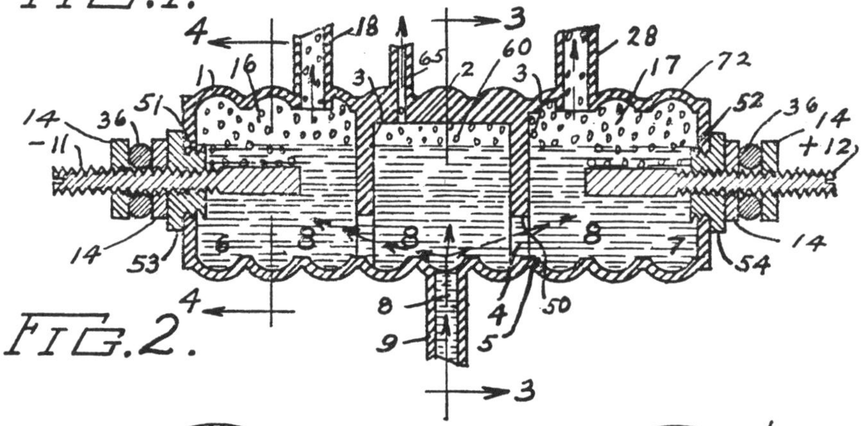

FIG. 2.

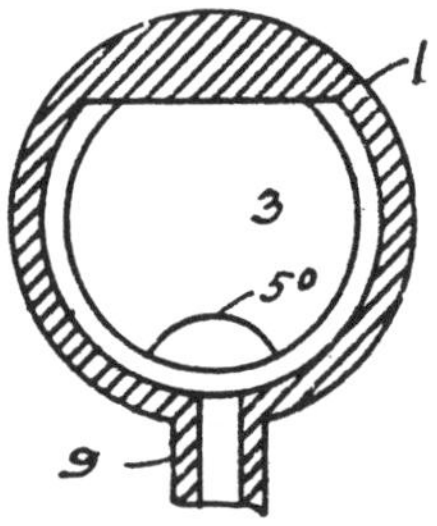

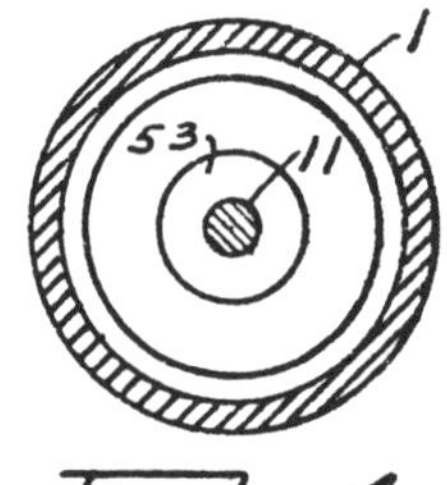

FIG. 4.

FIG. 3.

Inventor

BERNARD KLEIN.

By Howard J. Whelan.

Attorney

Patented Apr. 3, 1945 2,373,032

UNITED STATES PATENT OFFICE

2,373,032

ELECTROLYTIC DECOMPOSING GAS GENERATOR

Bernard Klein, Baltimore, Md.

Application June 29, 1942, Serial No. 448,991

3 Claims. (Cl. 204—278)

This invention refers to fuels for internal combustion engines and more particularly to a method and device to be used in decomposing or extracting gases from liquids such as water and then arranging them for use as an explosive mixture in engines of the internal combustion type. It has among its objects to provide a device that will separate combustible gases from the liquid compounds having same chemically combined in them, as for example: oxygen and hydrogen from water, through an electrolytic process. Another object is to have the device of such construction as to be readily adaptable for insertion or connection among the parts of a conventional internal combustion engine of the automobile, airplane, boat or other type without requiring radical or complicated arrangements for the purpose and capable of practical employment. A further object is to provide for the operation of the device in a convenient manner with adjustments permitting it to meet varied conditions encountered during operation. A further object is to provide a device that will automatically supply treated water or other liquid medium as needed for the device. A further object is to have the device adaptable for adjustment by remote control or at a distant location remote from the operator or user. It is a particular object of the invention to construct a self contained unit that can be applied to an internal combustion engine; and take relatively small space and be attached thereto in a compact and attractive manner.

Other objects will become apparent as the invention is more fully set forth.

There has been difficulty heretofore in employing the combustible gases of relatively inexpensive liquids, such as water, for power development, due to the cost of equipment, its cumbersome nature, and cost of operation. This invention is intended to relieve this situation by providing a simple manner for manufacturing explosive gases by decomposing same from liquids that are normally harmless and inexplosive.

A typical form of internal combustion engine furnished to propel an automobile, aircraft, boat or the like has fuel tank, carburetor, ignition. The fuel consists of gasoline or other hydrocarbons that runs from the tank to the carburetor, a current of air is drawn over the liquid fuel, and takes up the vapor and mixes with the latter to form the explosive mixture, to operate the internal combustion engine. This system has been found very satisfactory as long as a sufficient supply of gasoline or other fuel oil was available to meet its needs. Under present conditions the demand has increased and far exceeds the supply. Any additional system of fuel production that will be less costly and can use ingredients that can be found readily at hand and not heretofore used for the purpose herein described affords attractive possibilities. In this invention the fuel tank will be filled with water to which has been added an electrolyte, usually sulphur trioxide or sulphuric acid to increase its conductivity so an electric current will pass through it. The fluid is then passed from the tank to the generating device by a pump, vacuum tank, gravity or whatever system for the fuel supply the vehicle may be equipped with. The device is preferably placed between the fuel tank and engine and used with the existing carburetor. The device can be placed in the system conveniently, as it takes very little room, is of simple form, and has effective arrangement for attachment to the system. An important feature of the device is in the fact that it has arrangements for augmenting the supply of gas generated by the device. Another important feature of the device is that part of the gas generated may be used to replenish its respective auxiliary tube.

In the drawing which indicates a form of this invention:

Figure 1 is a diagrammatic view showing the various parts used to form a complete system embodying this invention,

Figure 2 is a longitudinal view taken through the generator,

Figure 3 is a sectional view taken along line 3—3 of Figure 2, and

Figure 4 is a sectional view taken along line 4—4 of Figure 2.

Similar reference characters refer to similar parts throughout the drawing.

In the drawing which indicates a water generator 1, by way of example, a source of electricity 35 is employed for converting the water into oxygen and hydrogen gases. The generator is preferably of hollow cylindrical form, with its upper wall 2 supporting partitions 3 dividing the interior into two compartments 6 and 7, respectively, and with an intermediate chamber 60. A space 4, is left between the lower portion 50 of the partition and the floor 5 of the generator as indicated in Figure 1. The liquid 8 to be processed is conducted to the bottom wall 5 into the generator through a pipe 9, connected with the liquid reservoir tank 10. The end portions of the generator have screw threaded holes 51 and 52, into which insulators 53 and 54 respectively screw.

The insulator 53, holds the negative cathode terminal 11, and insulator 54, the positive anode terminal 12. The insulators and terminals are water tight and are constructed to serve as stuffing boxes. The terminals 11 and 12 project into compartments 6 and 7 respectively, and contact the liquid residing therein. Electricity from a suitable source 35, feed the terminals when it is desired to decompose the liquid, which in this instance is water, into its component gases oxygen and hydrogen. The water is acidulated with sulphur trioxide (SO_3) or sulphuric acid (H_2SO_4) to improve its conductivity and electrolytic decomposition. The current passing into the water, when the circuit 36 is operating, causes hydrogen gas to liberate at the cathode terminal 11, and oxygen at the anode terminal 12. This is usually indicated by bubbles forming and arising from the terminals. The hydrogen gas collects in the compartment 6 and rises above the water into the space 16 and flows out of the outlet 18 into piping that conducts it to a cylinder when it is stored. Oxygen accumulating at terminal 12 passes into compartment 7 thence through the liquid into the space 17. From thence it passes into the piping 28. The piping is provided with check valves 19 and 34, fittings 20 and 30, carburetor 22, pipes 21, 23, 27, 33, 28 and 39, valves 25, 29 and 37 as indicated in the drawing. The internal combustion engine used for the operation is designated as 24. A by-pass connection with conventional accessories is indicated to eliminate the carburetor as may be required at any time. The hook-up comprises valves 63 and 61 and check valve 62 and pipes 64. The circuit 36 when its switch 46 is closed, causes electricity to flow through the acidulated water from anode 12 to cathode 11. This breaks up molecules of water into hydrogen and oxygen. The hydrogen arises from the cathode 11, into the space 16 in chamber or compartment 6, and the oxygen from the anode terminal 12 into the compartment 7; the partition 3, serving to keep them apart and segregated. The space 4 allows the water to move freely from one compartment of the generator to the other. The current also can pass through the water without obstruction by the partition. The partitions however mechanically prevent the hydrogen and oxygen mixing with each other in the generator. Any tendency of either hydrogen or oxygen to stray outside of their respective compartments is prevented by the intermediate chamber 60 which collects them. Such stray gases are removable through an outlet pipe 65, provided for the purpose. The vibration of the generator under any use in the vehicle mentioned, will not be sufficient to mix the gases by reason of the partitions 3 and the intermediate chamber 60, functioning as described. Pumps 66 or other conventional arrangement are utilized in the outlet piping 18 and 28, to force the hydrogen and oxygen respectively out of the compartments 6 and 7 through piping 21 and 39 to the mixer or carburetor 22 on their way to the engine 24, to be ignited by spark plugs 67 and ignition system 68. The hydrogen and oxygen are preferably combined through the medium of the spark and in doing so explode powerfully. The explosion causes the engine pistons to reciprocate and operate the engine in the usual manner. Low pressure explosions can be induced through a stoichiometric mixture of hydrogen and oxygen at a suitable ignition temperature. Ribs 72 provide expanding mechanical space and prevent the liquid from sloshing back and forth. In the operation of the device the tank 10, is filled with water to which has been preferably added sulphur trioxide or sulphuric acid to increase its conductivity. The tank is located high enough so the liquid can run down into the generator 1, by gravity, or supplied to the gas generator by the fuel pump or other lifting method which then exists in the automobile or other vehicle to which it is applied. The anode terminal 12 is connected to the positive side of an electric battery or other electric source, and the cathode 11 is connected to the negative side of the electric source. A switch 46 completes the electrical connections and may be placed on the instrument board convenient to the operator. When the electric source is turned on by switch 46, bubbles will form around the cathode and anode terminals and will arise through the solution and be collected in their respective chambers or spaces. The valve 37 is opened and pump 66 started and allows gas to flow through pipes 21 and check valve 19 into carburetor or other mixing device 22 which is opened or closed by lever 38. Oxygen flows from the space 17, through check valve 34, fitting 30, pipe 28, pump 66, valve 29 and pipe 39 into the carburetor or other mixing device 22 where the hydrogen is properly mixed with air 40 or oxygen to give proper combustion, after which the mixed gases are conveyed to the internal combustion engine where they are ignited similarly to present day use of gasoline or fuel oil in internal combustion engines. An auxiliary cylinder 26 of hydrogen is attached to pipe 27, valve 25 all of which connect with fitting 20. An auxiliary cylinder 31, of oxygen is attached to pipe 33, valve 32 and fitting 30. The tanks 26 and 31 are used to supplement the supply of gases developed by the generator, or may be used when it is desired to remove the generator from the lines for repairs etc. The generator can be constructed compactly and mounted close to the engine. Lock nuts 14 provide for adjustment of electrodes and for attachment of electric wires thereto.

The generator itself is preferably of insulating material such as any well known plastic, in order to increase the efficiency of the generator and prevent conducting of electricity to the walls where the action would be slightly short circuiting.

The ribbed or corrugated bottom surface 5 serves to create an interruption in the movement of the molecules or ions energized to a breaking point by the electric current flowing from the electrodes 12 and 11 respectively. The breaking point is that condition arising just prior to the separation of the water molecules (or other substances acted upon) into their atoms oxygen and hydrogen. The effect of this curvilinear and obstructing construction seems to be that of a mechanical disturbance or collision to the flow of the ions, causing them to collide more easily and induce separation of the atoms more quickly, than without it. This improves the effectiveness of the generator and increases the efficiency of the process. The walls of the tank 1 are given their corrugated form inside to reduce the sloshing of the water in the lower portions, whilst the upper surfaces thereof serve to entrap the gases and induce them to deposit their particles of loose water thereon and cause the latter to drip back into the main body of the water. This keeps the escaping gases relatively dry. The partitions

3 serve to cooperate with the corrugations in preventing sloshing and reducing the escape of moisture from the chamber 60 to the chambers 16 and 17. This also diminishes any tendency of the incoming water to appreciably disturb the level of the water in the electrode chambers. The chamber 60 acts as a cushion device.

While but one form of the invention is indicated in the drawing and described herein, it is not desired to limit this application to that particular form or in any other way, otherwise than limited by the descriptions and claims, as it is appreciated that other forms and constructions could be embodied that would employ the same principles and come within the scope of the appended claims.

Having thus described the invention what is claimed is:

1. A generator for automatic use and for decomposing water and segregating its component hydrogen and oxygen, comprising in combination a hollow cylindrical generator relatively long horizontally for holding water, having a plurality of orifices therein said generator having an internally corrugated form adapted to accumulate gases therein and drain same above and prevent sloshing of the water below, partition means in the interior of the generator extending through portion of same and arranged to divide same into compartments adapted to receive the hydrogen, oxygen and stray gases segregatively from each other, a plurality of electric terminals horizontally disposed and adjacent to the surface of the water projecting into the cylinder through some of said orifices, said terminals being arranged to contact the water in said cylinder for electrolyzing same when an electric current is passed from the terminal through the water, and to direct the hydrogen to flow to its compartment, the oxygen to its compartment, and stray gases to their compartment, means for removing the hydrogen, oxygen and stray gases separately, for conveying same away for use.

2. A generator for facilitating the decomposition and separation of hydrogen and oxygen resulting from the electrolysis of water, comprising in combination, a tank having its longitudinal dimension relatively long compared with its transverse dimension, and having its internal wall surface of corrugated form concentrically around same, so as to dampen any sloshing movement of water placed therein at a predetermined level to be electrolyzed and to induce entrapment of the said gases derived from the electrolysis in the portion above the water and facilitate the removal of moisture therefrom, partitions extending completely across the tank transversely with passages therethrough for the circulation of water therethrough but obstructing the gases from passing above same, said partitions being arranged to divide the tank into compartments for the hydrogen, oxygen and stray gases separate from each other, and the conducting of water to be processed thereto, electrodes in walls of the compartments and separate from each other to facilitate the transmittal of the respective gases to their respective compartments, and outlets from the chambers for removing the respective gases therefrom.

3. A generator for facilitating the decomposition and separation of hydrogen and oxygen resulting from the electrolysis of water, comprising in combination, a tank having its longitudinal dimension relatively long compared with its transverse dimension, and having its internal wall surface of corrugated form concentrically around same, so as to dampen any sloshing movement of water to be electrolyzed placed therein to a predetermined level and to induce entrapment of the said gases derived from the electrolysis in the portion above the water and facilitate the removal of moisture therefrom, partitions extending completely across the tank transversely with passages therethrough for the circulation of water therethrough but obstructing the gases from passing above same, said partitions being arranged to divide the tank into compartments for the hydrogen, oxygen and stray gases separate from each other, and the conducting of water to be processed thereto, electrodes in walls of the compartments and separate from each other to facilitate the transmittal of the respective gases to their respective compartments, and outlets from the chambers for removing the respective gases therefrom said compartment for the conducting of water to the tank being arranged with its partitions to contain a space above the water for the collection of air and other gases released therefrom independently separated from the oxygen and hydrogen so as to be segregated therefrom, an outlet in the last mentioned compartment for the removal of said air and gases therefrom, said electrodes being disposed within the tank so as to be wholly covered by the water at the predetermined level, and the water being conducted to the tank adjacent the bottom thereof and connections for the electrodes for tieing it into a circuit.

BERNARD KLEIN.

June 25, 1957 H. T. LUCAS 2,797,072

SUPPLEMENTAL CARBURETOR

Filed Sept. 20, 1954 2 Sheets-Sheet 1

Fig. 1

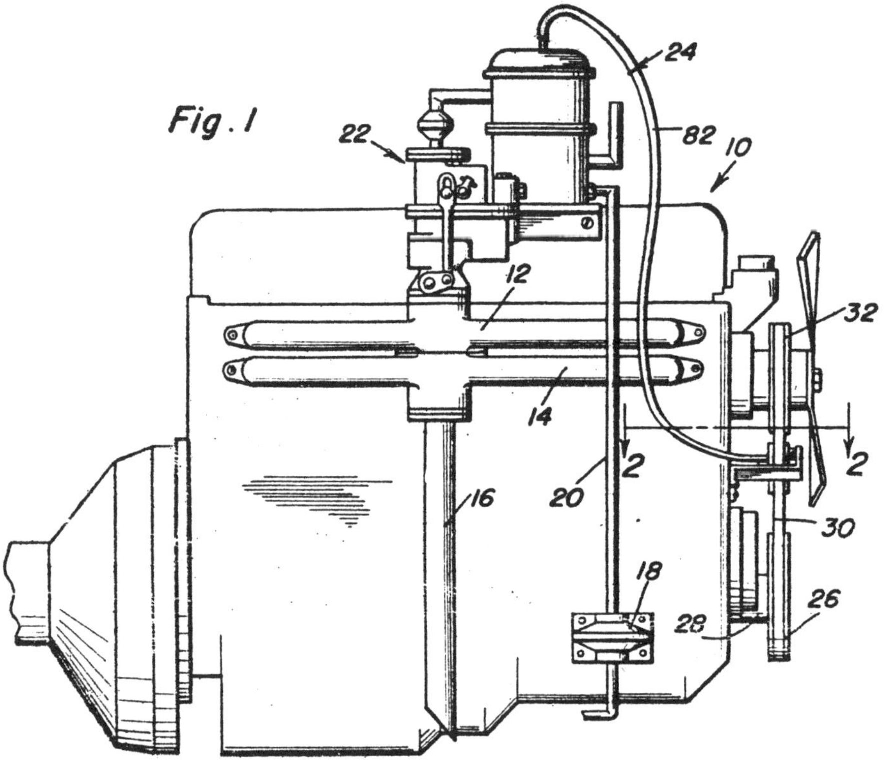

Fig. 2

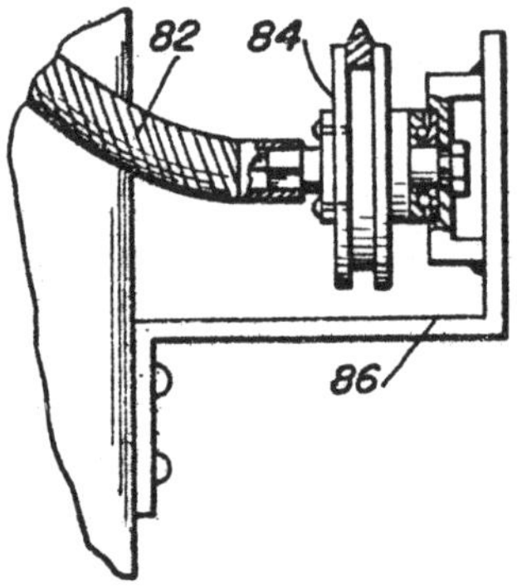

Harold T. Lucas

INVENTOR.

BY Clarence A. O'Brien and Harvey B. Jacobson Attorneys

June 25, 1957 H. T. LUCAS 2,797,072

SUPPLEMENTAL CARBURETOR

Filed Sept. 20, 1954 2 Sheets-Sheet 2

Fig.3

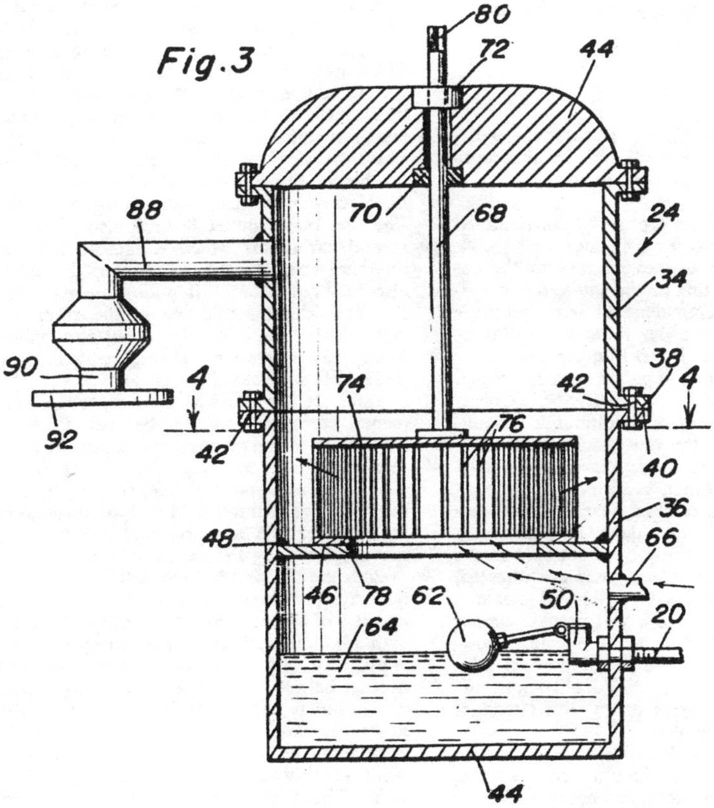

Fig. 4

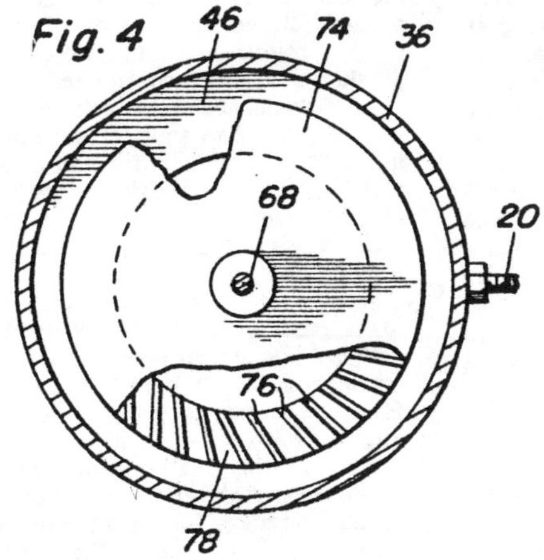

Fig. 5

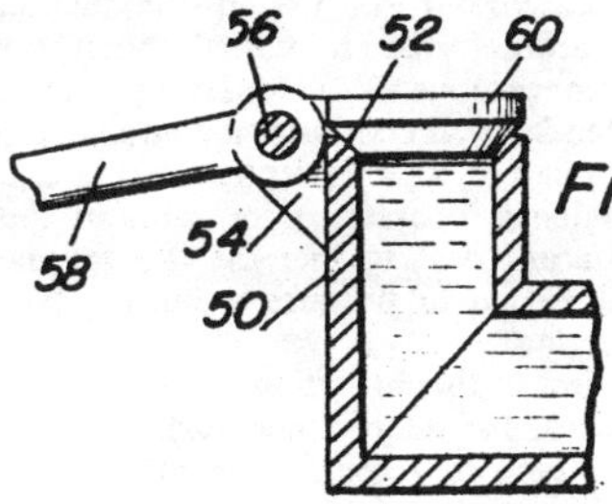

Harold T. Lucas
INVENTOR.

BY Clarence A. O'Brien
and Harvey B. Jacobson
Attorneys

United States Patent Office

2,797,072
Patented June 25, 1957

1

2,797,072

SUPPLEMENTAL CARBURETOR

Harold T. Lucas, Decatur, Ill., assignor of one-half to Ralph Anderson

Application September 20, 1954, Serial No. 456,965

1 Claim. (Cl. 261—24)

This invention relates generally to charge forming devices for internal combustion engines and pertains more particularly to a device to supplement the mixing action of a conventional carburetor.

A disadvantage accruing from the use of conventional carburetors for internal combustion engines is the difficulty of supplying a proper air-fuel mixture to the engine, particularly during starting of the same, this condition being aggravated as the ambient air temperature is lowered. The chief difficulty under these circumstances is that the raw gasoline which is fed into the carburetor does not sufficiently vaporize to supply a proper air-fuel mixture to all of the cylinders of the engine. Consequently, engine starting under these conditions is difficult, and it frequently occurs that the raw gasoline will form in pools or puddles in the manifold, sometimes to such an extent to cause the condition known as "flooding." Further, when an engine has reached operating temperature, to provide flexibility to the engine, it is necessary to heat the air-fuel mixture, usually by a hot spot in the intake manifold upon which exhaust gases are directed, in order to sufficiently volatilize the fuel and create a homogeneous air-fuel mixture which will provide proper combustion in all of the cylinders of the engine. However, due to the raised temperature of the air-fuel mixture as created by such intake manifold hot spots, the volumetric efficiency of the engine drops with attendant decrease in gas mileage and power.

It is therefore an object of this invention to provide improved means for supplying a homogeneous air-fuel mixture to the cylinders of an internal combustion engine.

It is a further object of this invention to provide a supplemental carburetor for use in conjunction with a conventional carburetor of an internal combustion engine to assure a homogeneous air-fuel mixture to the cylinders of an associated internal combustion engine and to supply such air-fuel mixture at a lower temperature than that which the engine is operated upon without the supplemental carburetor, so as to increase the volumetric efficiency of the engine to produce a more powerful and economical engine.

Another object of this invention is to provide a supplemental carburetor for internal combustion engines which incorporates a tank or receptacle member into which a predetermined amount of fuel is introduced and in which a fan is disposed to cause violent agitation of the air-fuel mixture therein.

These together with other objects and advantages which will become subsequently apparent reside in the details of construction and operation as more fully hereinafter described and claimed, reference being had to the accompanying drawings forming a part hereof, wherein like numerals refer to like parts throughout, and in which:

Figure 1 is a side elevational view of an internal combustion engine showing the invention in use thereon;

Figure 2 is an enlarged horizontal section taken substantially along the plane of section line 2—2 in Figure 1;

2

Figure 3 is an enlarged vertical section taken through the supplemental carburetor;

Figure 4 is a horizontal section taken substantially along the plane of section line 4—4 of Figure 3; and

Figure 5 is an enlarged sectional view through the float valve mechanism of the supplemental carburetor.

Referring now more particularly to the drawings, the reference numeral 10 indicates generally an internal combustion engine which incorporates the usual intake manifold assembly 12 and the exhaust manifold assembly 14, the latter having an exhaust pipe 16 leading therefrom. The engine also utilizes a conventional fuel pump assembly 18 and an outlet 20 leads therefrom for connection to the carburetor assembly 22 mounted upon the intake manifold 12, this carburetor being of entirely conventional construction and the details of which are not herein shown or described inasmuch as the conventional carburetor itself forms no part of this invention. The fuel pump outlet line 20 also leads to the supplemental carburetor which is indicated generally by the reference character 24. The internal combustion engine also includes a fan belt pulley member 26 attached to the forward projecting end of the crankshaft 28 and a belt member 30 is trained about this pulley and about a pulley 32 on the fan and water pump assembly for purposes well understood in this art.

Referring now more particularly to Figure 3, the supplemental carburetor 24 will be seen to consist of a receptacle or casing member formed of the upper and lower sections 34 and 36, respectively, each of which is provided with an annular flange, such as those indicated by the reference characters 38 and 40, through which fastening members 42 are projected for the purpose of removably securing the sections together.

Both the top and bottom sections of the receptacle or tank are preferably formed of cylindrical configuration, and it will be seen that the lower section 36 is provided with a bottom wall member 44 whereas the top section 34 is provided with a removable cover member 46, the connection between the upper section 34 and the cover 44 being similar to the connection between the upper and lower sections. Mounted within the lower section 36 is an annular baffle plate member 47, these two members being suitably secured together, as by welding 48 or the like. The fuel inlet line to the tank communicates therewith below the baffle plate 47 and is secured thereto in any suitable manner with the end of the line within the tank or receptacle terminating in an L-shaped fitting member 50 provided at its upper end with a valve seat portion 52 and with ear members 54 through which a pin 56 projects to pivotally secure the float rod member 58 thereto. The float rod terminates at one end in a valve disk member 60 which engages the previously described seat 52 to shut off the flow of fuel through the fitting 50, and the opposite end of the member 58 is provided with the float ball 62, the construction being obviously such as to maintain a predetermined level of fuel 64 within the bottom section 36 of the tank or receptacle 24.

An air inlet conduit 66 also communicates with the lower tank section 36 below the baffle plate 47 and is designed to direct air over the surface of the fuel 64 within the section 36. It is to be understood that suitable air filter means may be provided on the free end of the air inlet conduit 66, as is conventional.

An actuating shaft member 68 is rotatably received in the cover member 44 axially of the tank or receptacle, suitable bushings 70 and 72 being provided for this purpose. The lower end of the shaft has rigidly attached thereto a hub disk member 74 which is of smaller diameter than the internal diameter of the lower section 36 and peripherally disposed on this disk in depending

2,797,072

3

relation thereto are a series of impellers or vanes **76** such as to form what is commonly termed a squirrel cage fan. This fan is of the centrifugal type, and for the purpose of rigidifying the blades **76**, their lower ends are connected to a blade mounting ring or plate **78** which is of annular configuration such that its center opening aligns and matches with the center opening of the annular baffle plate **47**, as will be readily apparent from a study of Figures 3 and 4. It is to be understood that any suitable means may be provided for maintaining the shaft **68** and consequently the fan assembly in a predetermined longitudinal position and to prevent longitudinal shifting of the assemblies within the receptacle or tank.

The upper end of the shaft **68** is provided with a portion **80** of non-circular configuration in cross-section such as to be readily attached to a flexible cable assembly **82** which extends therefrom to a pulley member **84** mounted upon a Z-shaped bracket **86** secured to the engine **10**, see particularly Figures 1 and 2. The positioning of the bracket **86** is such as to engage the pulley **84** against the outside surface of the fan belt **30** and drive the flexible shaft or cable assembly **82** and consequently cause rotation of the fan assembly within the receptacle or tank. Adjacent the top end of the upper section **34** is an air outlet conduit **88** which is adapted for connection to the air horn of a conventional carburetor, such as that illustrated and previously described. For this purpose, the outlet conduit **88** may be provided with a downturned portion **90** terminating in a flange **92** for suitable connection to the carburetor air horn.

In operation, the fan assembly is rotated to act upon the fuel-air mixture which is occasioned by the evaporation of the fuel **64** as the air enters the air inlet conduit **66**, and the fan causes violent agitation of all the air introduced from the lower section to the upper section and out through the outlet conduit **88** to thoroughly mix and homogenize the air-fuel mixture such as to provide a proper mixture for the various cylinders of the associated internal combustion engine. At the same time, the temperature of the air-fuel mixture is materially reduced by the rapid evaporation of the fuel **64**, such as to lower the over-all temperature of the air-fuel mixture entering the internal combustion cylinders and effectively raise the volumetric efficiency thereof to increase the gas mileage and power of the engine.

It is to be noted that the assembly can be mounted directly within a gas tank of a vehicle or in the position shown in Figure 1, as desired. When mounted within the gas tank, the lower section **36** thereof is replaced by the gas tank itself, the fan serving to agitate the fumes directly from the gas tank and supply them to the carburetor assembly **22**, as will be readily apparent.

4

From the foregoing, the construction and operation of the device will be readily understood and further explanation is believed to be unnecessary. However, since numerous modifications and changes will readily occur to those skilled in the art, it is not desired to limit the invention to the exact construction shown and described, and accordingly, all suitable modifications and equivalents may be resorted to, falling within the scope of the appended claim.

What is claimed as new is as follows:

In combination with an internal combustion engine having an intake manifold and a principal carburetor disposed in communication therewith for supplying a combustible mixture thereto, said principal carburetor having an air intake throat, a supplemental carburetor mounted adjacent the principal carburetor, said supplemental carburetor comprising a receptacle having a bottom wall and a peripheral side wall, a flat baffle mounted in spaced relation to the bottom wall and having an enlarged central opening therein, an air inlet conduit communicating with said receptacle below said baffle plate, a fan mounted within said receptacle above said baffle plate, a fuel supply conduit communicating with the receptacle below said air inlet conduit, float control means operatively associated with said fuel conduit for maintaining the fuel level below the baffle plate at a predetermined level, a closure cover for said receptacle, and an air-fuel mixture outlet communicating with the receptacle above said baffle plate, said outlet connected to the intake throat of the principal carburetor for directing the air-fuel mixture discharged by the fan to the intake throat of the principal carburetor, said fan being of the centrifugal type and having the blades thereof disposed closely adjacent the upper surface of the baffle and having an inlet substantially equal in size to the baffle opening for pressurizing the receptacle above the baffle.

References Cited in the file of this patent

UNITED STATES PATENTS

760,061	Dubuis	May 17, 1904
1,000,518	Harris	Aug. 15, 1911
2,319,119	Feinberg	May 11, 1943
2,702,694	Featherston	Feb. 22, 1955

FOREIGN PATENTS

536,116	Great Britain	May 2, 1941

May 2, 1961 R. S. SHELTON **2,982,528**

VAPOR FUEL SYSTEM

Filed June 13, 1958 2 Sheets-Sheet 1

FIG. 1

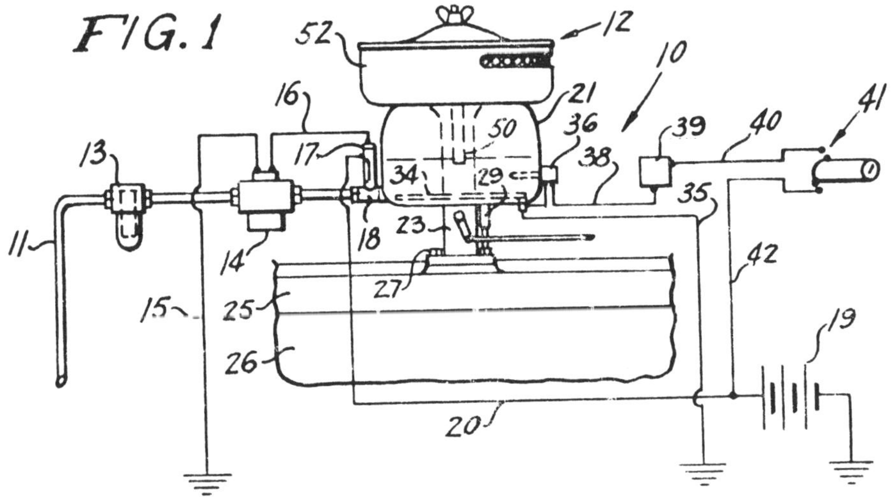

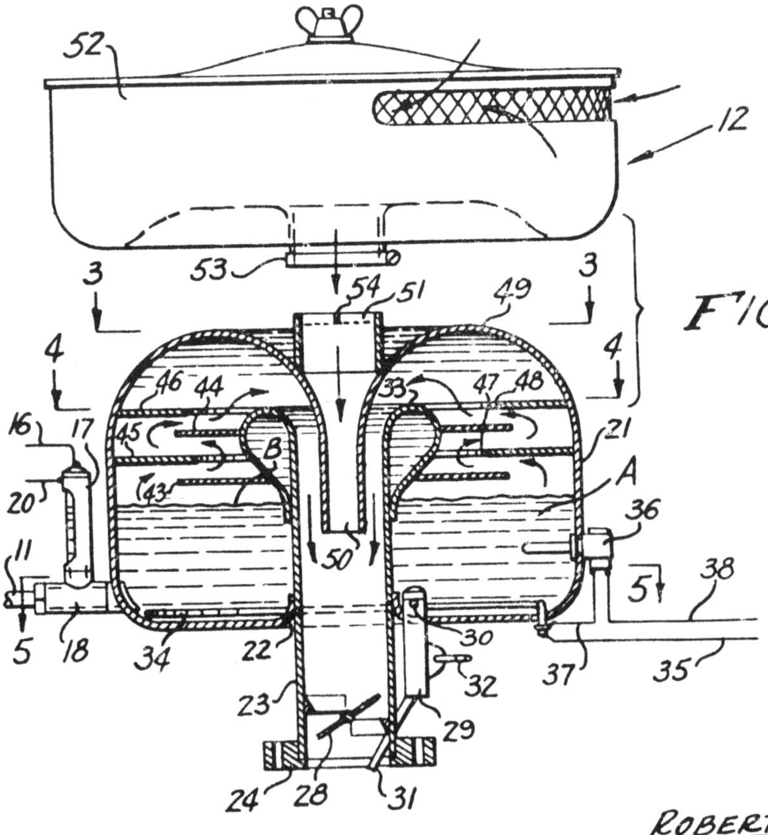

FIG. 2

INVENTOR.
ROBERT S. SHELTON
BY Patrick D. Beavers
ATTY.

May 2, 1961 | R. S. SHELTON | **2,982,528**

VAPOR FUEL SYSTEM

Filed June 13, 1958 | 2 Sheets-Sheet 2

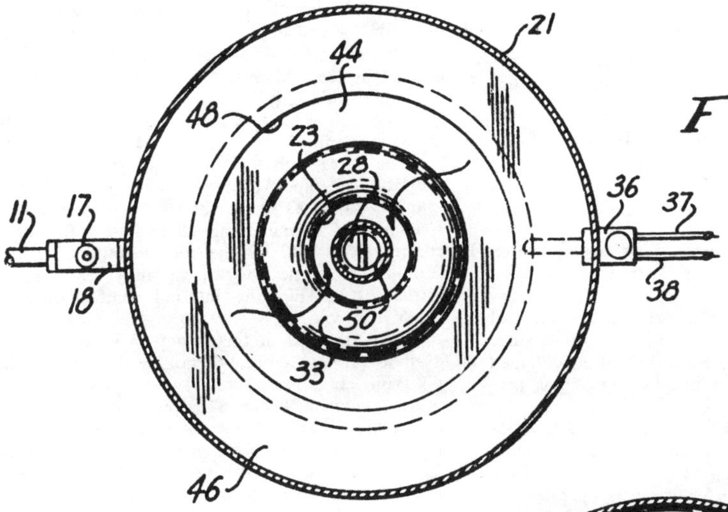

FIG. 4

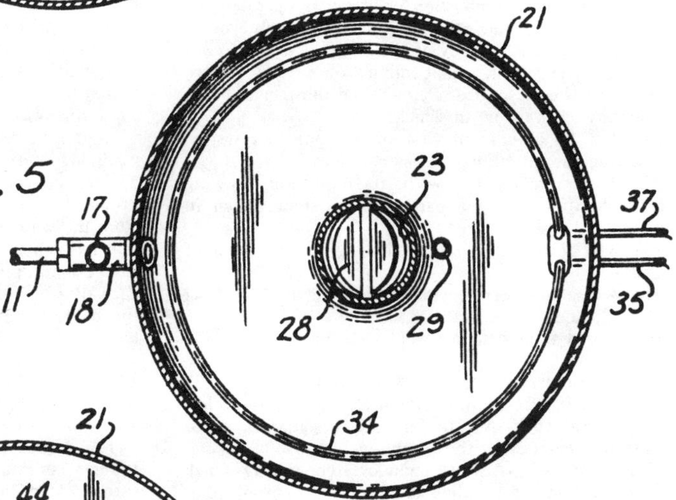

FIG. 5

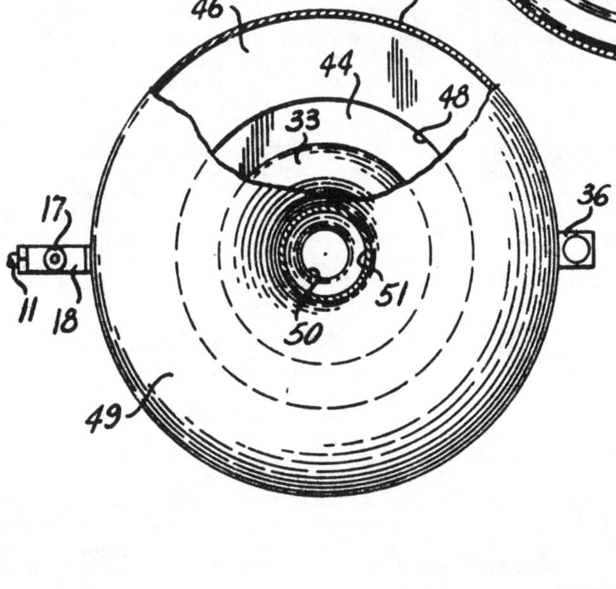

FIG. 3

INVENTOR.
ROBERT S. SHELTON
BY Patrick D. Beavers
ATTY.

United States Patent Office

2,982,528
Patented May 2, 1961

1

2,982,528

VAPOR FUEL SYSTEM

Robert S. Shelton, 39010 N. 12th St. E., Palmdale, Calif., assignor of forty percent to Maurice R. Shirley, Palmdale, Calif.

Filed June 13, 1958, Ser. No. 741,951

1 Claim. (Cl. 261—130)

This invention relates to improvements in vapor fuel systems that are to be used for internal combustion engines.

An object of this invention is to provide a vapor fuel system that will provide a great saving in gasoline, since approximately eight times the mileage that is obtained by the conventional internal combustion engine is provided by the use of such a system.

Another object of this invention is to provide a vapor fuel system that is provided with a reservior to contain liquid gasoline which is heated to provide vapors from which the internal combustion engine will operate.

With the above and other objects and advantages in view, the invention consists of the novel details of construction, arrangement and combination of parts more fully hereinafter described, claimed and illustrated in the accompanying drawing, in which:

Fig. 1 is an elevational view of a vapor fuel system embodying the invention;

Fig. 2 is an enlarged view, partly in section, showing the carburetor forming a part of the system shown in Fig. 1;

Fig. 3 is a transverse sectional view on line 3—3 of Fig. 2;

Fig. 4 is a transverse sectional view on the line 4—4 of Fig. 2; and

Fig. 5 is a transverse sectional view on the line 5—5 of Fig. 2.

Referring more in detail to the drawings, wherein like parts are designated by like reference numerals, the reference numeral **10** is used to generally designate a vapor fuel system embodying the invention.

The vapor fuel system **10** includes a conduit **11** which is connected to the gasoline tank for the internal combustion engine at one end and to a carburetor **12** at the opposite end. Interpolated in the conduit **11** is a gasoline filter **13** and an electric fuel pump **14**. A wire **15** grounds the pump **14** and a wire **16** connects the pump **14** to a gasoline gauge **18** on which is mounted a switch **17** which is connected to the conduit **11** when it is connected to the carburetor **12**. The gauge **17** is connected to a battery **19** of an internal combustion engine by a wire **20**.

Switch **18** is of conventional construction and is of the type disclosed in United States Patents Nos. 2,894,093; 2,825,895; and 2,749,401. The switch is so constructed that a float in position in the gauge is acted on by the liquid therein. As the liquid rises the float will disengage a pair of contacts mounted in the gauge to cut off electric pump **14**. As the float lowers due to the consumption of the liquid in the body the float will fall causing the contacts to be re-engaged to start pump **14** to replenish the liquid in the body.

The carburetor **12** includes a dome-shaped circular bowl or reservoir **21** which is provided with a centrally located flanged opening **22** whereby the reservoir **21** is mounted on a tubular throat **23**. An apertured collar **24** on the lower end of the throat **23** is positioned on the

2

intake manifold **25** of an internal combustion engine **26** and fastening elements **27** secure the collar to the manifold **25** in fixed relation thereto.

A vapor control butterfly valve **28** is pivotally mounted in the lower end of the throat **23** and the valve **28** controls the entrance of the vapor into the internal combustion engine **26** and thus controls the speed of the internal combustion engine **26**.

A raw gasoline pump **29** having an inlet **30** is mounted in the bottom of the reservoir **21** so that the inlet **30** communicates with the interior of the reservoir **21**. A spurt or feed pipe **31** connected to the pump **29** extends into the throat **23** so that by means of a linkage **32** that is connected to the pump **29** and to a linkage for the control valve **28** and the foot throttle of the internal combustion engine **26**, raw gas may be forced into the throat **23** to start the internal combustion engine **26** when it is cold.

The upper end of the throat **23** is turned over upon itself to provide a bulbous hollow portion **33** within the reservoir **21**.

An immersion heater **34** is positioned in the bottom of the reservoir **21** and a wire **35** grounds the heater **34**. A thermostat **36** is mounted in the wall of the reservoir **21** and extends into the reservoir **21**. A wire **37** connects the thermostat to the heater **34** and a wire **38** connects the thermostat **36** to a control **39** for the thermostat **36**. A wire **40** connects the control **39** to the ignition switch **41** and the ignition switch **41** is in turn connected to the wire **20** that is connected to the battery **19** by a wire **42**.

A pair of relatively spaced parallel perforated baffle plates **43** and **44** are connected to the bulbous portion **33** on the upper end of the throat **23** and a second pair of perforated baffle plates **45** and **46** extend inwardly from the wall of the reservoir **21** in parallel spaced relation to each other and to the baffle plates **43** and **44**.

The baffle plates are arranged in staggered relation to each other so that the baffle plate **45** is intermediate of the baffle plates **43** and **44** and the baffle plate **46** extends over the baffle plate **44**.

The baffle plate **45** is provided with a central opening **47** and the baffle plate **46** is provided with a central opening **48** that is of greater diameter than the opening **47** in the baffle plate **44**.

The domed top **49** of the reservoir **21** extends into a tubular air intake **50** that extends downwardly into the throat **23** and a mounting ring **51** is mounted on the exterior of the domed top **49** of the reservoir **21** in vertical alinement with the intake **50**. An air cleaner **52** is mounted on the mounting ring **51** by a coupling **53** as is the usual procedure and a spider **54** is mounted in the upper end of the mounting ring **51** to break up the air as it enters the ring **51** from the air cleaner **52**.

In operation and with the carburetor **12** mounted on the internal combustion engine in lieu of a conventional carburetor, the ignition switch **41** is turned on. Current from the battery **19** will cause the pump **14** to bring gasoline into the reservoir **21** until the switch **18** cuts off the pump **14** when the gasoline A has reached the level B in the reservoir **21**. The control **39** is adjusted so that the thermostat **36** will operate the heater **34** until the gasoline A has reached a temperature of 105° at which time the heater **34** will be cut off. When the gasoline A has reached the proper temperature, vapor therefrom will be available to follow the course indicated by the arrows in Fig. 2.

The internal combustion engine **26** is then started and if the foot control is actuated the pump **29** will cause raw gasoline to enter the intake manifold **25** until the vapors from the carburetor are drawn into the manifold **25** to cause the internal combustion engine **26** to operate.

2,982,528

3

As the gasoline A is consumed the pump **14** will again be operated and the heater **34** will be operated by the thermostat **36**. Thus the operation as described will continue as long as the internal combustion engine is operating and the ignition switch **41** is turned on. The reservoir **12** will hold from 2 to 3 quarts of gasoline and since only the vapors from the heated gasoline will cause the carburetor **12** to run the internal combustion engine **26**, the internal combustion engine will operate for a long time before more gasoline is drawn into the reservoir **21**.

The baffles **43**, **44**, **45** and **46** are arranged in staggered relation to prevent splashing of the gasoline within the carburetor. The level B of the gasoline A in the reservoir **21** is maintained constant by the switch **18** and with all elements properly sealed, the vapor fuel system **10** will efficiently operate the internal combustion engine **26**.

The valve **28**, controlling the entrance of vapors into the intake manifold **25**, controls the speed of the internal combustion engine in the same manner as the control valve in a conventional carburetor.

There has thus been described a vapor fuel system embodying the invention and it is believed that the structure and operation thereof will be apparent to those skilled in the art, it is also to be understood that changes in the minor details of construction, arrangement and combination of parts may be resorted to provided they fall within the spirit of the invention and the scope of the appended claim.

4

Having thus described the invention what is claimed as new and desired to be secured by Letters Patent is:

A carburetor for use in a vapor fuel system comprising a circular domed body, a mixing throat centrally of said body, means on one end of said throat for connecting said throat to an intake manifold of an internal combustion engine, the opposite end of said throat having a hollow bulbous portion thereon integral therewith, said body having a domed top, a tubular air intake integral with said top and positioned centrally of said mixing throat, a plurality of apertured baffles mounted in said body in staggered spaced relation to each other, means secured to said top centrally thereof in vertical alinement with said mixing throat for connecting an air cleaner thereto, a heater in said body for vaporizing fuel in said carburetor, a pump for supplying said carburetor with fuel, means for controlling the operation of said pump to maintain the fuel in said carburetor, a thermostat for controlling said heater, means for setting said thermostat to control the temperature of said heater and means for delivering raw gasoline into the intake manifold.

References Cited in the file of this patent

UNITED STATES PATENTS

1,065,370	Johnston et al.	June 24, 1913
1,739,818	Beals	Dec. 17, 1929
2,014,396	Rosier	Sept. 17, 1935

Aug. 21, 1962 P. A. SMITH 3,049,850

CARBURETER FOR INTERNAL COMBUSTION ENGINES

Filed June 25, 1959

Fig. 1.

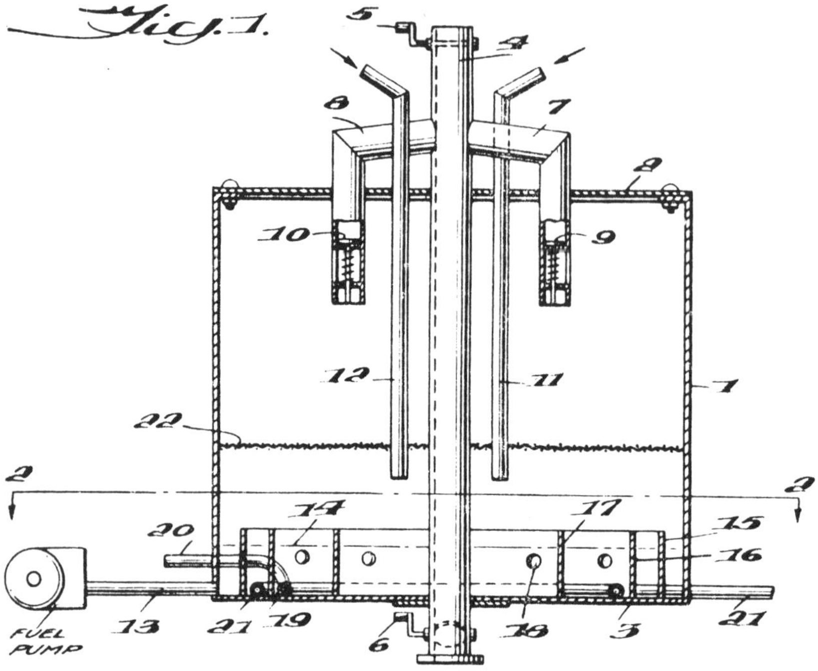

Fig. 2.

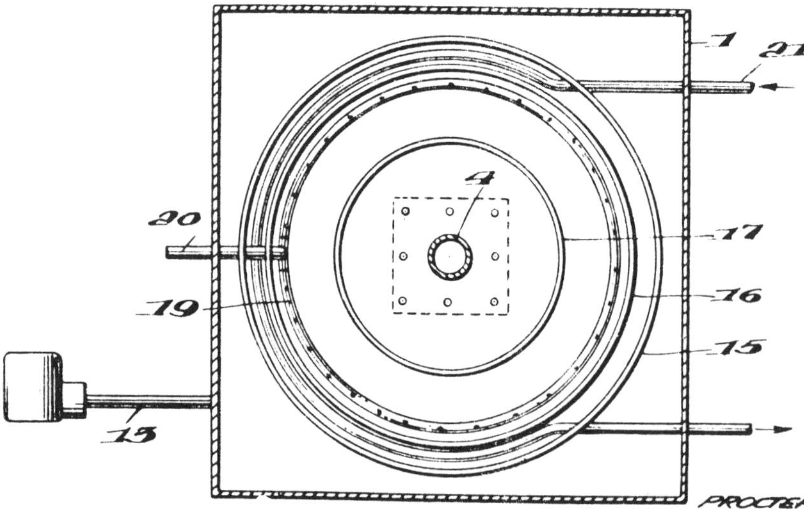

INVENTOR
PROCTER A. SMITH,

BY Lancaster and Taylor

ATTORNEYS

United States Patent Office

3,049,850
Patented Aug. 21, 1962

1

3,049,850
CARBURETER FOR INTERNAL COMBUSTION ENGINES
Procter A. Smith, Rte. 2, Beaver Dam, Va.
Filed June 25, 1959, Ser. No. 822,880
1 Claim. (Cl. 55—228)

This invention relates to a carbureter for internal combustion engines and more particularly to a device wherein the fuel is completely vaporized prior to being mixed with air.

Heretofore the usual type of carbureter utilizes a spray nozzle wherein the fuel is sprayed into a mixing chamber and the fine droplets of fuel are carried by the air into the intake manifold. This type of arrangement permits raw fuel to enter the cylinder and this raw fuel is not completely burned during the power stroke. This, of course, results in a wasting of fuel and furthermore the raw gas removes the lubricating film from the cylinder walls, thus producing greater wear on the engine parts.

According to the present invention there is provided means for completely vaporizing the gasoline prior to drawing the fuel mixture into the intake manifold. This results in a considerable savings in fuel and also permits the maintenance of a lubricating film on the cylinder walls due to the absence of raw fuel.

According to this invention a vaporizing chamber is provided into which the liquid gasoline is fed. A heating coil is provided in this chamber beneath the surface of the liquid gasoline so as to increase the temperature and thus facilitate vaporization. There is also provided agitation means in the form of an air feed pipe disposed beneath the surface of the liquid gasoline. This air feed pipe may be connected with any suitable pump which will force air through the liquid gasoline thus agitating and enhancing further vaporization. The ducts interconnecting the vaporization chamber with the intake manifold are provided with cut-off valves which automatically close in the event that the engine backfires. Valve means may also be provided to regulate the fuel air mixture.

An object of the present invention is to provide a carbureter for internal combustion engines in which the fuel is completely vaporized before being fed to the intake manifold.

Another object of the present invention is to provide a carbureter including means for agitating and vaporizing gasoline so that a sufficiently rich mixture of completely vaporized gasoline is fed to the cylinders of the internal combustion engine.

Other objects and many of the attendant advantages of the present invention will become apparent upon consideration of the following detailed specification in connection with the accompanying drawing wherein:

FIG. 1 is a sectional view in elevation of an embodiment of the present invention, and

FIG. 2 is a sectional plan view along the line 2—2 of FIG. 1.

Referring now more specifically to the drawing wherein like numerals indicate like parts throughout the several views there is shown at **1** in FIG. 1 a rectangular shaped container having a top plate **2** and bottom plate **3** secured thereto in any suitable manner. This container may be of any suitable size and shape, but preferably is of such dimensions that it can replace the carbureters presently in use and will fit beneath the engine hood.

An elongated duct **4** extends through the top plate and the bottom plate and this duct has a valve **5** at the upper end thereof and a valve **6** at the lower end. The upper end of the duct is connected with an air filter of any well known variety, and the valve **5** corresponds

2

to the choke on carbureters of known design. The lower end of the duct **4** is connected with the intake manifold of the internal combustion engine and the valve **6** corresponds to a throttle valve and is connected with the accelerator pedal.

The upper end of duct **4** has a pair of intake pipes **7** and **8** which extend through the top plate **2** of the vaporizing chamber and have the lower ends thereof open in the chamber. These intake pipes have valves **9** and **10** therein which are spring urged to the closed position. There is also provided air intake pipes **11** and **12** which extend into the chamber **1** to a point spaced from the bottom thereof.

A fuel feed pipe **13** is connected with the fuel pump and the fuel pump is intermittently controlled by any suitable mechanism (not shown) including a float for maintaining the fuel at a preselected level such as shown at **14** in FIG. 1. There may also be provided an overflow valve (not shown) for preventing the fuel level from rising to any substantial extent above the preselected level. A plurality of circular baffles such as shown at **15**, **16** and **17** may be arranged within the vaporizing chamber. These baffles may be provided with openings such as shown at **18** at the lower portions thereof so that the fuel level between the various baffles will be maintained the same. The baffles serve to prevent any substantial shifting of the liquid level when the vehicle turns a corner at high speed.

There is provided an air agitator in the form of a tubular pipe **19** having a plurality of apertures therein. This pipe is connected by a pipe **20** with an air pump of any suitable variety which serves to force air through the apertures into the liquid fuel.

The heating means for the liquid comprises a pipe **21** which passes around the lower portion of the container and is connected in the water cooling system of the engine. Thus, during operation of the engine warm water will pass through pipe **21** and serve to heat the fuel so as to facilitate vaporization.

A screen **22** surrounds the duct **4** and air intake pipes **11** and **12** as shown in FIG. 1. This screen may be of a highly porous material such as will permit the ready passage of the vaporized gases therethrough, but will form a barrier to the passage of raw gas therethrough.

The operation of the presently disclosed carbureter is as follows. Initially the vaporizing chamber **1** is filled with raw gas to a level **14** and the upper portion of the chamber is filled with the vaporized fumes. On starting the valve **5** is closed and a rich mixture is fed through the duct **4** to the intake manifold. The inner intake pipes **11** and **12** serve to replace the air passing out through the pipes **7** and **8**. The suction from the intake manifold causes the valves **9** and **10** to be opened but these valves will be closed as soon as this suction drops below a predetermined level. As the fuel vaporizes and the liquid level drops the fuel pump replenishes the supply so as to maintain the liquid level constant. Air is fed through pipe **20** so as to agitate the fuel and increase the vaporization thereof. Hot water passing through the pipe **21** further increases vaporization.

Thus it can be seen that the present invention provides means for insuring that only completely vaporized gasoline fumes enter the intake manifold. The device provides heating means and agitating means for facilitating the vaporization of the gas.

Obviously many modifications and variations of the present invention are possible in light of the above teachings.

What is claimed as new and desired to be secured by Letters Patent is:

In a device for mixing a fuel charge for supply to the intake manifold of an internal combustion engine com-

prising, in combination, a duct having one end thereof connected to the intake manifold and having the other end open to atmosphere, intake pipes having one end connected to said duct, a vaporization chamber, the other ends of said intake pipes communicating with the upper end of said vaporization chamber, spring actuated valve means in said intake pipes, said valve means being normally closed but adapted to open in response to reduced pressure in the intake manifold, a porous membrane dividing the vaporization chamber into upper and lower compartments, means for feeding liquid fuel to the vaporization chamber within the compartment, means for pumping air below the surface of the liquid in the vaporization chamber within the lower compartment, means comprising hot water conduits for supplying heat

to the liquid and a pair of valves disposed in said duct on opposite sides of the connection of said duct with the intake pipes to provide control over the supply of vaporized fuel and the ratio of fuel to air fed to the intake manifold.

References Cited in the file of this patent

UNITED STATES PATENTS

Re. 13,498	Bustard	Dec. 17, 1912
1,103,789	Macey	July 14, 1914
1,182,714	Schmidt	May 9, 1916
1,270,486	Brown	June 25, 1918
1,938,497	Pogue	Dec. 5, 1933
1,997,497	Pogue	Apr. 9, 1935
2,026,798	Pogue	Jan. 7, 1936

Dec. 4, 1962 G. L. HOLZAPFEL 3,067,020

CARBURETOR FOR LIQUEFIED GASEOUS FUELS

Filed March 16, 1959 3 Sheets-Sheet 1

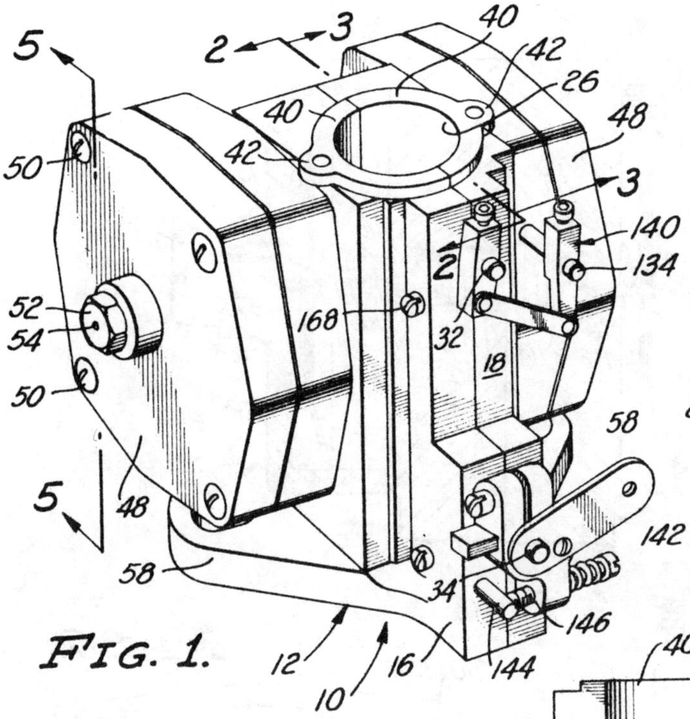

FIG. 1.

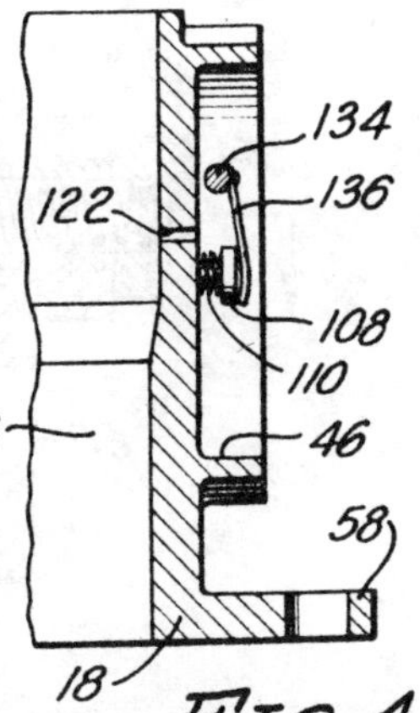

FIG. 4.

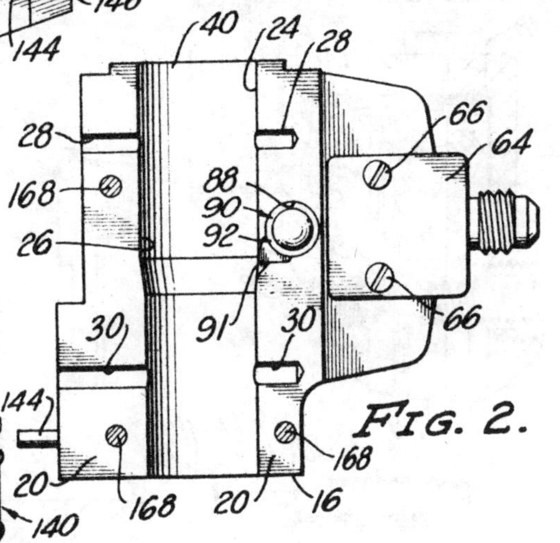

FIG. 2.

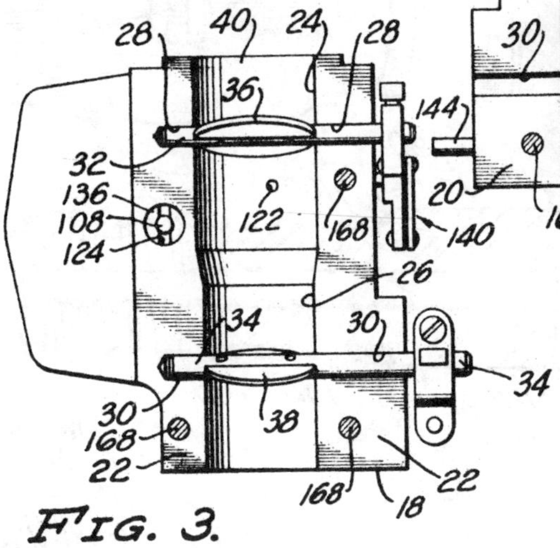

FIG. 3.

INVENTOR.
GEORGE L. HOLZAPFEL
BY
Thomas P. Mahoney
ATTORNEY

Dec. 4, 1962 G. L. HOLZAPFEL 3,067,020

CARBURETOR FOR LIQUEFIED GASEOUS FUELS

Filed March 16, 1959 3 Sheets-Sheet 2

FIG. 5.

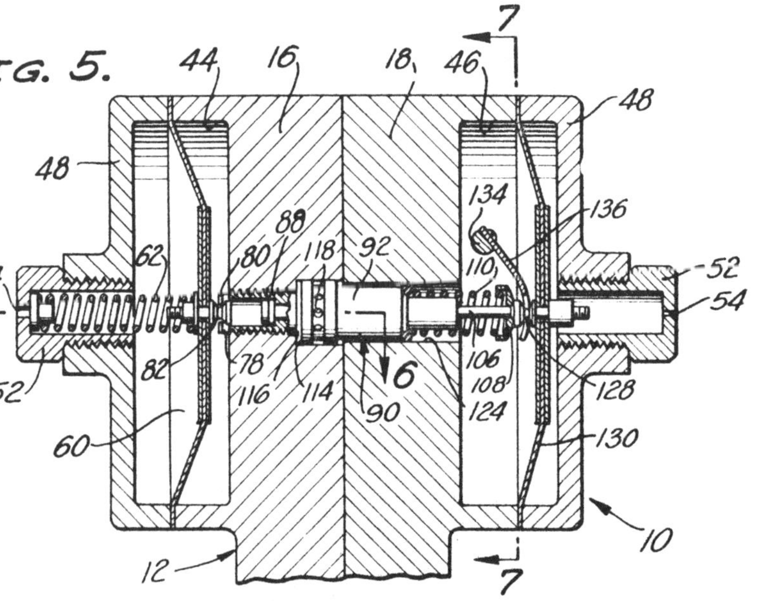

FIG. 6.

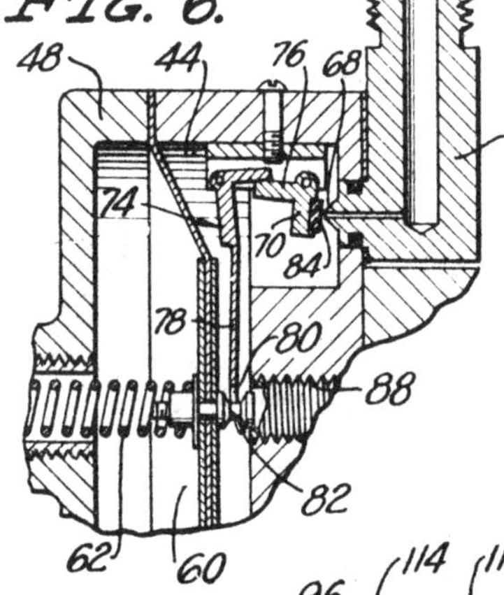

FIG. 7.

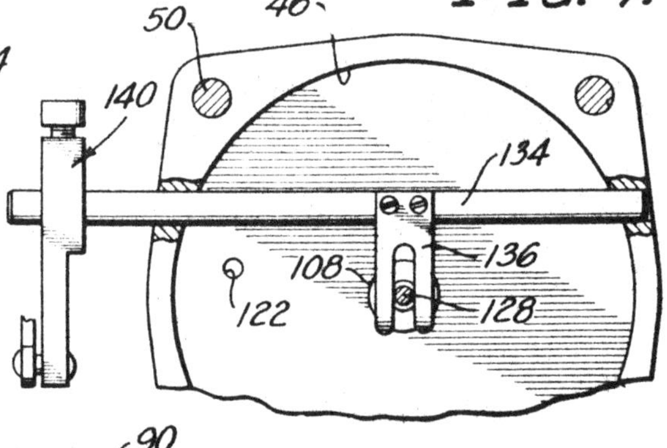

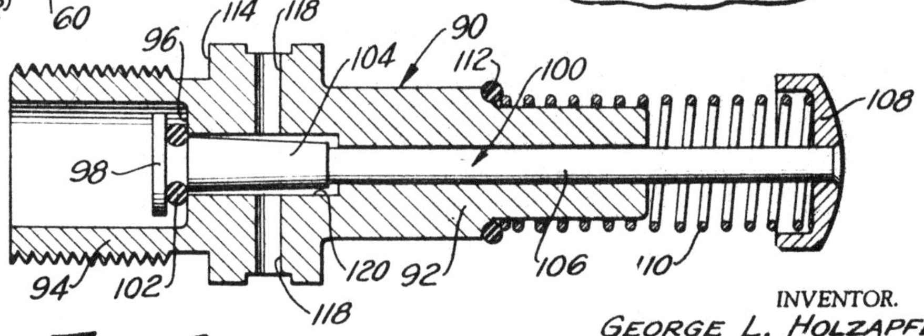

FIG. 8.

INVENTOR.
GEORGE L. HOLZAPFEL
BY
Thomas P. Mahoney
ATTORNEY

Dec. 4, 1962 G. L. HOLZAPFEL 3,067,020

CARBURETOR FOR LIQUEFIED GASEOUS FUELS

Filed March 16, 1959 3 Sheets-Sheet 3

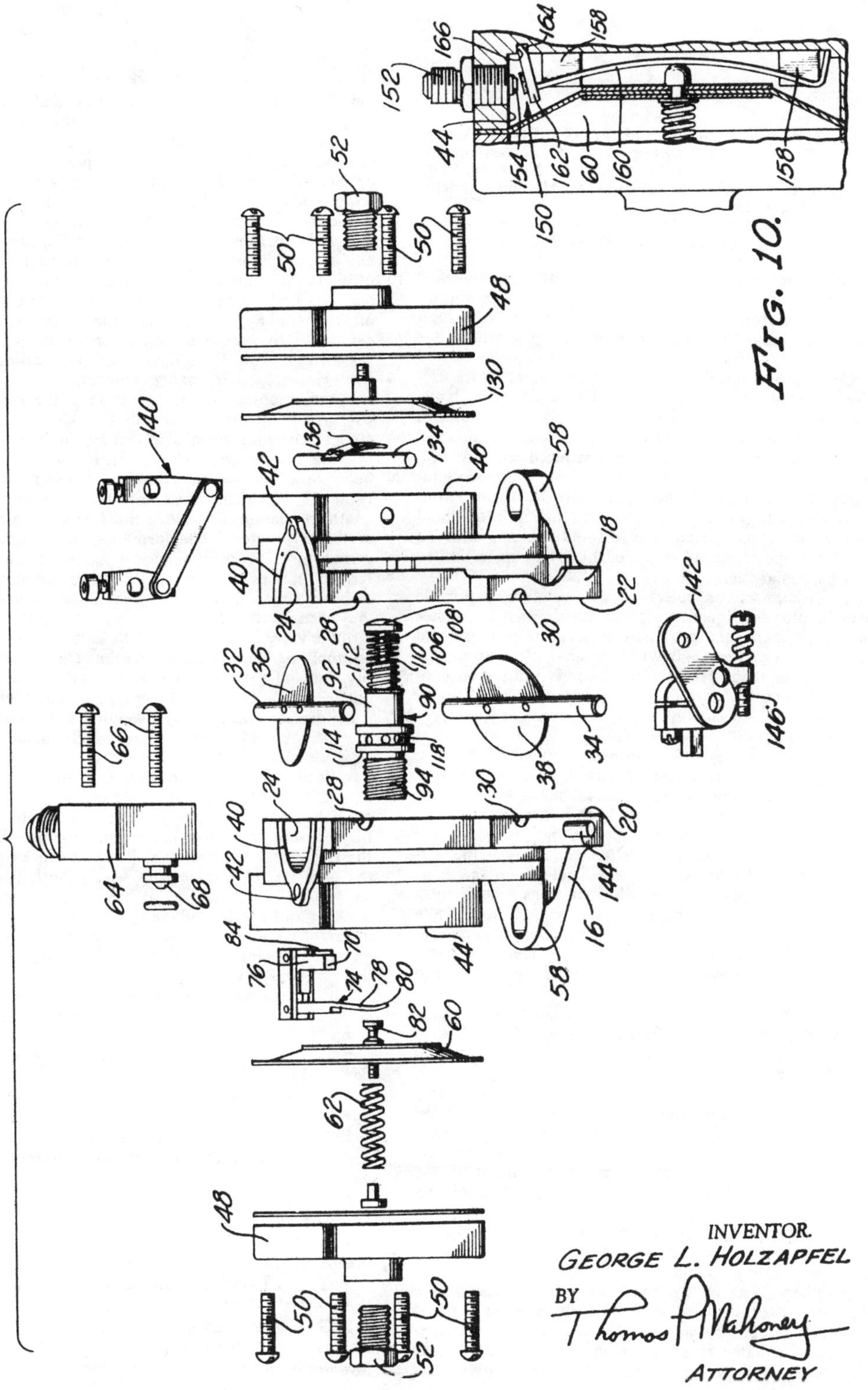

FIG. 9

FIG. 10.

INVENTOR.
GEORGE L. HOLZAPFEL
BY Thomas P. Mahoney
ATTORNEY

United States Patent Office

3,067,020
Patented Dec. 4, 1962

1

3,067,020

CARBURETOR FOR LIQUEFIED GASEOUS FUELS

George L. Holzapfel, Newbury Park, Calif., assignor of eighty percent to Concha Holzapfel, Newbury Park, and twenty percent to Thomas P. Mahoney, Pacific Palisades, Calif.

Filed Mar. 16, 1959, Ser. No. 799,529
13 Claims. (Cl. 48—184)

This invention relates to a carburetor for liquefied gaseous fuels and, more particularly, to a carburetor which is adapted to reduce the pressure of said fuels and to mix the fuel and air in correct proportions for optimum combustion.

Conventional liquefied gaseous fuel carburetors are usually quite bulky and are associated with relatively large, separately mounted primary and secondary regulators adapted to successively reduce the pressure of the liquefied gaseous fuel and to mix the fuel at reduced pressure with combustion air. Although the carburetor is primarily intended for use with liquefied gaseous fuels, it can be utilized with equal effectiveness in conjunction with liquid fuels such as gasoline and the like and it is not intended that the scope of the utilization of the invention be limited to a particular fuel.

It is, therefore, an object of my invention to provide a carburetor for liquefied gaseous fuels which eliminates the necessity for the provision of separate primary and secondary regulators and which is characterized by the fact that only one pressure regulator is utilized in conjunction with the fuel-air mixing chamber of the carburetor. An associated object of my invention is the provision of a carburetor of the aforementioned type wherein the pressure regulator is incorporated in the housing of the carburetor as a unitary part thereof and wherein the necessity for providing separate mountings for primary and secondary regulators is thus eliminated.

Another object of my invention is the provision of a carburetor for liquefied gaseous fuels which is of the self-starting type, that is, which immediately upon the energization of the engine with which it is associated by a crank, self-starter, or other means, will automatically deliver fuel to the fuel-air mixing chamber of the carburetor, thus eliminating the necessity for solenoid-type and other starters conventionally associated with liquefied gaseous fuel carburetors.

A further object of my invention is the provision of a liquefied gaseous fuel carburetor which incorporates, in conjunction with the pressure regulator thereof, a fuel outlet valve, said fuel outlet valve being adapted to feed fuel at relatively high pressure to the fuel-air mixing chamber of the carburetor, thus radicially improving the performance of the carburetor and eliminating the necessity of the provision of the customarily supplied secondary regulator. Therefore, in the carburetor of my invention the fuel is subjected to only one stage of regulation and the customary secondary regulator is eliminated with consequent economies in cost of manufacture and consequent improvement in ease of mounting the carburetor of my invention.

Another object of my invention is the provision of a carburetor for liquefied gaseous fuels wherein there is provided on the housing of the carburetor a pressure-energized actuator, said pressure-energized actuator being connected to both the aforementioned inlet valve and

2

the choke valve of the carburetor and said pressure actuator being also connected to the fuel-air mixing chamber of the carburetor. Therefore, when the pressure actuator is subjected to a pressure drop caused by energization of the engine with which the carburetor is associated, the actuator will automatically cause the opening of the choke valve and the concomitant opening of the fuel outlet valve, thus metering fuel to the fuel-air mixing chamber in direct proportion to the amount of air fed past the choke of the carburetor. In this manner, the fuel-air mixture can be precisely calculated for all the choke openings and the outlet valve can be calibrated to accurately meter the fuel in an accurate proportion with the amount of air passing into and through the fuel-air mixing chamber.

Another object of my invention is the provision of a carburetor of the aforementioned character which includes a housing formed in two symmetrical halves, said housing having opposed flat faces which are adapted to be secured in abutting relationship and having semi-circular faces which, when the flat faces are secured in abutting relationship, define the fuel-air mixing chamber of the carburetor. The formation of the housing of the carburetor in two symmetrical halves in the above described manner also permits the choke and butterfly valves to have their spindle extremities disposed in bearings in the opposed flat faces, thus permitting the choke and throttle valve to be mounted as unitary parts during the assembly of the carburetor and thus eliminating the necessity for the elaborate mounting expedients customarily utilized in carburetors of the type under discussion here.

A further object of my invention is the provision of a pressure regulator inlet valve which includes a simple actuator constituted by a flat spring, said flat or leaf spring having the inlet valve mounted upon one extremity thereof and associated with the inlet opening in said pressure regulator, said flat spring being operatively engaged by the pressure regulator diaphragm to cause displacement thereof and corresponding displacement or opening of the pressure inlet valve with which the flat spring is associated.

Other objects and advantages of my invention will be apparent from the following specification and the accompanying drawings which are for the purpose of illustration only and in which:

FIG. 1 is a perspective view showing a carburetor constructed in accordance with the teachings of my invention;

FIG. 2 is a side elevational view taken from the broken line 2—2 of FIG. 1;

FIG. 3 is a side elevational view taken from the broken line 3—3 of FIG. 1;

FIG. 4 is a fragmentary, sectional view showing the pressure actuator of the carburetor of my invention in part;

FIG. 5 is a vertical, sectional view taken on the broken line 5—5 of FIG. 1;

FIG. 6 is an enlarged, fragmentary, sectional view taken on the broken line 6—6 of FIG. 5;

FIG. 7 is a vertical, sectional view taken on the broken line 7—7 of FIG. 5;

FIG. 8 is an enlarged, longitudinal, sectional view of the fuel inlet valve incorporated in the carburetor of my invention;

FIG. 9 is an exploded view showing the component parts of the carburetor of my invention; and

FIG. 10 is an enlarged, fragmentary, sectional view showing an alternative form of fuel outlet valve for use in the carburetor of my invention.

Referring to the drawings and particularly to FIGS. 1–3 thereof, I show a carburetor **10** constructed in accordance with the teachings of my invention, said carburetor being embodied in a housing **12** which is formed from two symmetrically shaped parts, a first part **16** and a second part **18**. The first part **16** is formed with flat mating faces **20** adapted to engage similar mating faces **22** provided upon the second part **18** and both the first part **16** and the second part **18** incorporate semi-circularly formed walls **24** which, when the flat mating faces **20** and **22** are juxtaposed in abutting relationship, define a cylindrical fuel-air mixing chamber **26**.

The flat mating faces **20** and **22** are provided with bearing recesses **28** adjacent the upper extremities of the parts **16** and **18** as shown in FIGS. 2 and 3 of the drawings, while said flat faces **20** and **22** incorporate bearing recesses **30** adjacent the lower extremities of the aforementioned parts. Therefore, when the flat mating surfaces **20** and **22** of the first and second parts **16** and **18** of the carburetor housing **12** are juxtaposed in abutting relationship, the bearing recesses **28** and **30** cooperate to receive the spindles **32** and **34** of a choke and throttle valve **36** and **38**, respectively.

Thus, the opposite extremities of the spindles **32** and **34** of the choke and throttle valves **36** and **38**, respectively, can be deposited in the associated bearing recesses **28** and **30** and, in this manner, these valves can be easily and cheaply assembled as component parts of the carburetor structure. Semi-circular mounting rings **40** incorporating mounting lugs **42** are formed upon the upper extremities of the parts **16** and **18** to facilitate the mounting of an air cleaner or similar auxiliary device in conjunction with the carburetor **10**.

Each of the halves **16** and **18** of the housing **12** has a pressure chamber associated therewith, the first half **16** incorporating a first pressure chamber **44** and the second half **18** incorporating a second pressure chamber **46**, as best shown in FIGS. 4–7 of the drawings. Each of the pressure chambers **44** and **46** has a cover **48** operatively associated therewith and maintained in operative relationship with the housing **12** by means of screws or similar fasteners **50**. Adjustable fittings **52** are mounted in the covers **48** and incorporate orifices **54** establishing fluid communication between the interiors of the chambers **44** and **46** and the ambient air, for a purpose which will be described in greater detail below.

It will be noted that the lower extremity of each of the parts **16** and **18** of the housing **12** is provided with a mounting flange **58** to facilitate the securement of the carburetor **10** to the manifold or other portion of an associated engine, not shown.

The first pressure chamber **44** constitutes a pressure regulator and has a pressure responsive means constituted by a conventional diaphragm **60** mounted therein, said diaphragm being urgable in a right-hand direction, as best shown in FIGS. 5 and 6 of the drawings, by means of a spring **62** which has one extremity seated upon the center of the diaphragm and its opposite extremity seated in the associated fitting **52**. By rotation of the fitting **52** in the appropriate direction, the compressive effect of the spring **62** upon the diaphragm **60** may be varied to accomplish the desired pressure reduction through the pressure regulator chamber **44**.

A fuel inlet fitting **64** is mounted by means of screws **66** on the first part **16** of the housing **12** and has a valve seat **68** formed thereupon which is engageable by an inlet valve **70**, as best shown in FIG. 6 of the drawings, to permit or restrict the flow of liquefied gaseous fuel into the pressure regulator chamber **44**.

The energization of the inlet valve **70** is accomplished by a linkage **74** which, as best shown in FIGS. 5 and 6 of the drawings, includes a first bellcrank **76** and a second bellcrank **78** pivotally mounted with their juxtaposed legs in overlying and mutually engaging relationship. The lower extremity of the bellcrank **78** incorporates a bifurcated portion **80** which engages a yoke **82** on the diaphragm **60**.

A valve member **84** formed from rubber or the like is mounted on the bellcrank **76** and is urgable into engagement with the seat **68** on the fuel inlet fitting **64**. Since the spring **62** is normally compressed by the deflection of the diaphragm **60** in a left-hand direction, as viewed in FIGS. 5 and 6 of the drawings, due to the pressure in the chamber **44**, the bellcrank **78** urges the valve member **84** onto the seat **68**, thus preventing further flow of liquefied gaseous fuel into the pressure chamber **44**.

The pressure chamber **44** communicates, as best shown in FIGS. 2 and 5 of the drawings, with the mixing chamber **26** through a first bore **88**, said first bore terminating, as best shown in FIG. 2 of the drawings, in a downwardly and angularly inclined outlet portion **91** which communicates with the mixing chamber **26** to discharge liquefied gaseous fuel downwardly thereinto.

Mounted in the first bore **88** is an outlet fuel valve **90**, said outlet fuel valve being incorporated in an elongated housing **92** having a threaded extremity **94** adapted to engage a correspondingly threaded portion of the first bore **88** to retain said housing in said bore. The housing **92** incorporates a seat **96** against which the head **98** of a needle valve **100** is adapted to urge an O-ring **102**. The needle valve **100** includes a tapered metering portion **104**, as best shown in FIG. 8 of the drawings, and an elongated shank **106** formed integrally therewith.

A spring seat **108** is provided on the right-hand extremity of the elongated shank **106**, as best shown in FIGS. 5 and 8 of the drawings, said spring seat being adapted to support one extremity of a compression spring **110** whose other extremity is mounted upon a shoulder **112** of the valve housing **92**. The housing **92** is also provided with a shoulder **114** which, as best shown in FIG. 5 of the drawings, receives a sealing washer **116** which serves to seal the valve housing **92** against a corresponding portion of the carburetor housing **12**.

The valve housing **92** is provided wtih an outlet chamber **120** in which the tapered metering portion **104** of the needle valve **100** is located. A plurality of radial ports **118** communicates with said outlet chamber and thus permits fuel to be discharged laterally into right-angularly and downwardly oriented portion of the first bore **88**.

A second bore **122** is formed in the second part **18** of the carburetor housing **12** and establishes fluid communication between the fuel-air mixing chamber **26** and the second pressure chamber **46**. A third bore **124** is provided in the second part **18** of the carburetor housing **12** in alignment with the axially extending portion of the first bore and receives the portion of the valve housing **92** which projects beyond the axially extending portion of the first bore **88**, as best shown in FIG. 5 of the drawings.

Therefore, the spring seat **108** may be engaged by a yoke **128** mounted upon the pressure responsive diaphragm **130**, as best shown in FIG. 5 of the drawings, so that the left-hand movement of the diaphragm, **130**, as viewed in FIG. 5 of the drawings, will be accompanied by corresponding left-hand movement of the needle valve **100** urging the associated O-ring **102** off the seat **96** to permit flow of liquefied gaseous fuel under pressure into the fuel-air mixing chamber **26** of the carburetor, in a manner which will be described in greater detail below.

A spindle **134** has its opposite extremities journaled in the opposite walls of the second pressure chamber **46** and mounts a bifurcated lever **136** engageable with the yoke 128 on the pressure responsive diaphragm 130. The laterally projecting extremity of the spindle 134 is adapted to energize a linkage 140 operatively connecting the spindle 32 of the choke valve 36 to the pressure responsive diaphragm 130.

The outer extremity of the throttle valve spindle 34 has a throttle link 142 mounted thereupon which is engageable by any suitable means, such as a Bowden wire, to cause the opening of the throttle. A stop pin 144 is provided in the first part 16 of the carburetor housing 12 for engagement by a stop screw 146 and thus serves to limit rotation of the spindle 34 and thus rotation of the throttle 38 mounted thereupon.

An alternative construction 150 of the inlet valve is shown in FIG. 10 of the drawings as mounted within the first pressure chamber 44. In the alternative construction of the inlet valve 150, a simple fuel inlet fitting 152 is provided in a wall of the chamber 44 and has a valve seat 154 formed integrally therewith and projecting into the chamber 44. Mounted between spaced mounting blocks 158 is a flat or leaf spring actuator 160 which has a valve member 162 secured to an extremity thereof, one end of the valve member being beveled at 164 to permit it to pivot in a groove 166 in order that a rocking action may be imparted to the valve member 162 when the leaf spring 160 is deflected to the right by corresponding right-hand movement of the pressure responsive diaphragm 60, as viewed in FIG. 10 of the drawings.

The alternative valve construction 150 is considerably simpler than the construction of the inlet valve 70 previously disclosed hereinabove and greatly reduces the cost of fabricating the carburetor of my invention.

The component parts of the carburetor housing 12 are best fabricated by the utilization of conventional die-casting processes since the first and second parts 16 and 18 can be cast with the first and second pressure chambers 44 and 46 formed therein and the covers 48 can be of identical configuration. In the assembly of the carburetor 10 the spindles 32 and 34 of the choke valve 36 and throttle valve 38, respectively, are disposed in the respective bearing recesses 28 and 30 of the second part 18 of the housing 12 and the mating flat faces 20 and 22 of the parts 16 and 18 are brought into contiguity with each other after the valve housing 92 has its threaded portion 94 threaded into the correspondingly threaded portion of the first bore 88.

Of course, the other extremity of the valve housing 92 is inserted in the third bore 124 to permit the elongated shank 106 of the needle valve 100 to project into the second pressure chamber 46. Because of the fact that the housing 12 of the carburetor 10 is split on an axis which permits the mounting of the throttle and choke valves 36 and 38 and the outlet valve 90 in the simple manner described, the cost of producing the carburetor 10 of my invention is materially reduced.

After the two parts 16 and 18 of the carburetor housing have been assembled with the other components of the carburetor in the above described manner, screws 168 are utilized to maintain the two parts in operative relationship with each other. Subsequently, the inlet valve fitting 64 is secured to the first part 16 of the housing 12 by means of the screws 66 and the inlet valve 70 and the linkage 74 constituting a part thereof mounted in operative relationship with the first pressure chamber 44. The pressure responsive diaphragm 60 and the associated spring 62 are then mounted within the first pressure chamber 44 and the cover 48 secured in operative relationship with the pressure chamber by the utilization of the fastening screws 50.

In a similar manner, the spindle 134 is located within the second pressure chamber 46 and the bifurcated lever 136 affixed thereto, the pressure responsive diaphragm 130 being secured in juxtaposition thereto by means of the cover 48 and the associated mounting screws 50 therefor. After this is accomplished, the linkage 140 between the spindle 134 and the choke valve spindle 32 is installed and the throttle link 142 is also mounted upon the laterally extending portion of the throttle spindle 134. Therefore, it can be seen that once the component parts of the carburetor 10 of my invention have been assembled, the entire device is ready to be installed upon an engine by the utilization of the mounting flanges 58 and the elaborate starters and associated fuel regulators characteristic of prior art devices are eliminated. In addition, the simple assembly of the carburetor 10 materially reduces the cost of producing the same and results in a relatively small assembly which can be inserted in a much smaller space than prior art devices.

The assembly of the alternative inlet valve construction 150 in the first pressure chamber 44 merely involves the disposition of the leaf spring 160 between the spaced mounting blocks 158 and the location of the beveled portion 164 of the valve member 162 in the groove 166 in juxtaposition to the valve seat 154. The opposite bent extremity of the leaf spring 160 is located in juxtaposition to the opposite wall of the first pressure chamber 44. The assembly of the other components is similar to that described hereinabove.

In operation, the carburetor 10 is mounted upon the intake manifold of an internal combustion engine by the utilization of the mounting flanges 58. When so mounted, it constitutes a completely operative device independent of any auxiliary fittings, solenoid starters, or the like which are customarily utilized in conjunction with conventional gaseous fuel carburetors.

When the engine with which the carburetor 10 is associated is started by a manual or electrical means, the throttle valve 38 of the carburetor 10 is open and a negative pressure created within the fuel-air mixing chamber 26. When this occurs, a corresponding negative pressure is created within the second pressure chamber 46 through the second bore 122 causing deflection of the pressure responsive diaphragm 130 to the left, as viewed in FIG. 5 of the drawings. The deflection of the pressure responsive diaphragm 130 in the above described manner causes corresponding axial movement of the needle valve 100 to unseat the O-ring 102 from the seat 96 and permits flow of liquefied gaseous fuel under pressure from the first pressure chamber 44.

The flow of liquefied gaseous fuel from the chamber 44 permits right-hand movement of the pressure responsive diaphragm 60 causing right-hand deflection of the lever 78, as viewed in FIG. 6 of the drawings, and permitting the bellcrank 76 to rotate in a counterclockwise direction to unseat the valve 70 and permit flow of liquefied gaseous fuel through the fuel inlet fitting 64.

The liquefied gaseous fuel is thus subjected to a single pressure reduction stage as it flows through the pressure regulator constituted by the valve 70 and the associated first pressure chamber 44 and pressure responsive diaphragm 60 and flows from said first pressure chamber past the unseated O-ring 102 and the tapered metering portion 104 of the needle valve 100 to the radial ports 118 in the needle valve housing 92. It will be observed from FIG. 2 of the drawings that the ports 118 discharge the liquefied gaseous fuel under pressure downwardly through the angularly and downwardly oriented outlet port 91 so that the downward direction imparted to the liquefied gaseous fuel as it enters the fuel-air mixing chamber 26 tends to accelerate the flow of combustion air into the fuel-air mixing chamber 26 and to create an optimum fuel mixture. It will be noted that the tapered metering portion 104 of the needle valve 100 will be moved to the left as viewed in FIGS. 5 and 8 of the drawings as the throttle butterfly 38 is rotated to cause an increased negative pressure upon the pressure responsive diaphragm 130 to cause corresponding leftward movement of the needle valve 100.

The tapered metering portion 104 of the needle valve

100 can be calculated to meter the proper amount of fuel to the fuel-air mixing chamber 26 at all settings of the throttle 38 and it will be noted that the position of the choke butterfly 36 will correspond to that of the needle valve 100 at all times during the operation of the carburetor 10.

When the engine with which the carburetor 10 is shut down, the throttle butterfly 38 is, of course, returned to the closed position shown in FIG. 3 of the drawings and the air pressure in the chamber 26 returns to the ambient air pressure because of communication of the chamber 26 with the ambient air past the choke butterfly 36. When this occurs, the pressure in the second pressure chamber 46 returns to ambient and the compression spring 110 on the needle valve 100 causes simultaneous movement of the pressure responsive diaphragm 130 and the bifurcated lever 136 to the right, as viewed in FIG. 5 of the drawings, thus causing simultaneous impingement of the O-ring 102 of the needle valve 100 on its associated seat 96 and closure of the choke buterfly 36. Naturally, when the O-ring 102 seats in the above described manner, pressure in the first pressure chamber 44 on the right-hand side of the pressure responsive diaphragm 60, as viewed in FIGS. 5 and 6 of the drawings, will increase, causing deflection thereof to the left against the bias of the compression spring 62 and returning the valve 70 to seated position on the associated valve seat 68 to terminate the flow of liquefied gaseous fuel from the fuel source through the fuel inlet fitting 64.

With the alternative embodiment of the pressure regulator valve construction shown in FIG. 10 of the drawings, the valve member 162 is deflected from engagement with the valve seat 154 when a pressure drop occurs within the first chamber 44 due to the unseating of the O-ring 102 from its seat 96. The unseating of the O-ring 102 permits right-hand deflection of the first pressure responsive diaphragm 60 permitting the diaphragm to bend the leaf spring 160 at its center to rock the valve member 152 on its beveled extremity 164 in the groove 166. Conversely, when the needle valve 100 is closed to seat the O-ring 102 on the associated seat 96, pressure built up in the first pressure chamber 44 will cause deflection of the diaphragm 60 to the left, as viewed in FIG. 10 of the drawings, to permit the leaf spring 160 to return to the straight position and rock the valve member 152 into seated position on the associated seat 154.

I thus provide by my invention a liquefied gaseous fuel carburetor which can be utilized with liquefied gaseous fuels of relatively high pressures such as propane and butane, or can be utilized with pressurized natural gases of somewhat lower pressures. The carburetor of my invention eliminates the two pressure reduction stages conventionally utilized in conjunction with conventional liquefied gaseous fuel carburetors and is characterized also by the incorporation of the single stage pressure reducer or regulator in the body or housing of the carburetor itself. Furthermore, the fuel is fed at elevated pressures through a metering valve which coacts with the choke valve in response to pressure drops in the fuel-air mixing chamber to insure the proper fuel-air mixture for any throttle setting of the carburetor.

In addition, the carburetor of my invention is characterized by unique constructional features which facilitate the manufacture and assembly of the component parts of the carburetor, among them being the two-part split housing, the incorporation in the two parts of the housing of first and second pressure chambers, and the provision, in an alternative embodiment of the carburetor, of a simple, inexpensive and easily installed pressure regulator valve.

Other objects and advantages of the invention will have been made apparent by the detailed disclosure appearing hereinabove.

I claim:

1. In a carburetor, the combination: a housing having a fuel-air passage therein and first and second pressure chambers on sides thereof, said housing incorporating first and second bores respectively establishing fluid communication between said first and second chambers and said passage and a third bore in the wall of said second chamber aligned with said first bore; a fuel inlet communicating with said first chamber; first pressure responsive means in said first chamber; inlet valve means operatively connected to said first pressure responsive means for controlling fuel flow into said first chamber through said inlet; outlet valve means mounted in said first bore for controlling fuel flow into said passage, said outlet valve means having a portion extending into said second chamber through said third bore; second pressure responsive means in said second chamber operatively connected to said outlet valve means; a choke valve mounted in said passage above said first, second, and third bores; a throttle valve mounted in said passage below said first, second and third bores; and a linkage operatively connecting said second pressure responsive means to said choke valve to cause conjoint movement of said outlet valve and said choke valve.

2. In a carburetor, the combination of: a housing having a passage incorporating a fuel-air mixing chamber defined by a choke valve at the upper extremity of said passage and a throttle valve at the lower extremity thereof, said housing incorporating first and second pressure responsive members and first and second bores establishing fluid communication between said mixing chamber and said members; a fuel inlet valve in said housing operatively connected to said first pressure responsive member; and a fuel outlet valve in said first bore operatively connected to said second pressure responsive member, said second pressure responsive member also being operatively connected to said choke valve whereby a pressure drop across said mixing chamber will cause conjoint movement of said outlet valve and said choke valve.

3. In a carburetor, the combination of: a housing having a passage incorporating a fuel-air mixing chamber defined by a choke valve at the upper extremity of said passage and a throttle valve at the lower extremity thereof, said housing incorporating a pressure responsive member and fuel inlet means and first and second bores respectively establishing fluid communication between said fuel inlet means and said pressure responsive member and said mixing chamber; and a fuel outlet valve mounted in said first bore in communication with said fuel inlet means and operatively connected to said pressure responsive member, said pressure responsive member also being operatively connected to said choke valve whereby a pressure drop across said second bore will cause conjoint movement of said choke and outlet valves.

4. In a carburetor, the combination of: a housing consisting of mating halves having opposed flat surfaces engaging one another and spaced surfaces of semi-circular configuration between said flat surfaces defining a fuel-air mixing chamber having upper and lower extremities in said housing; a choke valve mounted in the upper extremity of said chamber having a spindle with its extremities secured between said mating halves; and a throttle valve located at the lower extremity of said mixing chamber having a mounting spindle whose opposite extremities are secured between the opposed flat surfaces of said mating halves, said mating halves having opposed bores communicating with said chamber for the reception and control of a fuel outlet valve.

5. In a carburetor, the combination of: a housing consisting of mating halves having opposed flat surfaces engaging one another and spaced surfaces of semi-circular configuration between said flat surfaces defining a fuel-air mixing chamber in said housing, said housing having pressure responsive means and fuel inlet means thereupon and first and second bores therein respectively establishing fluid communication between said mixing chamber

and said fuel inlet means and said pressure responsive means; a choke valve mounted in the upper extremity of said chamber having a spindle with its extremities secured between said mating halves; a throttle valve located at the lower extremity of said mixing chamber having a mounting spindle whose opposite extremities are secured between the opposed flat surfaces of said mating halves; and a fluid outlet valve located in said first bore in communication with said fuel inlet means and operatively connected to said pressure responsive means and said choke valve whereby a pressure drop across said second bore will cause corresponding and conjoint opening of said outlet valve and said choke valve.

6. In a carburetor, the combination of: a housing consisting of mating halves having opposed flat surfaces engaging one another and spaced surfaces of semi-circular configuration between said flat surfaces defining a fuel-air mixing chamber in said housing, said halves each incorporating pressure responsive means and a bore establishing fluid communication between said mixing chamber and said pressure responsive means; a choke valve mounted in the upper extremity of said chamber having a spindle with its extremities secured between said mating halves; a throttle valve located at the lower extremity of said mixing chamber having a mounting spindle whose opposite extremities are secured between the opposed flat surfaces of said mating halves; and a fuel outlet valve mounted in one of said bores for feeding fuel into said mixing chamber, said fuel outlet valve being fed fuel through said one of said pressure responsive means and being operatively connected to the other of said pressure responsive means and said choke whereby a pressure drop in said mixing chamber will energize said second pressure responsive means to cause conjoint movement of said outlet and choke valves.

7. In a carburetor, the combination of: a housing having a fuel-air mixing chamber therein and first and second chambers formed upon opposite sides thereof, said housing incorporating a first bore establishing fluid communication between said first chamber and said mixing chamber and a second bore establishing fluid communication between said second chamber and said mixing chamber and a third bore axially aligned with said first bore and extending into said second chamber; a choke valve at the upper extremity of said mixing chamber; a throttle valve at the lower extremity of said mixing chamber; pressure responsive means in said second chamber responsive to a pressure drop across said second bore; and a fuel outlet valve mounted in said first and third bores, said fuel outlet valve being operatively connected to said pressure responsive means and said pressure responsive means being operatively connected to said choke valve whereby pressure drop induced movement of said pressure responsive means will cause conjoint movement of said choke and outlet valves.

8. In a carburetor, the combination of: a housing having a fuel-air mixing chamber therein and a pressure chamber and a fuel inlet thereupon, said housing incorporating a first bore establishing fluid communication between said fuel inlet and said mixing chamber, a second bore establishing fluid communication between said pressure chamber and said mixing chamber and a third bore axially aligned with said first bore and extending into said pressure chamber; a throttle valve at the lower extremity of said mixing chamber; pressure responsive means in said pressure chamber responsive to a pressure drop across said second bore; and a fuel outlet valve mounted in said first and third bores, said fuel outlet valve being operatively connected to said pressure responsive means and said pressure responsive means being operatively connected to said choke valve whereby pressure drop induced movement of said pressure responsive means will cause conjoint movement of said choke and outlet valves.

9. In a carburetor, the combination of: a housing having a fuel-air mixing chamber therein, said mixing chamber having upper and lower extremities and, said housing having first and second chambers formed therein and a fuel inlet communicating with said first chamber, said housing incorporating a first bore establishing fluid communication between said first chamber and said mixing chamber, a second bore establishing fluid communication between said second chamber and said mixing chamber and a third bore axially aligned with said first bore and extending into said second chamber; a choke valve at the upper extremity of said mixing chamber; a throttle valve at the lower extremity of said mixing chamber; first and second pressure responsive means in said first and second chambers; a fuel inlet valve operatively connected to said first pressure responsive means for controlling flow of fuel into said first chamber; and a fuel outlet valve mounted in said first and third bores, said fuel outlet valve being operatively connected to said first and second pressure responsive means and said second pressure responsive means being operatively connected to said choke valve whereby pressure drop induced movement of said first and second pressure responsive means will cause conjoint movement of said choke and outlet valves.

10. In a carburetor, the combination of: a housing having a mixing chamber and first and second pressure chambers formed therein on opposite sides thereof, said housing incorporating a first bore between said first chamber and said mixing chamber and second and third bores communicating with said second chamber, said housing having a fuel inlet into said first chamber; first and second pressure responsive means in said first and second chambers, respectively; a fuel inlet valve at said fuel inlet including an inlet valve member and a flat spring actuator therefor in said first chamber, said first pressure responsive means being engageable with said flat spring actuator to move said inlet valve member; a choke valve at one extremity of said mixing chamber; a throttle valve at the other extremity thereof; and an outlet valve mounted in said first and third bores and being operatively connected to said second pressure responsive means and said second pressure responsive means being operatively connected to said choke valve.

11. In a carburetor, the combination of: a housing having a mixing chamber and first and second pressure chambers formed therein on opposite sides thereof, said housing incorporating a first bore between said first chamber and said mixing chamber and second and third bores communicating with said second chamber, said housing having a fuel inlet into said first chamber; first and second pressure responsive means in said first and second chambers, respectively; a fuel inlet valve at said fuel inlet including an inlet valve member and a flat spring actuator therefor in said first chamber, said first pressure responsive means being engageable with said flat spring actuator to move said inlet valve member; a choke valve at one extremity of said mixing chamber; a throttle valve at the other extremity thereof; and an outlet valve mounted in said first and third bores and being operatively connected to said second pressure responsive means and said second pressure responsive means being operatively connected to said choke valve and said outlet valve having a spring thereupon urging said outlet valve and choke valve into closed positions.

12. In a carburetor, the combination of: a housing having a pressure regulator on one side and a pressure actuator on the other side, said pressure regulator including a pressure responsive diaphragm, a fuel inlet and an inlet valve consisting of a spring urging a valve member on said fuel inlet, said pressure actuator including a pressure responsive diaphragm, said housing having a mixing chamber and a first bore between said pressure regulator and said mixing chamber and second and third bores communicating with said pressure actuator; a choke valve at one extremity of said mixing chamber; a throttle valve at the other extremity of said mixing chamber; and an outlet valve mounted in said first and third bores, said

outlet valve being operatively connected to said pressure actuator and said pressure actuator being connected to said choke valve.

13. In a carburetor, the combination of: a housing incorporating a fuel-air mixing chamber, a fuel inlet, pressure responsive means on one side of said mixing chamber and a first bore between said fuel mixing chamber and said fuel inlet, a second bore between said mixing chamber and said pressure responsive means and a third bore communicating with said pressure responsive means; a choke valve at one extremity of said chamber operatively connected to said pressure responsive means; and a fuel outlet valve located in said first and third bores and

operatively connected to said pressure responsive means whereby a pressure drop across said second bore will cause said pressure responsive means to urge said choke and outlet valves into open positions.

References Cited in the file of this patent

UNITED STATES PATENTS

2,597,335	Jones	May 20, 1952
2,717,827	Best	Sept. 13, 1955
2,831,758	Warner	Apr. 22, 1958
2,855,283	Schumacher	Oct. 7, 1958
2,957,492	Volpin	Oct. 25, 1960

Jan. 4, 1966 F. W. WELLS ET AL 3,227,427

CARBURETOR SYSTEM

Filed Feb. 20, 1963

Fig.1.

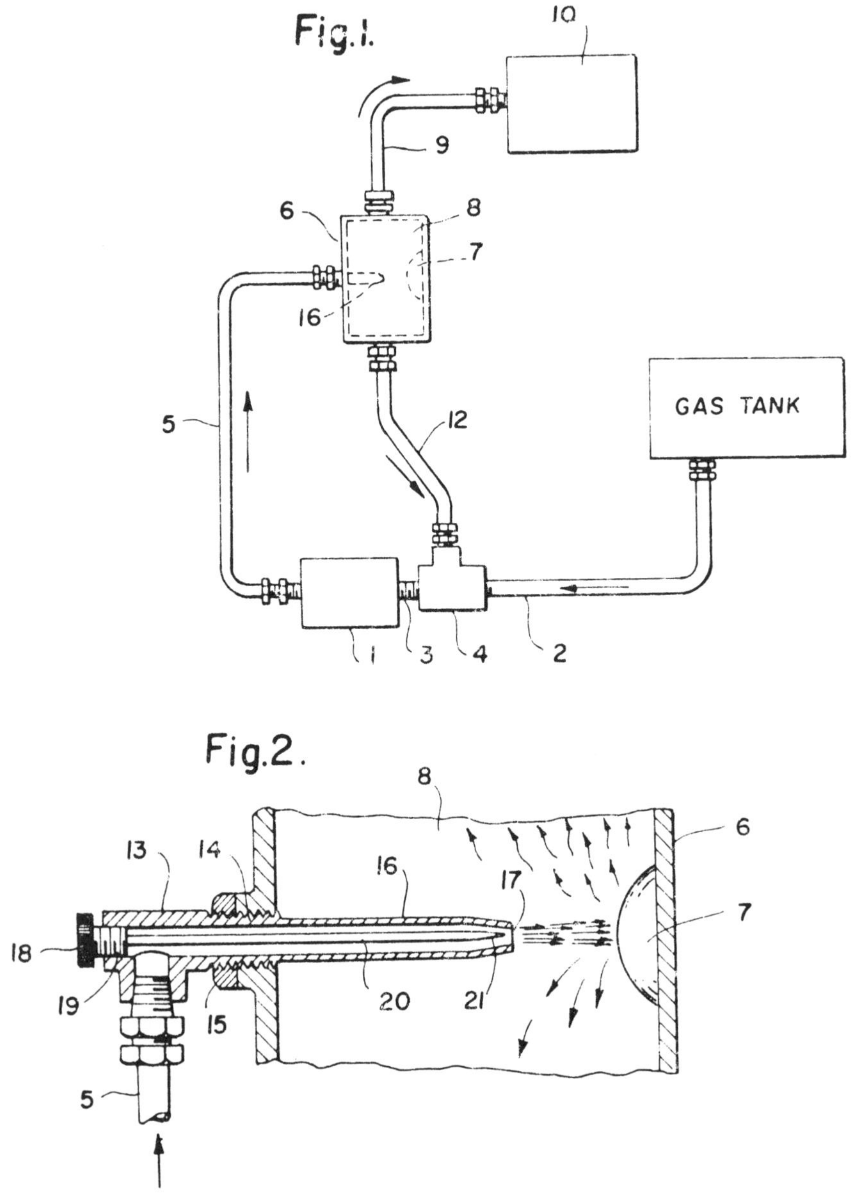

INVENTORS
FRANCIS W. WELLS and
ALBERT E. PHILLIPS
BY
William J. Ruano
their ATTORNEY

United States Patent Office

3,227,427

Patented Jan. 4, 1966

1

3,227,427

CARBURETOR SYSTEM

Francis W. Wells, Pittsburgh, and Albert E. Phillips, McKees Rocks, Pa., assignors of thirty percent to William J. Russo, Pittsburgh, Pa.

Filed Feb. 20, 1963, Ser. No. 259,946

5 Claims. (Cl. 261—20)

This invention relates to a carburetor system for increasing mileage of an automobile, truck or other gas-engine driven conveyance.

Various attempts have been made in the past to increase the efficiency of gasoline engines for vehicles, but these have not been altogether successful or widely adopted since they require complicating the carburetor construction and greatly increasing the expense thereof as well as providing only a very small increase in efficiency that does not warrant the increased cost of the equipment.

Another disadvantage of prior systems involving atomizing gasoline in the carburetor is that droplets of gasoline pass through without complete atomization, therefore not providing optimum efficiency.

An object of our invention is to provide a system for increasing gas mileage for an automobile and which is devoid of the above named disadvantages.

A more specific object of our invention is to provide, between the fuel pump and existing carburetor of an automobile, a loop circuit including an atomizing chamber for atomizing gasoline before entering the carburetor and which recirculates droplets which are not sufficiently atomized through the loop circuit so as to very greatly increase the gasoline mileage or efficiency of the vehicle.

Another object of the invention is to considerably reduce harmful exhaust fumes from the engine by reducing the amount of raw gasoline injected therein, thereby minimizing or substantially avoiding air pollution which has plagued many areas of the country.

Other objects and advantages will become more apparent from a study of the following description taken with the accompanying drawing wherein:

FIG. 1 shows a gasoline atomizing system for an automobile carburetor embodying the principles of our invention; and,

FIG. 2 is an enlarged, fragmentary, vertical cross-sectional view of a modification of the atomizing nozzle inside the atomizing chamber 6 of FIG. 1.

Referring more particularly to FIG. 1 of the drawing, numeral 1 denotes a fuel pump connected by pipe 3 and a standard T 4 to pipe 2 leading to a gas tank (not shown). The delivery connection end of the fuel pump 1 leads to a conduit 5 connected to a nozzle 16. Nozzle 16 has a tapered end 17 (see FIG. 2) for squirting, under high velocity, a stream of gasoline onto a spherical segment 7 which causes the particles to deflect and atomize, as indicated by the arrows. Most of the atomized gasoline is sucked into the upper portion 8 of the atomizing chamber 6 and will be drawn through conduit 9 into a vacuum portion of carburetor 10 connected to the engine (not shown).

However, some of the gasoline squirted by nozzle 16 and atomized by the spherical segment 7 will not be sufficiently small in size so as to be drawn upwardly, therefore, will drain downwardly through drain-back line or conduit 12, particularly as a consequence of suction therein created by pump 1, through T 4 and fuel pump 1, thence again flowing through conduit 5 to nozzle 16, and thus is recirculated. So long as the particle size of the gasoline is too large to be drawn upwardly through conduit 9, the droplets will be recirculated over and over again through pipe 12, T 4 and conduit 5 until they atomize sufficiently to be drawn upwardly into the carburetor 10.

2

FIG. 1 shows a nozzle 16 with a tapered end portion, shaped somewhat as shown in FIG. 2, but without a needle valve 20. Such arrangement is for the purpose of assuring that the nozzle will always be open and permit at least a predetermined flow of gasoline at all times.

FIG. 2 shows a modification of the nozzle shown in FIG. 1, which comprises the addition of a needle valve 20 having a tapered end 21 which may be adjustably projected into the opening of the nozzle so as to variably restrict the area of the nozzle opening, merely by adjustably turning the knurled head 18 so as to effect greater screwing action at threads 19. It should be noted that even when head 18 is screwed to the limit against nozzle body 13, the end 21 will not completely close the opening at 17. This is to assure that the device will be fool-proof and prevent the possibility of either stopping or unduly restricting the flow of gasoline through the nozzle.

Efficiencies ranging from a 50 percent increase to a 100 percent increase over existing efficiencies have been obtained by the present invention, not only because of the effective atomizing of the gasoline in chamber 6 even before entering the carburetor, but particularly by virtue of recirculation of the heavier particles of the gasoline, over and over again, until completely atomized before conducting them through the carburetor.

Thus it will be seen that we have provided an amazing and almost unbelievable increase in mileage of automobiles by virtue of our novel atomizing system involving thorough atomization and recirculation; furthermore, we have provided an atomizing system that can be added to existing, standard automobile carburetors without the necessity of replacing the carburetor, which system involves relatively simple and inexpensive standard parts; furthermore, we have provided a system for considerably reducing the amount of raw gas injected into the engine to an extent as to sharply reduce and practically eliminate harmful exhaust fumes that would otherwise contaminate the atmosphere, therefore, solving the air pollution problem existing in many areas.

While we have illustrated and described several embodiments of our invention, it will be understood that these are by way of illustration only, and that various changes and modifications may be made within the contemplation of our invention and within the scope of the following claims.

We claim:

1. A carburetor system for a gasoline engine driven conveyance, comprising, in combination, a complete and independent carburetor for connection to the engine, a separate atomizing chamber disposed at a lower level than said carburetor, a fuel inlet pipe connecting the top portion of said atomizing chamber to a vacuum portion of said carburetor, said atomizing chamber including a nozzle and a confronting surface for deflecting and atomizing gasoline squirted thereon by said nozzle, a drain line leading downwardly from the bottom portion of said atomizing chamber, a source of gasoline located below said atomizing chamber and connected to said drain-line, and a fuel pump serially connected between said drain line and said nozzle so that squirted droplets which are not atomized will drain from the bottom of said chamber through said drain-line and will be recirculated back to the nozzle until effectively atomized, said atomizing chamber being devoid of an air supply.

2. A system as recited in claim 1 together with an adjustable needle in said nozzle for adjustably throttling gasoline flowing through said nozzle.

3. A system as recited in claim 1 wherein said confronting surface is a spherical segment in substantially horizontal alignment with said nozzle substantially centrally of said atomizing chamber.

4. A system as recited in claim 1, together with a needle valve for adjustably throttling the opening of said nozzle

3,227,427

3

and being so arranged that when fully throttled, there is still a predetermined flow of gasoline through said nozzle.

5. A carburetor system for a gasoline engine driven conveyance, comprising, in combination, a complete and independent carburetor for connection to the engine, a separate atomizing chamber disposed at a lower level than said carburetor, a fuel inlet pipe connecting the top portion of said atomizing chamber to a vacuum portion of said carburetor, said atomizing chamber including a nozzle and a confronting arcuate surface for deflecting and atomizing gasoline squirted thereon by said nozzle, a gasoline tank located below said atomizing chamber, a T having one leg connected to said gasoline tank, a drain line connected between the bottom of said atomizing chamber and a second leg of said T, and a fuel pump connected between the third leg of said T and said nozzle for recirculating over and over again squirted droplets in said atomizing chamber which have drained through said drain line, whereby gasoline in said fuel inlet pipe is effectively atomized, said atomizing chamber being devoid of an air supply.

4

References Cited by the Examiner

UNITED STATES PATENTS

890,970	6/1908	Durr	261—36
980,946	1/1911	Heermans.	
992,260	5/1911	Rush.	
1,585,094	5/1926	Hendricks.	
1,997,497	4/1935	Pogue	261—36
2,702,694	2/1955	Featherston	261—23
2,892,692	6/1959	Anderson	123—133 X

FOREIGN PATENTS

808,823	11/1936	France.	
1,111,376	10/1955	France.	
22,259	1912	Great Britain.	

ROBERT F. BURNETT, *Primary Examiner.*

HARRY B. THORNTON, *Examiner.*

T. R. MILES, *Assistant Examiner.*

Sept. 20, 1966 | C. MIÑOZA | 3,273,983

FUEL VAPORIZER CARBURETOR

Filed Nov. 20, 1961 | 2 Sheets-Sheet 1

FIG. 1.

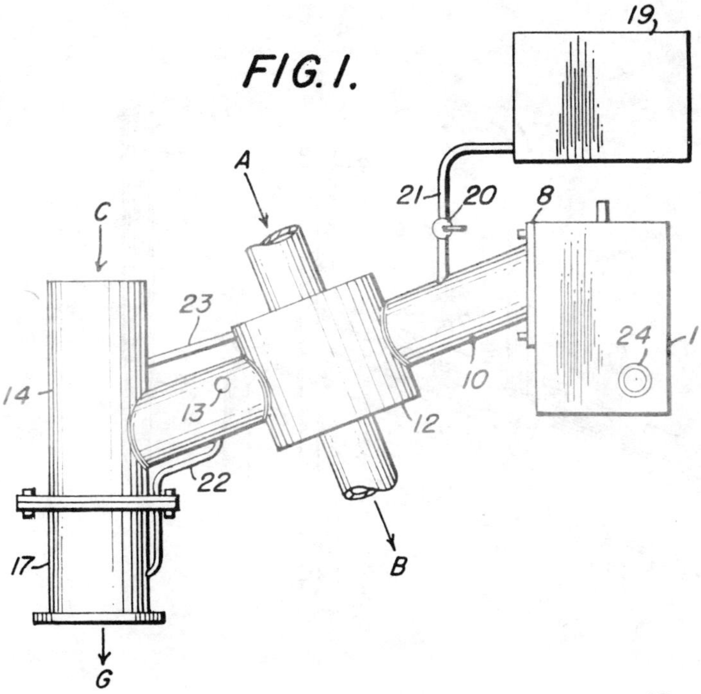

FIG. 2.

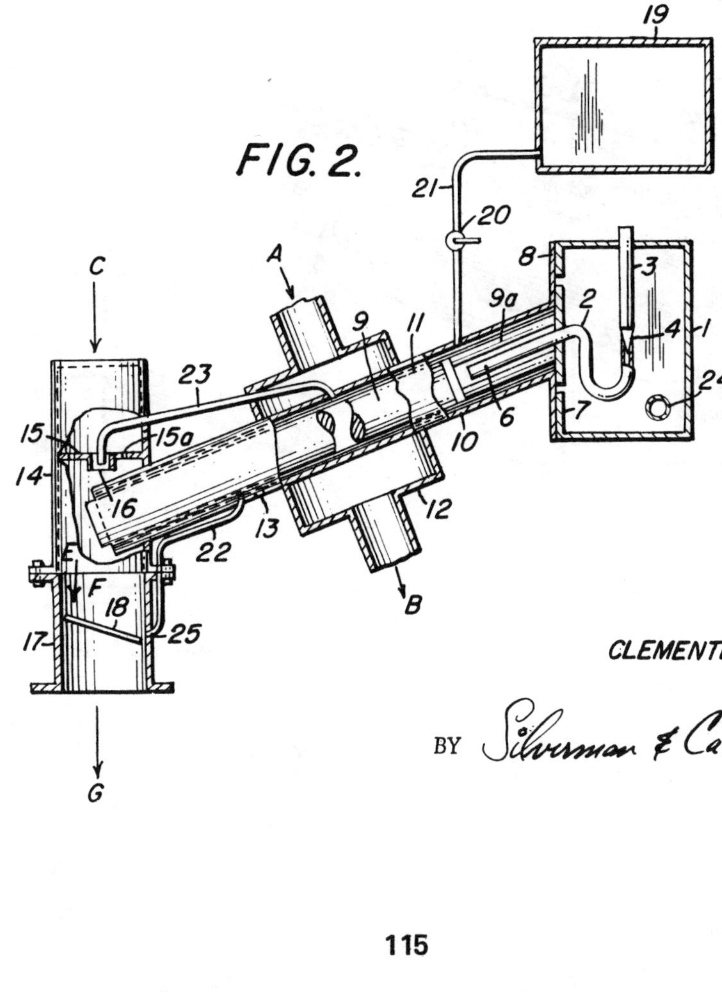

INVENTOR
CLEMENTE MIÑOZA

BY Silverman & Cass

ATTORNEYS

Sept. 20, 1966 C. MIÑOZA 3,273,983

FUEL VAPORIZER CARBURETOR

Filed Nov. 20, 1961 2 Sheets-Sheet 2

FIG. 3.

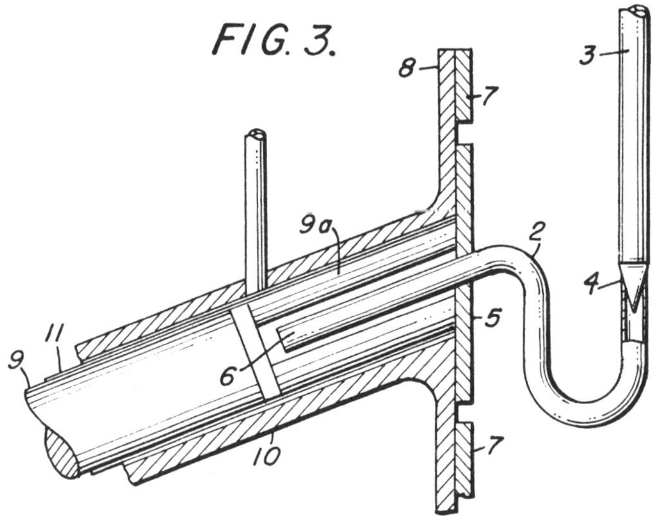

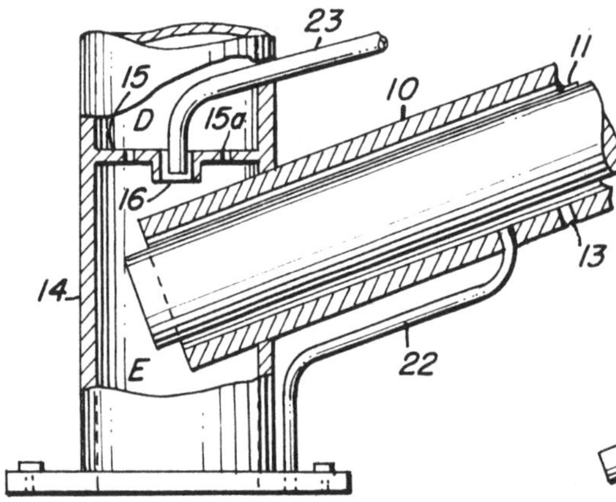

FIG. 5.

FIG. 4.

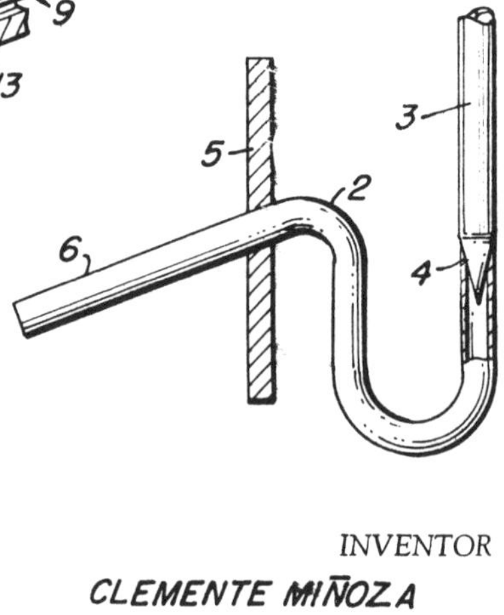

INVENTOR
CLEMENTE MIÑOZA

BY Silverman & Cass
ATTORNEYS

United States Patent Office

3,273,983
Patented Sept. 20, 1966

1

3,273,983
FUEL VAPORIZER CARBURETOR

Clemente Miñoza, Diliman, Quezon City, Philippines, assignor of one-fifth to Felipe LL. Santillan, Quezon City, Philippines

Filed Nov. 20, 1961, Ser. No. 154,107
Claims priority, application Republic of the Philippines, Jan. 27, 1961, 3,449
12 Claims. (Cl. 48—180)

This invention relates in general to carburetion and fuel induction in internal combustion engines and more specifically to an improved and novel carburetor for vaporizing liquid fuels like kerosene, diesel oil, gasoline and alcohol; mixing the resulting fuel vapors with the proportioned amount of air; and finally feeding the metered combustible fuel-air mixture into the combustion chamber.

It is a common knowledge that the formation of a fuel-air mixture consists of vaporizing the fuel and mixing the finely divided particles of fuel with air. In conventional jet carburetors, vaporization of liquid fuel is accomplished by applying the liquid fuel through nozzles or jets into a stream of moving air. The induction stroke of the engine reduces the pressure in the cylinder and reduces the pressure gradient from atmospheric pressure to the lower pressure that exists in the carburetor, manifold and engine cylinder. This pressure drop causes the air to flow through the induction system and the fuel to be sprayed from the fuel nozzle or jet. The pressure drop at the exit of the fuel nozzle or jet is accentuated by a restricted or venturi section which increases the air velocity and pressured drop at this point. The fuel leaves the nozzle or jet more or less in a stream of droplets of various sizes. During this process, vaporization from the surfaces of the drop and particles takes place, causing the disappearance of the finest particles and the reduction in sizes of the others. The ideal situation would be to have all the particles vaporized and uniformly distributed before the mixture enters the engine cylinders. Since very little vaporization of the fuel occurs in the carburetor, further vaporization is usually obtained in the induction system of the engine. In this process even the most volatile fuel like gasoline is not completely vaporized, consequently, there is always liquid fuel in the form of fine mist present in the mixture as it enters the engine cylinder. This liquid fuel in the form of fine mist is the principal source of waste of fuel in the jet carburetor, since it cannot impart power to the engine because it only burns but does not explode.

The primary object of the present invention is to provide for a fuel vaporizer-carburetor which will completely vaporize the liquid fuel before being mixed with the air and fed to the engine cylinder.

Another object of this invention is to provide for a novel fuel vaporizer-carburetor which preheats and completely vaporizes the liquid fuel into highly explosive vapors thereby effecting a complete combustion in the engine cylinder and a high thermal efficiency of the engine.

A further object of this invention is to provide for an efficient and effective fuel vaporizer-carburetor which is simpler in construction and operation and cheaper in cost of production than the conventional jet carburetors and fuel oil atomizers.

Still another object of this invention is to provide for a

2

unique fuel vaporizer-carburetor which could efficiently replace the conventional jet carburetors for volatile fuels and the conventional diesel fuel atomizers and injectors.

Still another object of this invention is to provide for a unique fuel vaporizer-carburetor which would reduce the noise of the engine and thereby eliminate the use of the usual muffler or silencer.

Further objects, novel features of construction and arrangement of parts will appear from the following description when read in the light of the accompanying drawings, wherein are disclosed a preferred embodiment of my invention.

According to the present invention a fuel vaporizer-carburetor for use with an internal combustion engine, consists of an evaporator tube assembly heated by the engine's exhaust gases and adapted to continuously vaporize the fuel supplied by a siphon from an auxiliary fuel tank, and to feed the resulting vapor into a mixer tube assembly which is connected to the inlet manifold of the engine and adapted to admit air and mix it in correct proportions with the vaporized fuel issuing from the evaporator tube assembly and allow a metered amount of the resulting fuel vapor and air mixture into the engine cylinder through the inlet manifold of the engine.

In carrying out the purpose of the present invention, briefly, the principle of operation is as follows: The liquid fuel in the auxiliary fuel tank flows through a siphon tube and drops on the upper end of a declining metal rod core which extends throughout the length of the evaporator tube and is wrapped laterally around with metal sheet sleeve provided with holes at the upper portion. A stream of air flows at high speed through the holes at the lower portion of this evaporator tube, out of the said tube and then into the mixer tube at the suction stroke of the engine. This stream of air has a suction effect inside the evaporator tube. The combined action of gravity and the said stream of air forces the liquid fuel to flow through the holes at the upper portion of the metal sheet sleeve and down through the clearance between the inside surface of the evaporator tube and the outside surface of the metal sheet sleeve. The evaporator tube is being heated by the exhaust gases of the engine, which flow through the heat exchanger, so that when the liquid fuel comes in contact with the hot surface of the evaporator tube the fuel vaporizes, and flows in a vapor state into the mixer tube, where the vapor mixes with the air that flows through the holes in the baffle plate of the mixer tube. The fuel vapor and air mixture flows down the engine inlet manifold in metered amounts through the engine throttle valve.

In the accompanying drawings:

FIGURE 1 is the elevation of the whole fuel vaporizer-carburetor assembly showing the relative locations of the different parts and indicating its position relative to the induction flow C–G and the exhaust flow A–B of the engine.

FIGURE 2 is the sectional view of the fuel vaporizer-carburetor of FIGURE 1, showing the inside details of the different parts.

FIGURE 3 is a broken sectional view of the evaporator tube showing the relative position of the metal sheet sleeve and the metal rod core and showing also the position of the siphon tube with the needle valve, at the upper end of the evaporator tube.

FIGURE 4 is the elevation of the siphon tube with the needle valve, showing the attachment of the siphon tube to its flange, and

FIGURE 5 is the sectional view of the upright mixer tube showing in detail the baffle plate with holes inside the mixer tube and how the lower end of the evaporator tube of FIGURE 3 is held in place.

In the following description reference numbers are used to indicate corresponding or similar parts in all of the views of the accompanying drawings and throughout the discussion, corresponding numerals refer to the same or similar parts.

In the preferred form of the invention as shown in the drawings, an auxiliary fuel tank **1**, FIGS. 1 and 2, is connected to the fuel pump (not shown) of the engine. The level of the fuel in the tank **1** is kept constantly above the upper bend of the siphon tube **2** by means of a float valve (not shown). The fuel tank flange **7**, FIGS. 2 and 3, is bolted (bolts not shown) to the evaporator tube flange **8**, with gasket (not shown) in between the two flanges making the joint thereat airtight.

The siphon tube **2**, FIGS. 2, 3 and 4, is a metal tube with small inside diameter, opened at both ends and bent in the form as shown in the drawings. This siphon tube **2** is fixed to the flange **5**, FIGS. 2, 3 and 4. The evaporator tube flange **8** and the siphon tube flange **5** are bolted (bolts not shown) together so that the lower portion of the siphon tube **2**, with the mouth **4** FIGS. 2, 3 and 4 is kept upright and the discharge end **6** of the siphon tube is kept in place in the axial hole of the metal rod core **9*a*** inside the evaporator tube **10**. In between the flange **8** and flange **5**, a gasket (not shown) is placed to make the joint of the two flanges airtight.

The needle valve **3**, FIGS. 2, 3 and 4, opens the siphon mouth **4**, FIGS. 2, 3 and 4 which is always submerged under the liquid fuel, to allow the flow of the fuel through the siphon tube **2** when the engine is running and closes the same to stop the fuel flow when the engine is not operating. This needle valve **3** is always kept closed by means of a spring (not shown) when the engine is not running. The end of needle valve **3**, outside the tank **1**, as shown in FIGS. 1 and 2, is connected to one end of a lever (not shown) and the other end of the lever is connected by means of a chain (not shown) with a turnbuckle (not shown) to the plunger of a solenoid (not shown or an electromagnet (not shown). The fulcrum of the lever is between the two ends of the lever so that when the plunger (not shown) of the solenoid is pulled downward the needle valve **3** is lifted upward hence the siphon mouth **4** is opened. When the said plunger is released the needle valve **3** is pushed downward by the action of the spring (not shown) so that the siphon mouth **4** is closed. The purpose of the turnbuckle is to increase or decrease the tension on the chain (not shown) so as to vary the opening of the siphon mouth **4**, FIGS. 2, 3 and 4. Increasing the tension on the said chain increases the opening of the siphon mouth **4**, while decreasing the tension on the chain decreases the opening of the siphon mouth **4**. Thus the amount of fuel flowing through the siphon tube **2** is controlled by the tension of the chain, because the said tension governs the opening of the siphon mouth **4**. The opening of the siphon mouth **4** affects the fuel economy and the idling speed of the engine. If the opening of the siphon mouth **4** is too big, the idling speed of the engine will be high and not constant and the fuel economy of the engine will be low. The tension on the said chain should be adjusted by turning the turnbuckle clockwise or counter clockwise, until the idling speed of the engine is low and steady. When the proper idling speed of the engine is obtained the turnbuckle should not be turned any more so that the opening of the siphon mouth **4** will remain constant throughout the running operation of the engine. The coils of the electromagnet or solenoid are electrically connected to the battery of the engine with a switch in the connection. When the said switch is turned on, the solenoid is energized, its plunger pulls the said chain and lever thus the needle valve **3** is opened. When the switch is turned off the

needle valve **3** will close by the action of the spring (not shown). This method of controlling the needle valve is a combination of electrical and mechanical means.

The needle valve can also be controlled by mechanical means alone. In the above mentioned method, the plunger of the electromagnet or solenoid may be replaced by a rod which could be pushed or pulled by the operator of the engine; pulling the rod opens the needle valve and pushing same closes it. Another method of controlling the needle valve is to use a device which operates like a pneumatic windshield wiper instead of the electro-magnet or solenoid. The needle valve, by means of a lever and chain with a turnbuckle is connected to the diaphragm of the pneumatic device which is connected to the suction manifold of the engine in such a way that when the engine is cranked or is running the needle valve is opened and when the engine is stopped the needle valve is closed by the action of the spring.

The evaporator tube **10**, FIGS. 1, 2 and 3, opened at both ends and with a flange on the upper end is a metal tube of uniform inside diameter throughout its length. The lower end of the evaporator tube **10** is inserted into a hole at the side of the upright mixer tube **14**, FIGS. 1, 2 and 5, in such a way that the evaporator tube is inclined with the horizontal as shown. The outside surface of the lower end of the evaporator tube **10** is welded to the mixer tube **14** so that the evaporator tube **10** and the mixer tube **14** are integral. At the lower portion of the evaporator tube **10** are primary air holes **13**, FIGS. 1, 2 and 5. These primary air holes may be enclosed by an air tight casing (not shown) which may be connected by a tube to the air cleaner of the engine, so that the primary air passing through these air holes **13** will be clean. Near the middle portion of the evaporator tube **10** is the heat exchanger **12**, FIGS. 1 and 2. Inside and extending throughout the length of the evaporator tube **10** is a metal rod core which is divided into two sections: the lower section **9** and upper section **9*a***, FIGS. 2 and 3. The entire length of said metal rod core is wrapped laterally around with metal sheet sleeve **11**, FIGS. 2, 3 and 5, which has holes in its upper portion. The lower section **9** which is a solid rod is wrapped tightly with the metal sheet sleeve **11**, FIGS. 2, 3 and 5, so that the fuel will not be able to pass over the outside surface of the metal rod core **9** while the upper section **9*a*** with an axial hole throughout its length fits loosely inside the upper portion of the metal sheet sleeve **11**. There is a small clearance between the inside surface of the evaporator tube **10**, and the outside surface of the metal sleeve **11** wherein the fuel will be able to pass. The discharge end **6** of the siphon tube **2**, FIGS. 2 and 3, fits in the axial hole of the upper section **9*a*** of the metal rod core. The fuel from the discharge end **6** of the siphon tube **2**, FIGS. 2 and 3, will flow through the junction of the upper section **9*a*** and lower section **9** of the metal rod core and out into the clearance space between the inside surface of the evaporator tube **10** and the outside surface of the metal sheet sleeve **11**, through the holes in the upper portion of the metal sheet sleeve **11**.

In order to heat the evaporator tube with the exhaust gases of the engine, the evaporator tube is provided with a heat exchanger **12**, FIGS. 1 and 2, which is located at the middle portion of said evaporator tube **10**. The main part of the heat exchanger is a short tube with its inside diameter bigger than the outside diameter of the evaporator tube. The evaporator tube passes through the inside of the short tube and the ends of the short tube are welded to the outside surface of the evaporator tube so that the joints thereat are airtight. The sides of this short tube are provided with two pipes opposite each other, marked A and B in FIGS. 1 and 2. Pipe A (partially shown) is connected from the exhaust manifold of the engine, while pipe B (also partly shown) leads to the atmosphere. The hot exhaust gases of the engine enter through A, FIGS. 1 and 2, surrounding that part of the evaporator tube **10** enclosed inside the short tube of the heat exchanger **12**, imparting some heat thereto, and come out of B, hence the evaporator tube is heated. With this heat exchanger **12**, FIGS. 1 and 2, the engine will run with very little noise or sound even without the usual muffler or silencer which can now be eliminated.

The mixer tube **14**, FIGS. 1, 2 and 5 is a vertical tube of uniform inside diameter throughout its length. The lower end of the mixer tube **14** provided with flange is bolted to the upper flange of the adapter **17**, FIGS. 1 and 2, while the lower flange of said adapter **17** is attached to the engine inlet manifold (not shown) so that the mixer tube **14** is kept upright. The upper end of said mixer tube **14** is connected to the air cleaner (not shown) of the engine. The lower portion of the evaporator tube **10**, FIGS. 1, 2 and 5 is inside the mixer tube **14**, FIGS. 1, 2 and 5 as described in the previous paragraph. The baffle plate **15**, FIGS. 2 and 5, which is a circular metal sheet, is welded to the inside surface in the upper portion of the mixer tube **14** as shown in FIGS. 2 and 5, above the lower end of the evaporator tube **10**, FIGS. 2 and 5. The plane of the said baffle plate **15** is perpendicular to the vertical axis of the mixer tube **14**, as shown. This baffle plate **15**, divides the mixer tube **14** into two compartments, namely, the upper compartment D and the lower compartment E, FIGS. 2 and 5. The upper compartment D contains fresh air while the lower compartment E contains the mixture of fuel vapor and air. This baffle plate **15**, FIGS. 2 and 5 is provided with venturi throat **16** and supplementary air holes **15*a***, FIGS. 2 and 5. The lower end of the high speed pipe **23**, FIGS. 1, 2 and 5, is inside the venturi throat **16**. One of these supplementary air holes **15*a***, is provided with a needle valve (not shown) so that the opening of this hole could be varied so as to vary the amount of fresh air flowing into the mixer tube **14**. Another of the supplementary air holes **15*a*** may be provided with a flap valve (not shown) so that at high engine speed this flap valve will open to admit more fresh air into the mixer tube **14** in order to keep the fuel-air ratio more or less constant.

The adapter **17**, FIGS. 1 and 2 is a tube of uniform inside diameter. The upper and lower ends of the said adapter are each provided with a flange. The upper flange is connected to the lower end of the mixer tube **14**, FIGS. 1 and 2 while the lower flange is attached to the inlet manifold (not shown) of the engine. Inside the adapter **17**, is a throttle valve **18**, which controls the flow of the fuel-air mixture into the cylinder of the engine (not shown). The throttle valve **18** as shown in FIG. 2, which is a circular plate is set inside the adapter tube **17** such that its plane is inclined with the vertical axis of said adapter tube, when the engine is idling. The throttle valve **18** is attached to a spindle (not shown) which passes through the center of the said valve as a diametral axis thereof, so that the valve is opened or closed by turning the said spindle clockwise or counter clockwise. By means of a spring (not shown) connected to the spindle, the throttle valve **18** is always set in the position as shown in FIG. 2, when the engine is idling or at standstill. The spindle (not shown) which is perpendicular to the vertical axis of the mixer tube **14**, is provided with a set screw so that the position of the throttle valve **18**, for engine idling speed could be varied. To increase the valve opening, the valve **18**, FIG. 2, is turned so that its lower edge will go down while its upper edge will go up. The valve **18**, FIG. 2, is fully opened when it is in a vertical position. The lower end of the idling pipe **22**, FIGS. 1, 2 and 5, extends into the inside surface of the adapter **17**, at the airtight joint **25** thereat as shown in FIG. 2. This adapter **17** may be also called the base of the carburetor.

The high speed pipe **23**, FIGS. 1, 2 and 5, is a tube with small inside diameter. Its upper end connects with an airtight joint with the upper inner surface of the evaporator tube **10**, near the middle portion thereof and inside the heat exchanger **12**, FIGS. 1 and 2. The lower end of

this high speed pipe **23** is in the venturi throat **16** as shown in FIGS. 2 and 5. This venturi throat **16** passes through the baffle plate **15** as shown, so that the fuel vapor from the evaporator tube **10**, FIGS. 1, 2 and 5, could pass through said high speed pipe **23**, into the mixer tube **14** as shown, and then into the cylinder of the engine (not shown).

When the throttle valve **18**, FIG. 2, is widely opened, the velocity of the air passing through the venturi throat **16**, FIGS. 2 and 5 will be very high, thus, a low pressure is created at the lower end of the high speed pipe **23** at said venturi throat **16** so that the fuel vapor in the evaporator tube **10** will flow into the cylinder of the engine through the said high speed pipe **23**. With this high speed pipe **23** which may be also called accelerating pipe, the engine will accelerate easily.

The idling pipe **22**, shown in FIGS. 1, 2 and 5, is similarly a tube with small inside diameter. Its upper end is connected with an airtight joint with the lower inner surface of the evaporator tube **10**, below the primary air holes **13** as shown in FIGS. 2 and 5. The lower end of this idling pipe **22** extends into the inside surface of the adapter **17**, FIGS. 1 and 2, at the airtight joint **25** thereat as shown, slightly above the lower edge of throttle valve **18** at its position shown in FIG. 2, set for idling speed of the engine. At this idling position of the throttle valve **18**, there is small clearance between its lower edge and the inside surface of said adapter **17**, FIGS. 1 and 2, thus the speed of the mixture of fuel vapor and air passing through this small clearance is very high, so that the pressure at the lower end **25** of said idling pipe **22**, FIGS. 1 and 2, will be very low, producing a suction effect thereat. The said idling pipe **22**, helps to suck fuel from the evaporator tube **10**, FIGS. 1, 2, 3 and 5, at the idling speed of the engine. The air passing through the air holes **13**, FIGS. 1, 2 and 5, will also suck fuel from the siphon tube **2** into the evaporator tube **10**, FIGS. 1, 2, 3 and 5.

Without this idling pipe **22**, FIGS. 1, 2 and 5, at high engine speed, when the throttle valve **18**, FIG. 2, is instantly closed up to idling position as shown, the engine will stop. The explanation is as follows. At high engine speed, the throttle valve **18**, FIG. 2, is wide open. Air entering through the air holes **13**, FIGS. 1, 2 and 5, together with the fuel vapor, will flow rapidly through the lower end of the evaporator tube **10**, FIGS. 1, 2, 3 and 5, passing the throttle valve **18**, FIG. 2, and into the cylinder of the engine (not shown). When the throttle valve **18** is instantly closed, the flow of air through the lower end of the evaporator tube **10** will be stopped, causing a rise of pressure in the evaporator tube **10** and hence the flow of fuel through the discharge end **6** of the siphon tube **2**, FIGS. 2, 3 and 4, will be stopped so that the engine will also stop.

With this idling pipe **22**, FIGS. 1 2 and 5, the engine will not stop when the throttle valve **18**, FIG. 2 is instantly closed up to idling position, because when said throttle valve **18** is at idling position there is at once a suction effect in the idling pipe **22**, so that the flow of fuel into the evaporator tube **10**, FIGS. 2, 3 and 5 will not stop, hence the engine will continue to run. This most important function of said idling pipe **22**, to prevent the engine from stopping when the throttle valve **18** is closed up to idling position is necessary because at this state, the high speed pipe **23**, FIGS. 1, 2 and 5, has no more suction effect in the evaporator tube **10**, because the speed of the air through the venturi throat **16**, FIGS. 2 and 5, is very low.

This idling pipe **22**, FIGS. 1, 2 and 5, may be provided with a needle valve (not shown) at its lower end **25**, FIG. 2, thereat, in order to vary its opening so that the idling speed of the engine could be controlled.

The parts of the present invention are assembled and joined together as shown in FIGS. 1 and 2. It is necessary that all the joints, internal and external, are per-fectly airtight. In a mobile engine, the present fuel vaporizer carburetor is preferably installed in such a way that the vertical plane containing the axes of the upright mixer tube **14**, the inclined evaporator tube **10**, and the siphon tube **2**, as shown in FIG. 2, should be at right angles to the direction of motion of the vehicle. The auxiliary fuel tank **1**, is connected to the discharge of the engine fuel pump (not shown). The inlet opening A of the heat exchanger **12** is connected by a pipe (not shown) to the outlet of the engine exhaust manifold (not shown) while opening B is connected by a pipe (not shown) that leads to the atmosphere. The upper opening C of the mixer tube **14** connects from the air cleaner (not shown) of the engine, and the lower flange G of adapter **17** is attached to the engine inlet manifold (not shown).

The detailed operation of the present invention is described as follows, explained in the light of FIG. 2. If *gasoline* is used, the starting tank **19**, starting valve **20**, pipe **21** and the fuel tank heater tube **24** (FIGS. 1 and 2) are not used. The fuel tank **1** is filled with gasoline up to a level above the upper bend of the siphon tube **2**. The ignition switch (not shown) of the engine and the switch (not shown) of the needle valve **3**, are turned on and with the needle valve **3** open, the engine is cranked. The gasoline will flow into the siphon tube **2** and out through the discharge end **6** since the fuel in the fuel tank **1** is higher than the discharge end **6**. At the suction stroke of the piston (not shown) of the engine, air enters the primary air holes **13** and flows at high speed through the small clearance between the inside surface of the evaporator tube **10** and the outside surface of the metal sheet sleeve **11**, into compartment E of the mixer tube **14**. This air may be called primary air. This high speed primary air creates a partial vacuum in the evaporator tube **10**, so that the gasoline in the evaporator tube **10** will partially vaporize. The fuel vapor passes through the high speed pipe **23**, while the liquid fuel flows down through the clearance between the outside surface of the metal sheet sleeve **11** and the inside surface of the evaporator tube **10**. The primary air licks the liquid fuel as it flows down the evaporator tube **10**. This action of the primary air will vaporize the liquid fuel so that the primary air is charged with fuel vapor. This charged primary air flows from the evaporator tube **10**, in a direction inclined with the horizontal and veers to a direction perpendicular to the horizontal. This change of direction makes the charged primary air turbulent. Fresh air from the air cleaner (not shown) of the engine, flows through the small supplementary air holes **15***a*, of the baffle plate **15** into the big compartment E, so that said fresh air will become turbulent. This turbulent fresh air, the charged primary air, and the fuel vapor from the high speed pipe **23** will mix homogeneously in compartment E to form an explosive mixture. This explosive mixture will flow into compartment F and out of the adapter **17** through opening G into the cylinder (not shown) of the engine. Then the mixture will explode in the cylinder of the engine and the engine will run. The hot exhaust gases of the engine entering the heat exchanger **12** through opening A, impart some of their heat to the evaporator tube **10**. The evaporator tube **10** becomes hot, so that the fuel vaporizes when it comes in contact therewith. Some of the fuel vapor passes through the high speed pipe **23** and the rest flows down the evaporator tube.

Vaporization of the liquid fuel will be more or less complete, because a small amount of fuel at every suction stroke of the engine is under high temperature but at low pressure inside the evaporator tube. When the fuel is completely vaporized, the combustion inside the cylinder of the engine is nearly complete so that the exhaust gases of the engine will be more or less free from carbon monoxide and hydrocarbons.

When the hot exhaust gases of the engine are in the heat exchanger **12**, they collide, come in contact with, and

impart heat to the surface of that portion of the evaporator tube **10** inside the heat exchanger **12**. The exhaust gases coming out of the heat exchanger **12** will then be cooled and have low speed, hence, will not be noisy. Therefore with this heat exchanger **12**, the usual muffler or silencer of the engine may be eliminated.

When the engine is idling, the throttle valve **18** is at the position shown in FIG. 2. The speed of the fuel-air mixture passing through the clearance between the lower edge of the throttle valve **18** and the inside surface of the adapter **17** will be very high. There is suction effect in the idling pipe **22**, so that some of the fuel vapor flows through the said pipe, and the rest of the fuel vapor flows through the lower end of the evaporator tube **10**. The fuel vapor will not flow through the high speed pipe **23**, since there is very little suction effect in the said pipe because the speed of the air passing through the venturi throat **16** is very low. When the throttle valve **18** is opened a little wider, the amount of fresh air passing through the supplementary air holes **15***a* of the baffle plate **15** will increase. The speed of the air passing through the venturi throat **16** and through the primary air holes **13** will be very high so there will be a big suction effect inside the evaporator tube **10**. The flow of fuel from the siphon tube **2** will increase and the amount of fuel-air mixture fed into the cylinder of the engine will increase; consequently, the engine speed will increase. But when the throttle valve **18** is opened too wide, the clearance between the lower edge of the throttle valve **18** and the inside surface of the adapter **17** will be very big, hence the speed of the fuel-air mixture passing through this said clearance will be very low so that there will be no more suction effect in the idling pipe **22**. Thus the throttle valve **18** controls the speed of the engine and allows a richer fuel-air mixture during idling.

When heavy fuel like *diesel oil* is used, then the starting tank **19**, starting valve **20**, pipe **21** and the preheating tube **24** as shown in FIGS. 1 and 2, are used. The preheating tube **24** is connected by a tube (not shown) to the exhaust manifold (not shown) of the engine. This said tube **24** is like the tube of a fire tube boiler. The hot exhaust gases of the engine flow inside the heating tube **24** while the fuel surrounds the outside surface so that the fuel in the auxiliary tank **1**, FIGS. 1 and 2 is thereby preheated. The heat exchanger **12** and the tube (not shown) leading from the exhaust manifold (not shown) of the engine to the heat exchanger are all heat insulated when heavy fuel is used.

The starting tank **19** is filled with gasoline while the auxiliary fuel tank **1** is filled with diesel oil. The starting valve **20** is opened and gasoline will flow through the pipe **21** and into the evaporator tube **10**. The needle valve **3** is closed so no diesel oil will flow into the evaporator tube **10**. The engine is cranked and it will start and run for a while with gasoline as fuel, as discussed before. When the evaporator tube **10** and the auxiliary fuel tank **1** are hot, the needle valve **3** is opened and the starting valve **20** is closed. The engine will now continue to run with diesel oil as fuel.

It is believed that the advantages and features of the invention have been fully explained in connection with the above specification of the drawings and that those skilled in this art will understand the manner of construction of the invention and operation and use thereof. It is desired to point out that a considerable variation in many of the details may be made without in any way departing from the spirit or scope of the invention as defined in the appended claims.

What I claim is:

1. In a fuel carburetion system for internal combustion engines in which liquid fuel from a fuel feeder assembly is vaporized and mixed with air in a mixer tube assembly in predetermined ratios of air to fuel and said mixtures thereafter are introduced into the inlet manifold of the engine; in combination: an evaporator tube assembly disposed inclined and in communicating relation between the fuel feeder assembly and the mixer tube assembly, said evaporator tube assembly having an inlet and an outlet, said inlet in communication with said fuel feeder assembly to receive liquid fuel therefrom and said outlet being interior of said mixer tube assembly to discharge vaporized fuel therefrom, a heat exchanger assembly surrounding said evaporator tube between the inlet and outlet ends theerof, said heat exchanger assembly constructed and arranged for receiving exhaust gases from said engine for heating said evaporator tube, a first conduit means communicating between the interior of the heated evaporator tube and said mixer tube assembly, a second conduit means communicating between the interior of said evaporator tube assembly at a location adjacent said heat exchanger assembly and said mixer tube assembly, means in the mixer assembly for selectively passing vaporized fuel through one or the other of said first and second conduit means into said mixer assembly, and air inlet means in communication with the inerior of said tube at a location between the second conduit means and the heat exchanger assembly whereby to permit introduction of air into the evaporator tube assembly during the operation of the engine.

2. The system as claimed in claim **1** in which said evaporator tube assembly comprises an elongate tubular member, a solid core member having an axial bore extending along a fraction of the length of said core and opening to one end thereof, said core member being disposed within said elongate tubular member, said core member having an axial passageway opening to the said inlet and a sleeve having a perforate portion adjacent one end thereof, said sleeve being arranged with the perforate portion loosely wrapped around that length of the core member coextensive with said axial passageway and the remainder of the sleeve tightly wrapped around the remaining portion of said core member.

3. The system as claimed in claim **1** in which said mixer tube assembly comprises a hollow cylindrical vertically disposed member having a pair of co-axial compartments, said compartments being defined by a baffle plate and a venturi provided in said baffle plate, said first conduit means terminating in the venturi and said evaporator tube terminating interior of the lower one of said compartments.

4. The system as claimed in claim **3** in which said second conduit means terminates interior of said lower compartment and below the outlet of the evaporator tube.

5. The system as claimed in claim **1** in which said fuel feeder assembly comprises an auxiliary fuel tank and a siphon tube having an entry and disposed interior of said tank and a discharge end disposed interior of said evaporator tube inlet, and needle valve means for said entry end.

6. In a fuel carburetor system for internal combustion engines and said system including fuel vaporizing means; a mixer tube assembly for mixing and feeding to the cylinders of said engines a selectively proportioned fuel-air mixture, said mixer tube assembly comprising a vertically arranged tubular member having an air inlet at its upper end and outlet means for connecting said tube to the inlet manifold of the engine at the other end, baffle means disposed within said tube and spaced from the upper end thereof to define upper and lower chambers, said baffle means including a venturi throat and at least one air passageway, first and second conduit means communicating between said venturi throat and fuel vaporizing means and between said lower chamber and said fuel vaporizing means respectively for introduction of said vaporized fuel into the respective chambers, and throttle means in said lower chamber for varying the internal pressure within said chamber to vary the passage of vaporized fuel through one or the other of said first and second conduit means.

7. In a fuel vaporizer carburetor of the type described, the combination comprising a fuel feeder assembly, an inclined evaporator tube assembly having upper and lower ends, the upper end being sealingly engaged with the fuel feeder assembly, and a vertically arranged mixer tube assembly, said lower end of said evaporator tube assembly terminating within said mixer tube assembly, heat exchanger means surrounding said evaporated tube assembly between the ends thereof and communicating with said engine to receive the exhaust gases therefrom, means to continuously feed fuel to and through said evaporator tube assembly to the engine in properly proportioned mixtures of air and fuel, said means comprising, a fuel feeder assembly including a liquid fuel reservoir and a siphon means communicating between said reservoir and said evaporator tube assembly to discharge a quantity of liquid fuel into the upper end of said evaporator tube assembly, vaporizing means within said evaporator tube assembly to vaporize said liquid fuel, and means for discharging said vaporized fuel into said mixer assembly, said last mentioned means comprising air inlet means provided in said evaporator tube assembly exterior of said mixer tube but closely adjacent the lower end of said evaporator tube assembly, and said mixer tube assembly being connected to the engine inlet manifold and comprising means for introducing additional air into said mixing tube assembly, plate means including a venturi throat for accelerating said additional air, a mixing chamber for proportionally mixing vaporized fuel with said additional air and throttle means for varying the pressure gradient in the mixing chamber whereby to meter a predetermined quantity of fuel-air mixture into the inlet manifold of the engine and further to operate said first mentioned means.

8. The combination as claimed in claim **7** in which said evaporator tube assembly includes an outer tubular shell member, an inner core member concentrically disposed therein, said shell and core members defining an elongate passageway extending the length of said members, said air inlet means formed in said outer shell member between the heat exchange means and the mixer tube.

9. The combination as claimed in claim **7** in which said evaporator tube assembly comprises an elongate hollow cylindrical member open at both ends, a concentric passageway interior of the tube and extending the length thereof, said first conduit means comprising an exterior accelerator pipe, one end thereof passing through said heat exchanger means to communicate with said passageway, the other end thereof communicating to the interior of the mixer assembly at the venturi throat.

10. The combination as claimed in claim **7** in which said evaporator tube assembly comprises an elongate hollow cylindrical member open at both ends, a concentric passageway interior of the tube and extending the length thereof, first conduit means comprising, an exterior accelerator pipe, one end thereof passing through the heat exchanger means and communicating with said passageway, the other end thereof communicating to the interior of the mixer assembly at the venturi throat for passing vaporized fuel from the heated portion of the evaporator tube into the mixer tube at the venturi throat, and a second conduit means comprising an idling pipe having one end thereof communicating with the lower portion of said cylindrical member and the opposite end thereof communicating with the said mixing chamber of said mixer tube assembly to provide an extra rich fuel-air mixture when the engine is idling.

11. The combination as claimed in claim **7** in which said siphon means is provided with a valve for opening and closing same.

12. The combination as claimed in claim **8** in which said inner core is provided with an axial recess opening to the siphon means, and a sleeve disposed between said core and said tubular shell member, said sleeve being tightly engaged with said core along the length thereof

3,273,983

11

but loosely engaged with that portion of the core carrying said recess.

References Cited by the Examiner

UNITED STATES PATENTS

1,950,806	3/1934	Mathes	123—133 X
2,140,071	12/1938	Cleem	123—133

12

FOREIGN PATENTS

878,382	9/1961	Great Britain.
349,054	6/1937	Italy.

MORRIS O. WOLK, *Primary Examiner.*

GEORGE D. MITCHELL, *Examiner.*

J. S. COVRONEK, *Assistant Examiner.*

Dec. 27, 1966 H. SCHWARTZ **3,294,381**

CARBURETOR

Filed March 18, 1963

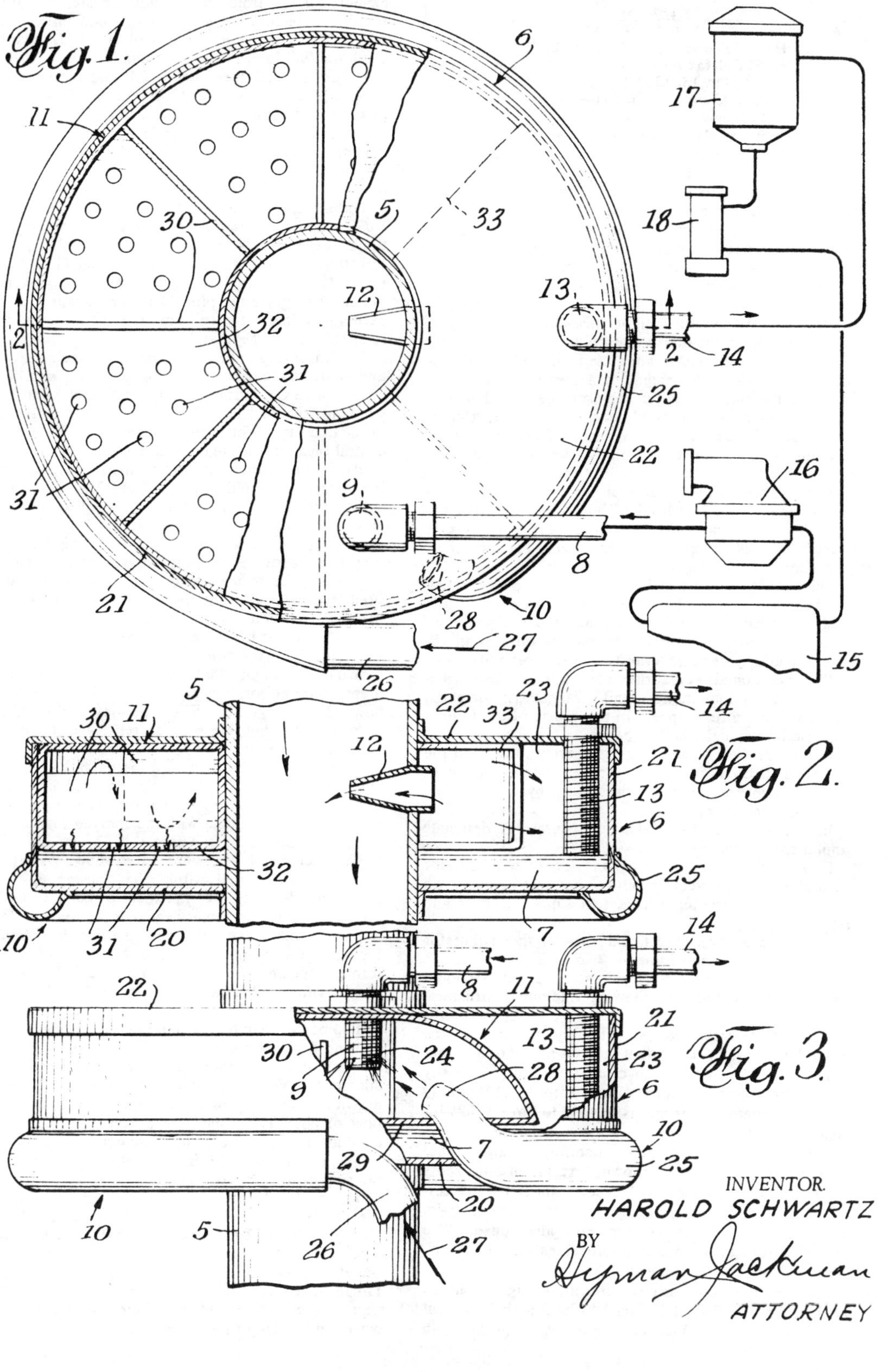

INVENTOR.
HAROLD SCHWARTZ
BY
Hyman Jackman
ATTORNEY

United States Patent Office 3,294,381
Patented Dec. 27, 1966

1

3,294,381

CARBURETOR

Harold Schwartz, Baldwin Park, Calif.
(12260 E. Colefax Ave., Aurora, Colo. 80010)

Filed Mar. 18, 1963, Ser. No. 265,885
2 Claims. (Cl. 261—145)

This invention relates to a carburetor construction.

An object of the present invention is to provide a carburetor in which the fuel, before the same is combined with air and fed to an internal combustion engine, is treated by the hot exhaust fumes of said engine to increase its burning efficiency in the cylinders of said engine.

Another object of the invention is to provide a carburetor, as above characterized, that circulates the fumes-laden fuel in a manner to free the same of inordinately large globules of fuel, thereby insuring that only finely divided and pre-heated fuel of mist-like consistency is directed to the engine intake manifold.

The present carburetor, when used for feeding the cylinders of a six-cylinder engine of a popular automobile, improved the miles per gallon performance of said engine, during ordinary driving conditions, over 200%, a common grade of gasoline being used, the efficiency achieved resulting from the mentioned pre-heating of the fuel and because the same is kept under low pressure imposed by suction applied to the carburetor for the particular purpose of maintaining the level of fuel constant during engine operation. Such low pressure in the carburetor causes increased vaporization of the fuel in the carburetor and increased efficiency of operation.

This invention also has for its objects to provide such means that are positive in operation, convenient in use, easily installed in a working position and easily disconnected therefrom, economical of manufacture, relatively simple, and of general superiority and serviceability.

The invention also comprises novel details of construction and novel combinations and arrangements of parts, which will more fully appear in the course of the following description, and which is based on the accompanying drawing. However, said drawings merely shows and the following description merely describes, one embodiment of the present invention, which is given by way of illustration or example only.

In the drawing, like reference characters designate similar parts in the several views.

FIG. 1 is a partly broken plan view of a carburetor according to the present invention, the same being shown in operative association with a fuel supply, feeding and return system.

FIG. 2 is a vertical sectional view of the carburetor as taken on the plane of line 2—2 of FIG. 1.

FIG. 3 is a partial side elevation and partial sectional view of said carburetor showing additional structural details.

The present carburetor is preferably mounted on the usual downdraft air tube 5 that receives a flow of air from atmosphere by way of an air cleaner, said tube 5, in any usual way, being provided with a throttle or butterfly valve that controls flow and is formed to have a flow-increasing Venturi passage. The above common features of the fuel feed to the intake manifold of an internal combustion engine are not shown; first, because the same are old and well known and, second, the same are disclosed in applicant's pending application, Serial No. 182,420, bearing the same title, now abandoned. The present carburetor embodies improvements over the disclosure of the earlier application.

The present carburetor comprises a housing 6 mounted on said air tube 5, and designed to hold a shallow pool of fuel 7, a fuel inlet 8 terminating in a spray nozzle 9, an

2

exhaust gas manifold 10 to conduct heated exhaust gases for discharge into the spray of fuel emanating from the nozzle 9 and for heating said pool of fuel 7, means 11 to scrub the fuel-fumes mixture to eliminate large droplets of fuel from said mixture (which fall into the pool 7 therebeneath), a nozzle tube 12 receptive of said scrubbed mixture and passing the same under Venturi action into air tube 5 combined with air in said tube for passage to the intake manifold of an engine, and a pickup pipe 13 connected to an outlet 14 for drawing excess fuel from the pool 7 during operation of the carburetor.

The system connected to said carburetor is shown in FIG. 1, and comprises a fuel tank 15, a generally conventional fuel pump 16 for drawing fuel from said tank and directing the same to the inlet 8, a fuel filter 17, and a pump 18 connected in series between said tank and the outlet 14 to place the pipe 13 under suction and to draw excess gasoline in the carburetor back to the tank 15 for re-circulation to the inlet 8.

The carburetor housing 6 may be circular, as shown, and quite flat compared to the diametral size, so as to have a large flat bottom 20 that, with the cylindrical wall 21, holds the fuel pool 7. A cover 22 encloses the top of the housing. The bottom 20 and cover 22 have aligned central openings through which the down draft tube 5 extends, said pipe thereby forming the interior of the housing to have an annular inner space 23.

The fuel inlet 8 is separably secured to the cover 22 with the spray nozzle 9 thereof extending through said cover. While shown with spray-emitting holes 24 arranged to provide a spray around the nozzle 7, the nozzle may be formed so the spray is directional as desired, for most efficient interengagement of the sprayed fuel with the heating gases supplied by the manifold 10.

Said manifold is shown as a pipe 25 that has an end 26 extending from the conventional heat riser chamber) not shown) of the engine that is being supplied with fuel by the present carburetor, the arrow 27 indicating exhaust gas flow into the pipe 25. Said pipe may encircle the lower portion of the housing 6 to, thereby, heat the pool 7 by transfer of heat through the wall of the housing. Said manifold pipe is shown with a discharge end 28 that extends into the housing in an inward and upward direction toward the nozzle 9 so that the exhaust gases flowing in said pipe intermingle with the sprayed fuel to heat the same as the fuel leaves the nozzle.

The fuel-scrubbing means 11 is shown as an arcuate chamber 29 disposed within the housing 6, the same being provided with a complement of baffle walls 30 that causes the flow of fumes-heated misty fuel to follow a serpentine path that intercepts the heavier globules of fuel so the same may run down the faces of said walls, through openings 31 in the bottom wall 32 of the scrubbing chamber 29, and into the pool 7. Thus, only the finer globules of the heated fuel emanate from the open end 33 of the chamber 29 into the interior space 23 of the housing 6 above the level of the pool 7. This fuel, by the suction action thereon through the nozzle tube 12 due to the Venturi action in the downdraft tube 5, is drawn into the air flow in the latter tube to carburize said flow of air. The same then proceeds to the engine for combustion therein in the usual manner.

The pickup pipe 13 is also shown as carried by the housing cover 22 and may be adjusted so its lower open end is so spaced from the housing bottom 20 as to regulate the depth of the pool 7, which is preferably below the bottom wall 32 of the scrubbing chamber 29. Since this pipe is subject to the suction of the pump 18 through the outlet 14 and filter 17, the level of the pool 7 is maintained by a return to the tank 15 of excess fuel for subsequent pre-heating and scrubbing, as above described, when fed by pump 16 to the nozzle 9.

3

It will be seen that the surface of the pool 7 is subject not only to the Venturi action in tube 5, but also to the suction of the pump 18 as the same draws fuel back to the supply tank. Thus, the surface of the pool is under pressure somewhat less than atmospheric. Hence, the vaporization of the pool surface is increased, the resulting vapors combining with the flow from the scrubbing chamber to the downdraft tube 5.

While the foregoing has illustrated and described what is now contemplated to be the best mode of carrying out the invention, the construction is, of course, subject to modification without departing from the spirit and scope of the invention. Therefore, it is not desired to restrict the invention to the paricular form of construction illustrated and described, but to cover all modifications that may fall within the scope of the appended claims.

Having thus described the invention, what is claimed and desired to be secured by Letters Patent is:

1. A carburetor for an internal combustion engine having an exhaust and an intake manifold, and a downdraft tube leading to said intake manifold, comprising:

(a) a housing having a flat, horizontal bottom and adapted to hold a shallow pool of fuel, said housing encircling said downdraft tube and mounted thereon,

(b) an arcuate scrubbing chamber in said housing open at one end and closed at its other end to the interior of the housing and having a top and bottom spaced above the housing bottom provided with holes open to the pool of fuel,

(c) a complement of radial baffles in said scrubbing chamber in arcuate arrangement and mounted vertically between the top and bottom of said scrubbing chamber,

(d) a liquid fuel inlet terminating in a spray nozzle discharging into said scrubbing chamber adjacent its closed end,

(e) means to conduct exhaust gases of said engine into said scrubbing chamber, said means terminating in a discharge for said exhaust gases directed toward said nozzle to commingle with the sprayed fuel emanating therefrom, thereby to produce a flow of

4

fumes-laden finely-divided fuel that includes larger globules and means positioning said baffles so the fumes flow in a serpentine path toward the open end of the chamber around and in contact with said baffles so the larger globules are intercepted and run down the faces of the baffles and discharged through the holes in the chamber bottom and into the mentioned pool of fuel in the housing,

(f) a pickup pipe for drawing excess fuel from the pool with its pickup end located in spaced relation to the housing bottom but closer thereto than is the bottom of the scrubbing chamber, thereby maintaining the level of the pool below said chamber bottom,

(g) the remainder of the fumes-laden fuel discharging from said open end of the scrubbing chamber into the housing, and

(h) a nozzle tube intersecting the downdraft tube open to said fumes-laden fuel in the housing and directed to discharge the same into a flow of air in said downdraft tube to provide a combustible fuel mixture.

2. A carburetor according to claim **1** in which the means for conducting the exhaust gases comprises a pipe in heat-exchange engagement with the housing to heat the fuel forming the pool therein to increase vaporization of the surface of said pool.

References Cited by the Examiner

UNITED STATES PATENTS

1,134,633	4/1915	Noll et al.	55—444
1,997,497	4/1935	Pogue	261—151
2,531,661	11/1950	Arvastsson	261—133

FOREIGN PATENTS

688,874	5/1930	France.
219,020	1926	Great Britain.

HARRY B. THORNTON, *Primary Examiner.*

HERBERT L. MARTIN, RONALD R. WEAVER, *Examiners.*

Aug. 29, 1967 R. E. WILLIAMS **3,338,223**

CARBURETORS

Filed May 26, 1966

FIG. 1

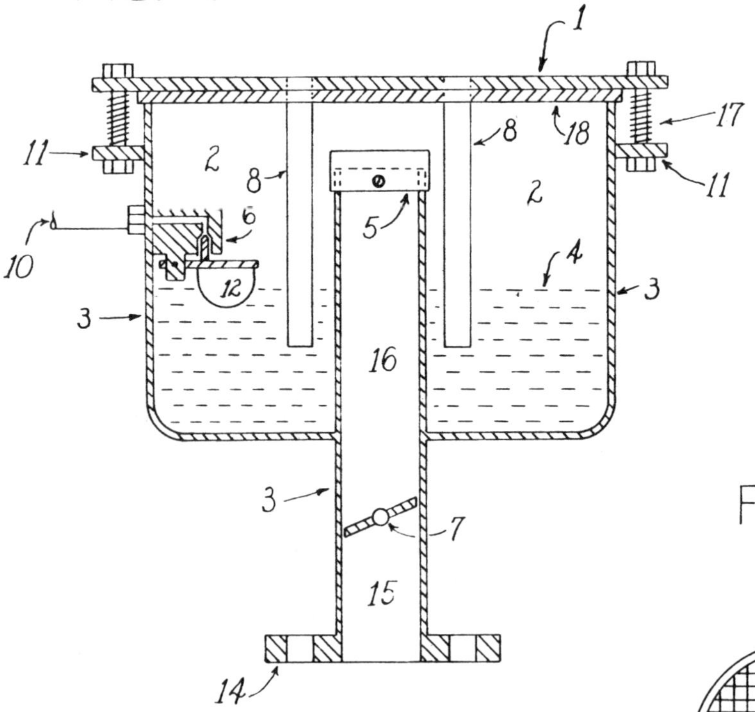

FIG. 3

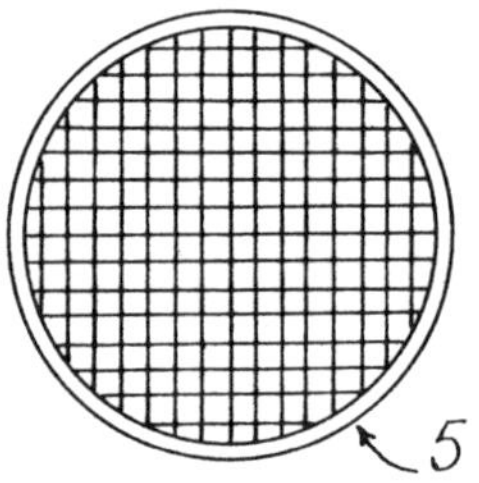

FIG. 2

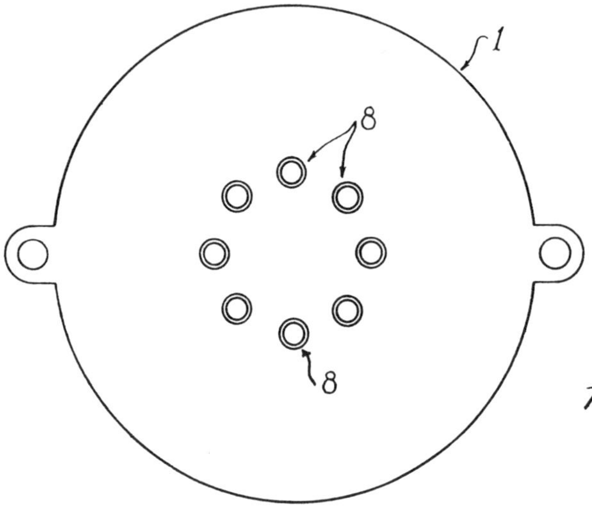

INVENTOR.

ROBERT E. WILLIAMS

United States Patent Office

3,338,223
Patented Aug. 29, 1967

1

3,338,223
CARBURETORS
Robert E. Williams, Waynesburg, Pa.
(1607 Lower Silver Lake Road, Topeka, Kans. 66608)
Filed May 26, 1966, Ser. No. 553,071
1 Claim. (Cl. 123—134)

This invention relates to carburetors for internal combustion engines and more specifically to a device whereby the air is mixed with the fuel in a highly combustible and economical manner.

Aside from the conventional jet type carburetor which sprays the fuel directly into the air, other types of carburetors have all employed means for vaporizing the fuel by various methods; such as, injecting compressed air into the fuel; heating the fuel and collecting the vapors, etc.

This invention is new and different in that it converts raw fuel and air into a highly vaporized and combustible mixture by the action of the air passing through the fuel without the assistance of heat, compressed air or other mechanical or chemical means.

It is widely known among those skilled in the art that the conventional jet type carburetor produces fair performance over a wide speed range of the engine but also due to the inefficient droplets of gasoline which are not converted to vapor, there is the disadvantage that much of the fuel is never converted into enough of a vapor to burn properly. As these larger droplets of fuel are unburned they cause unnecessary wear on the cylinder walls due to their tendency to "wash off" any oil film which should be present for proper lubrication of the cylinder.

An object of my invention is to produce a highly combustible vapor whereby there will be no appreciable waste of fuel nor unnecessary wear of the engine due to unburned fuel particles.

Another object of this invention is to provide a highly efficient means of converting gasoline into a vaporous state with air without the use of complicated linkages, venturis, or jets.

Another object of my invention is that due to the simplicity of design it can be easily and economically manufactured. The mechanical simplicity assures the carburetor of trouble free performance and longevity of operation.

Still a further object of my invention is that the air which is drawn into the fuel does not have to be filtered. Any foreign matter in the air such as dust, dirt, and most impurities remain suspended in the gasoline and are not drawn into the vapor. Any large accumulation of foreign matter is easily removed by taking off the top of the carburetor and cleaning the residue from the body.

And still another object of this invention is that should a fire or back fire occur within the intake manifold of the engine, the flame produced is not able to travel past the flame arrester and into the fuel which is held in the body of the carburetor.

Still a further object of my invention is to provide a carburetor which can easily be adapted to engines which employ any type of a carburetor; whether side draft, down draft, or up draft.

This invention and its operation can easily be understood by reference to the attached drawing of which:

FIGURE 1 is a cross sectional side view of the entire carburetor assembly.

FIGURE 2 is a top view of the cover plate and attached air tubes.

FIGURE 3 is a top view of the fine mesh wire screen or flame arrester which presents a fire from entering the carburetor vapor area.

Referring now to FIGURE 1; as the throttle valve **7** is moved out of the closed position as shown, the low pressure created by the intake cycle of an internal combustion

2

engine is admitted into area **16**, past the flame arrester **5**, and into the vapor chamber **2**. This low pressure causes atmospheric air to be drawn into the carburetor body through the air tubes **8** which are attached to the cover plate **1**. In order that the air be admitted into the vapor chamber **2**, it must first pass through the fuel **4** in the form of bubbles. These air bubbles mix with the gasoline as they rise from the bottom of the air tubes **8** to the top of the fuel level. The vapor created by this action is then drawn past the flame arrester **5**, into area **16**, thence past the throttle valve **7** and into area **15** which is bolted to the intake manifold of the engine by the base of the carburetor **14**. Upon leaving area **15** the fuel air vapor passes into the cylinders by way of the intake manifold (not shown).

The fuel level **4** is held at a constant level within the body **3** by means of a conventional float valve **6** and float **12**. The fuel is admitted into the valve **6** by a small diameter fuel line **10**. Pressure is applied to the fuel in line **10** by means of an external fuel pump (not shown) or by a gravity feed system (not shown).

FIGURE 2 shows the top of the cover plate **1** which is attached to the body **3** by bolts **17**. The bolts **17** are secured to the body **3** by protruding lugs **11** which are welded or cast onto the body. An air-tight gasket **18** is located between the cover plate **1** and the body **3** to insure an airtight seal between the atmosphere and the vapor chamber **2**.

The body **3** and the cover plate **1** are made of steel or of a metal which is easily cast. The air tubes **8** are made of brass and secured to the cover plate **1** by means of brazing or any suitable manner to assure an airtight seal.

A flame arrester **5** is shown in FIGURE 3. This safety device is secured by means of a screw to the top of the pipe extending up through the fuel. The inside of the pipe forms areas **15** and **16** plus it acts as a means of attachment for the throttle valve **7**. This pipe is an integral part of the body **3**.

Should a fire occur within the engine or the intake manifold it cannot reach the vapor chamber **2** due to the fine mesh wire screen (flame arrester **5**). Flame or fire will not pass through a screen of an extremely fine mesh; however, the fuel air vapors are readily and easily admitted without restriction.

It is possible that minor changes in size, detail and arrangements coming within the field of invention may be adopted in practice.

Having thus described my invention what I claim is:

A carburetor for supplying a fuel air mixture to be burned comprising enclosed chamber means for holding a supply of liquid fuel, said chamber means including a bottom wall having a central opening therein, an upstanding peripheral side wall, and a removable top wall having a plurality of inlet openings therein, means for supplying liquid fuel to said chamber, means for regulating the fuel supplied to said chamber means for maintaining a selected upper level of liquid fuel above said bottom wall, a fuel air conduit for delivering a fuel air mixture to be burned and projecting upwardly into said chamber means through said central opening in said bottom wall, said conduit having an open upper end spaced above said selected upper level of liquid fuel in said chamber means and below said top wall and an open lower end adapted for connection with the intake manifold of an internal combustion engine, throttle valve means in said conduit below said chamber means for controlling the flow of fuel air mixture to said engine, screen means mounted across the upper open end of said conduit, and a plurality of spaced apart, air supply tubes for directing air into the liquid fuel in said chamber means, said tubes being equally spaced around said conduit and supported by said top wall, each of said tubes including an open upper

3,338,223

3

end in communication with the atmosphere through one of said inlet openings in said top wall and an open lower end spaced beneath the selected upper level of fuel in said chamber means whereby combustion air is drawn downwardly through said tube from the atmosphere into said liquid fuel and rise upwardly therein before flowing downwardly through said conduit for combustion.

References Cited

UNITED STATES PATENTS

434,695	8/1890	Barrett et al.	
662,304	11/1900	Reenstierna	261—124 X

4

1,156,924	10/1915	Nichols	261—124 X
2,338,044	12/1943	Lanser	261—72 X
2,658,489	11/1953	Putt	261—121 X

FOREIGN PATENTS

356,405	1/1938	Italy.

HARRY B. THORNTON, *Primary Examiner.*

RONALD R. WEAVER, *Examiner.*

Feb. 24, 1970 F. E. GERRARD 3,496,919

CARBURETOR AND SYSTEM FOR SUPPLYING VAPORIZED FUEL TO COMBUSTION ENGINES

Filed Sept. 5, 1967

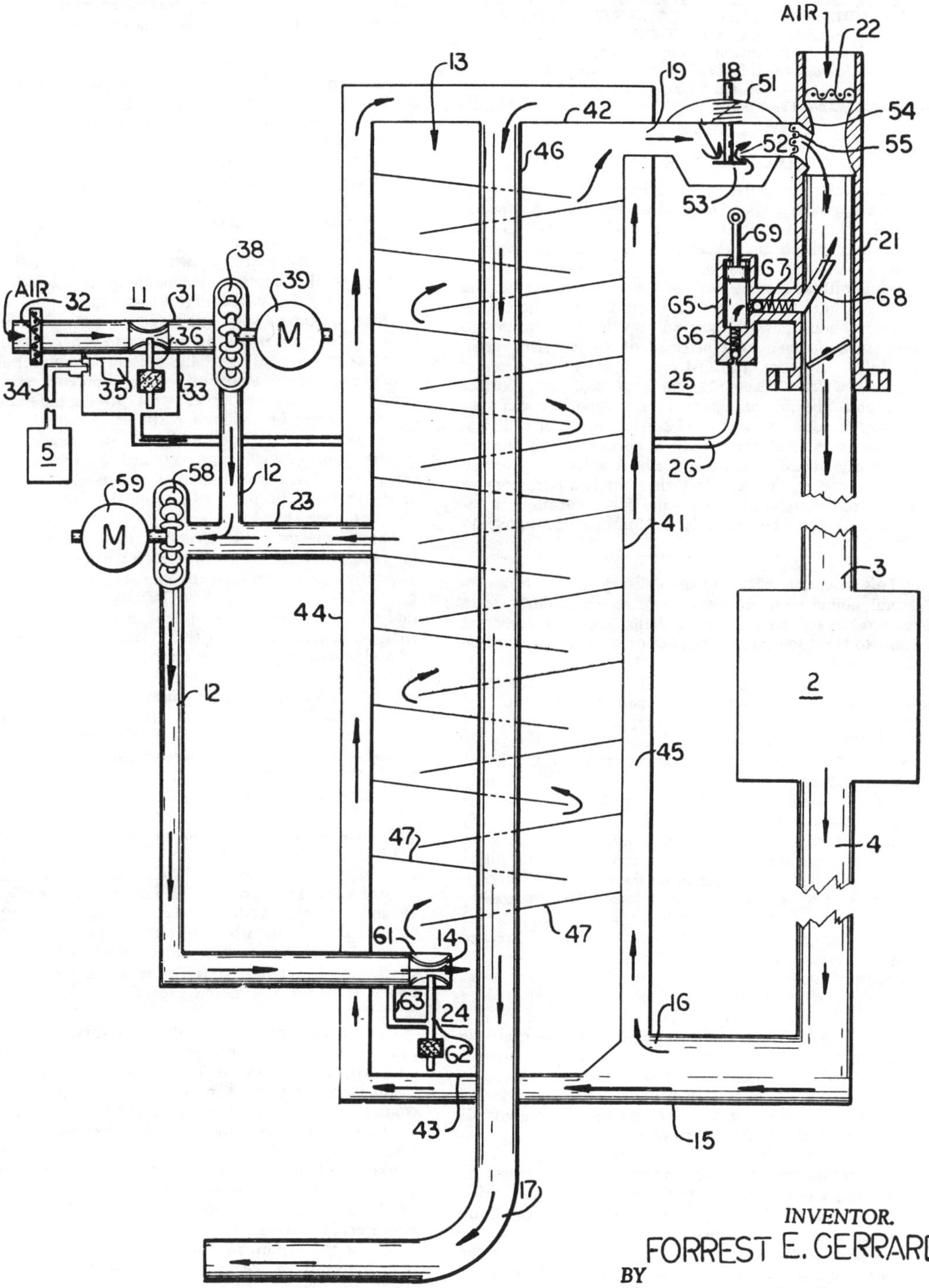

INVENTOR.

FORREST E. GERRARD

BY

McGrew & Edwards

ATTORNEYS

United States Patent Office

3,496,919
Patented Feb. 24, 1970

1

3,496,919

CARBURETOR AND SYSTEM FOR SUPPLYING VAPORIZED FUEL TO COMBUSTION ENGINES

Forrest E. Gerrard, 1070 S. Dudley St., Lakewood, Colo. 80226

Filed Sept. 5, 1967, Ser. No. 665,400
Int. Cl. F02m 31/04
U.S. Cl. 123—122 4 Claims

ABSTRACT OF THE DISCLOSURE

Carburetor for delivering fuel from a fuel tank to an engine combustion chamber in vaporized form. A vaporizing chamber has an inlet and outlet and internal surfaces in contact with a heat exchanger. The heat exchanger is connected to the exhaust gas conduit of the engine. A first conduit with an atomizer and impeller is connected to the inlet for introducing a fuel and air mixture thereto. A second conduit having a second impeller is connected to the first conduit at a point intermediate the vaporizer chamber inlet and outlet. The first conduit has an atomizer at the inlet which is disposed in the bottom of the vaporizer chamber. The interior space of the vaporizer chamber above the point of the second conduit being of sufficient volume to store all the vaporized portion of the mixture at the established pressures therein. A valve is connected to the outlet of the vaporizer chamber and is responsive to intake pressure of the engine for controlling withdrawal of the vaporized mixture through the outlet.

This invention relates to the delivery of fuel to an internal combustion engine and more particularly to a carburetor and system for supplying fuel in a vaporized form to the combustion chamber of an internal combustion engine.

Gasoline which is commonly in use as fuel for automobile engines is a liquid comprising a number of hydrocarbons having boiling points ranging from about 90° F. to 425° F. Those portions of the gasoline having the lower boiling point are commonly referred to as the lighter end or lighter fractions and those of the higher boiling point the heavier end or heavier fractions. Carburetors presently in use deliver much of the fuel and particularly that of the heavier fraction boiling point into the cylinder of the combustion chamber in an unvaporized condition and during the heat of combustion there is insufficient oxygen present to burn the fuel to release all of the energy stored in the fuel. As a result, a large portion of the energy contained in the fuel is not used and is passed out the exhaust pipe and wasted and the surrounding air is contaminated with these products of combustion.

In the practice of the present invention, both the lighter and heavier fractions of the fuel are vaporized by first passing them in heat exchange relation to an available heat supply before delivery to the combustion chamber and independently of the section of the engine cylinder so that a maximum amount of available energy is released from the gasoline and a lesser amount of combustible matter is released to the atmosphere.

An object of this invention is to improve the efficiency and operating cost of an internal combustion engine by providing a novel carburetor arrangement for delivering fuel in a vaporized condition to the engine in a condition to be substantially consumed therein.

Another object of this invention is to provide for the vaporization of both lighter and heavier fractions of the gasoline before it enters the combustion chamber.

Still another object of this invention is to provide simple, efficient and economical means for reducing the amount of pollutants which are discharged to the atmosphere from an internal combustion engine.

2

A further object of this invention is to provide a carburetor means for delivering fuel to an internal combustion engine which may be easily adapted to existing carburetion systems of current standard makes of present types of automobile engines.

Other objects, advantages and capabilities of the present invention will become more apparent as the following description proceeds taken in conjunction with the drawing in which:

The drawing is a schematic diagram of an arrangement embodying features of the present invention shown as coupled in flow relationship between a fuel tank and the throttle body and associated combustion chamber of an internal combustion engine.

Referring now to the drawing, a combustion chamber of an internal combustion engine designated **2** with a fuel intake **3** and exhaust discharge **4** and a fuel supply or tank **5** which are operatively associated with the arrangement of the present invention are illustrated schematically for purposes of explanation of the present invention but it is understood such components form no part thereof. This engine may be of various known types for different uses but the present invention is particularly suitable for use in conjunction with automobile engines because of the greatly improved efficiency attained by releasing more energy stored in the fuel and the reduction of the amount of discharge of products of combustion into the atmosphere attained thereby.

The arrangement illustrated in the drawing embodying features of the present invention will first be described generally by progressing from the fuel tank **5** to the engine **2** with the direction of flow indicated by arrows. The fuel supplied from the fuel tank **5** is first atomized or changed in an atomizer **11** to a mixture of small particles of fuel suspended in air in an induced air flow which is directed as an incoming fuel mixture along an intake conduit **12** through a fuel intake **14** into a vaporizing chamber **13**. In the vaporizing chamber **13** the fuel mixture progresses upwardly in a sinuous flow in heat exchange relation with surfaces heated by the exhaust gases of the engine **2** which are conducted from the exhaust of the engine through a conduit **15**, into a heat intake **16** passage extending lengthwise of chamber **13** and out a central heat discharge passage **17** so that portions of the fuel mixture absorb heat and are vaporized as they progress upwardly in chamber **13**. A discharge control valve **18** is coupled between the vapor discharge **19** and a throttle body **21** having an air intake **22** coupled in flow relationship with the intake of the engine combustion chamber which selectively regulates the discharge of the vaporized fuel discharged from chamber **13** in accordance with the pressure in the combustion chamber.

Unvaporized portions of the fuel mixture are withdrawn from the vaporizing chamber in a recirculation cycle through a conduit **23** coupled between an intermediate portion of the vaporizing chamber and the intake conduit **12** and are returned or recirculated in an induced air flow to mix with the incoming fuel mixture in the intake conduit **12** prior to being delivered to the fuel intake **14** of the vaporizing chamber. Liquefied portions of the fuel which drain and settle in the bottom portion of the vaporizing chamber are reatomized in a second atomizer **24** associated with the induced air flow at the intake of the vaporizing chamber. A primer or pump **25** pumps fuel from the fuel tank through a line **26** into the throttle body **21** and engine combustion chamber **2** as required for furnishing additional fuel for starting and acceleration of the engine.

More specifically, the atomizer **11** includes a venturi tube **31** having a screened air intake **32** and a fuel reser-

3,496,919

3

voir **33** into which fuel is delivered from the tank by a line **34**. A float **35** is associated with the discharge of this line for controlling the level of fuel in the reservoir. The fuel pipe **36** having restricted orifices in its upper end is disposed in an upright position in the reservoir **33** and extends from a submerged position in the fuel of the reservoir up through into the venturi tube so that when an induced air flow is produced in the tube fuel will be drawn up through the pipe and into the tube and discharged as minute particles of fuel suspended in air in an induced air flow in the venturi tube.

For producing an induced air flow in the venturi tube **31** and intake conduit **12** there is provided an impeller **38** disposed in a casing coupled in flow relation between the venturi tube **31** and conduit **12** which is mounted on the shaft of a drive **39**, preferably an electric motor. Rotation of the impeller **38** at a constant speed produces a static pressure in intake conduit **12** and an air flow along conduit **12** for introducing the fuel mixture into the fuel intake **14** of the vaporizing chamber.

The vaporizing chamber **13** preferably will be disposed in an upright position and in practice preferably will be mounted adjoining the engine. As shown, chamber **13** includes upright sidewalls **41** and top and bottom walls **42** and **43** formed of a heat conductive material. The fuel intake **14** is disposed in a lower portion of the sidewall and the vapor discharge **19** in an upper portion of the opposite sidewall. An outer housing or jacket **44** fully encloses the inner chamber walls and is disposed in spaced relation thereto to form a heat passage **45** around the inner walls in flow communication with a tubular member **46** which terminates in the heat discharge **17** at its lower end. Tubular member **46** is disposed centrally in the inner chamber side wall and as shown forms an inner wall of the chamber and is of a heat conductive material so that the heated exhaust gases will circulate in heat exchange relation with the chamber walls along the bottom wall **43**, upwardly around the sidewalls **41**, over the top wall **42** and then downwardly through the central tubular member **46** and out the exhaust discharge **17** as shown by arrows.

The interior surface of the sidewalls is provided with a plurality of baffle members **47** which are secured at one sidewall and incline downwardly therefrom and terminate in spaced relation to the opposing sidewall to form passages which permit escape of elevating gases. These baffle members are supported from opposing sidewalls in an alternating pattern and are preferably imperforate so as to define a sinuous flow path for the fuel mixture as it progresses upwardly toward the vapor discharge **19**, thereby providing a flow path of substantial length in a condensed space. In addition, the unvaporized fuel which is elevated and then liquefies at some stage in its progressive movement upwardly will drain down along these baffles, and eventually collects at the bottom of the chamber. In this manner, the heat absorbed by the walls **41**, **42**, and **43** and from the tubular member **46** will be transferred to the incoming fuel mixture which is forced into the chamber under pressure and will absorb heat so as to vaporize and then rise to the top of the chamber. This vaporized fuel is then stored in an upper portion of the chamber until being drawn through valve **18**.

While the vaporizing chamber **13** has been described with a degree of particularity it is understood that other forms may be provided. The exterior dimension may be changed to fit in available space under the hood of a particular vehicle and the heat exchange passages and baffle arrangement may be changed in shaping, dimension, and arrangement so long as there is provided a heat exchange flow path of substantial length in a condensed space to vaporize the incoming fuel, recirculation of unvaporized portions and a collection space for liquid portions therein which may be reintroduced in an atomized condition for vaporization thereof.

The control valve **18** may be of various pressure responsive types and as shown includes a housing **51** having inner walls defining a port **52** and a spring biased valve member **53** which will open and close the port in accordance with the suction or pressure at its discharge side produced by the reciprocating action of the piston in the cylinder of the combustion chamber **2**.

4

The throttle body **21** has a venturi portion **54** having a screened aperture **55** in flow relation coupled to the valve **18** so that the suction of the engine chamber will draw in an amount of vaporized fuel and air from intake **22** determined by the setting of a valve into the throttle body **21** and then the combustion chamber **2** in accordance with the requirements of the engine.

The recirculation conduit **23** is disposed in flow communication with an intermediate portion of the chamber and preferably in the side walls and one at approximately in equally spaced relation between the fuel intake **14** and the vapor discharge **19** as shown. An impeller **58** is mounted in a casing coupled in flow relation with conduits **12** and **23** and is driven preferably by an electric motor **59** which will draw in an induced air flow unvaporized portions of the fuel from the chamber which has risen upwardly therein. This impeller is connected as shown in intake line **12** so that the unvaporized portions will recirculate and mix with the incoming fuel mixture from the atomizer **11** and deliver this mixture to the fuel intake **14**.

The second atomizer **24** includes a venturi section **61** formed as a part of the discharge end of the intake conduit **12** having a fuel pipe **62** which has a restricted orifice in the venturi secton and extends downwardly into the bottom of the chamber and associated pipe **63** in flow communication with the conduit **12** upstream of the pipe **62** and the pipe **63** so that the induced air flow produced in the intake conduit **12** by impeller **38** and **58** will produce a venturi effect so as to draw the fuel in liquid form up into the venturi section and atomize or change it to a mixture of small particles of fuel suspended in air in an induced air flow for revaporization thereof.

The primer or pump **25** preferably will be some type of pressure-type fuel pump and as shown for purposes of description includes a housing **65** with a ball valve **66** on the intake side and a ball valve **67** on the discharge side with a fuel pipe **68** projecting into the throttle body **21**. A reciprocating plunger **69** is mounted within a passage in the housing between the two valves which will reciprocate so as to draw fuel through line **26** from the reservoir supplied by the fuel tank and discharge it into the air flow in the throttle body **21**. This pump will operate by a suitable mechanical motive means independently of the other portion of the system to pump fuel into the combustion chamber for starting and acceleration purposes.

While electric motors have been described as a preferred form of drive for impellers **38** and **58** another arrangement is by using a belt or gear driven from the engine **2** to which fuel is being supplied.

From the foregoing it is apparent that the impeller **58** will induce an air flow in the form of a forced draft that will circulate unvaporized portions of the fuel mixture around and around in a cyclic manner through the lower portion of the chamber **13**, conduit **23** and downstream portion of conduit **12** as indicated by the direction of arrows independent of the suction of the combustion chamber so that the fuel will absorb additional heat during each cycle and eventually become vaporized so as to then pass to the upper portion of the chamber which is of sufficient capacity to store the vaporized fuel portion therein. Atomizer **24** reatomizes fuel that has condensed to a liquid state and has flowed back down to the bottom of the chamber.

The upper portion of the chamber **13** above intake conduit **12** stores the vaporized portion of the fuel at pressures in the chambers established by the air being forced into the chamber. Impeller **38** running at a constant speed will deliver a preselected amount of air to the chamber and thereby establish a pressure in the chamber for producing an induced flow of the vaporized portions of the fuel upwardly into the upper storage portion. This stored vaporized fuel is withdrawn or drawn off by the engine intake through the regulator valve.

I claim:

1. A carburetor system for delivering fuel to an engine combustion chamber comprising a heat exchange chamber having an external shape to fit available space adjoining an engine of a self-propelled vehicle, said chamber having inner surfaces for directing an incoming fuel and air mixture through the chamber interior from an intake portion to an outlet portion, and means maintaining said surfaces in heat exchange relation with exhaust gases of the engine, atomizer means inclusive of a first impeller having rotary drive for injecting a fuel mixture suspended in air under pressure into said intake portion of the chamber, said mixture moving in an induced flow from said intake portion to said outlet portion in heat exchange relation with said inner surfaces for vaporizing at least a major portion of the mixture to a gaseous state, conduit means having a second rotary impeller for cyclic withdrawal of unvaporized portions of said mixture at a point intermediate the intake portion and the discharge portion for return into the induced flow entering the intake portion so as to be subjected to more heat exchange action, said conduit means having a second atomizing means at its discharge end in the chamber and actuated by the air flow produced by a least said second impeller for reatomizing fuel in a liquid state collecting in the chamber, the interior space of said chamber above the point of recycle withdrawal being of sufficient volume to store all the vaporized portion of the mixture at the established pressures therein, and valve means responsive to the intake pressure of the engine for controlling the withdrawal of the vaporized mixture through said outlet portion to the combustion chamber of the engine.

2. A carburetor system as defined in claim 1, in which the chamber has exterior heat-conductive walls maintained in contact with exhaust gases passing from the engine, and the return flow of recirculating unvaporized portions are subjected to more heat exchange in their repeated passage through the chamber with the induced flow.

3. A carburetor system as defined in claim 1, means maintaining the vaporized portion of the mixture above the point of withdrawal of unvaporized portions in vapor state by continuing heat exchange action.

4. A carburetor system for delivering fuel to an engine

combustion chamber comprising a heat exchange chamber having an external shape to fit available space adjoining an engine of a self-propelled vehicle, said chamber having inner surfaces for directing an incoming fuel and air mixture through the chamber interior from a lower intake portion to an upper outlet portion, and means maintaining said surfaces in heat exchange relation with exhaust gases of the engine, atomizer means inclusive of a first impeller having a rotary drive for injecting a fuel mixture suspended in air under pressure into said lower intake portion of the chamber, said mixture moving in an induced flow from said lower intake portion to said upper outlet portion in heat exchange relation with said inner surfaces for vaporizing at least a major portion of the mixture to a gaseous state, conduit means having a second rotary impeller for cyclic withdrawal of unvaporized portions of said mixture at a point intermediate the intake portion and the discharge portion for return into the induced flow at the intake portion subjected to said heat exchange action, said conduit means having a second atomizing means at its discharge end disposed in the bottom of the chamber and actuated by the air flow produced by at least said second impeller for reatomizing fuel in a liquid state collecting at the bottom of the chamber, the interior space of said chamber above the point of recycle withdrawal being of sufficient volume to store all the vaporized portion of the mixture at the established pressures therein, and valve means responsive to the intake pressure of the engine for controlling the withdrawal of the vaporized mixture through said outlet portion to the combustion chamber of the engine.

References Cited

UNITED STATES PATENTS

1,140,064	5/1915	Rakestraw.
1,198,013	9/1916	Dempsey.
1,733,792	10/1929	Good.
1,734,723	11/1929	Gildehaus.
2,323,639	7/1943	Anderson.

FOREIGN PATENTS

688,874	5/1930	France.

RONALD R. WEAVER, Primary Examiner

U.S. Cl. X.R.

48—144; 123—131; 261—24, 34, 145

United States Patent

[11] 3,630,698

[72] Inventor Joseph H. Baldwin
Route 4, Box 305-A-2, Waycross, Ga. 31501

[21] Appl. No. 4,509
[22] Filed Jan. 21, 1970
[45] Patented Dec. 28, 1971

[54] FUEL SYSTEM
5 Claims, 3 Drawing Figs.

[52] U.S. Cl. .... 48/180 R, 48/180 C, 123/119, 261/70, 261/72, 261/119, 261/DIG. 38

[51] Int. Cl. F02m 21/04

[50] Field of Search .... 48/180 M, 180 P, 180 R, 180 S, 180 B, 102 A, 103; 123/119; 261/119, 36.1, 115, 72, 70

[56] References Cited

UNITED STATES PATENTS

2,856,169	10/1958	Mustain	48/180 X
2,957,759	10/1960	Jettinghoff	48/180
3,081,160	3/1963	Ensign	48/180 X
3,171,467	3/1965	Featherston	261/36.1
3,227,427	1/1966	Wells et al.	261/115 X
3,279,770	10/1966	Parker, Jr.	261/36.1
3,433,608	3/1969	Epifanio, Sr.	48/180
3,464,803	9/1969	Kimberley	48/180
3,486,524	12/1969	Brooks	261/72 X

Primary Examiner—Morris O. Wolk
Assistant Examiner—R. E. Serwin
Attorney—Browdy and Neimark

ABSTRACT: A fuel system constructed to furnish gaseous fuel, rather than liquid fuel, to an internal combustion engine. The delivery system converts liquid fuel to its gaseous state by lowering the pressure on the surface of the liquid, thereby causing the liquid to boil thereby providing the vapor which is fed to a venturi and thence into the intake manifold.

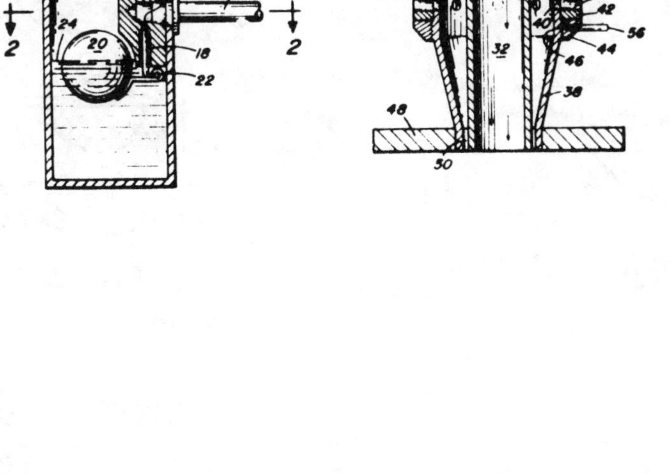

PATENTED DEC 28 1971 3,630,698

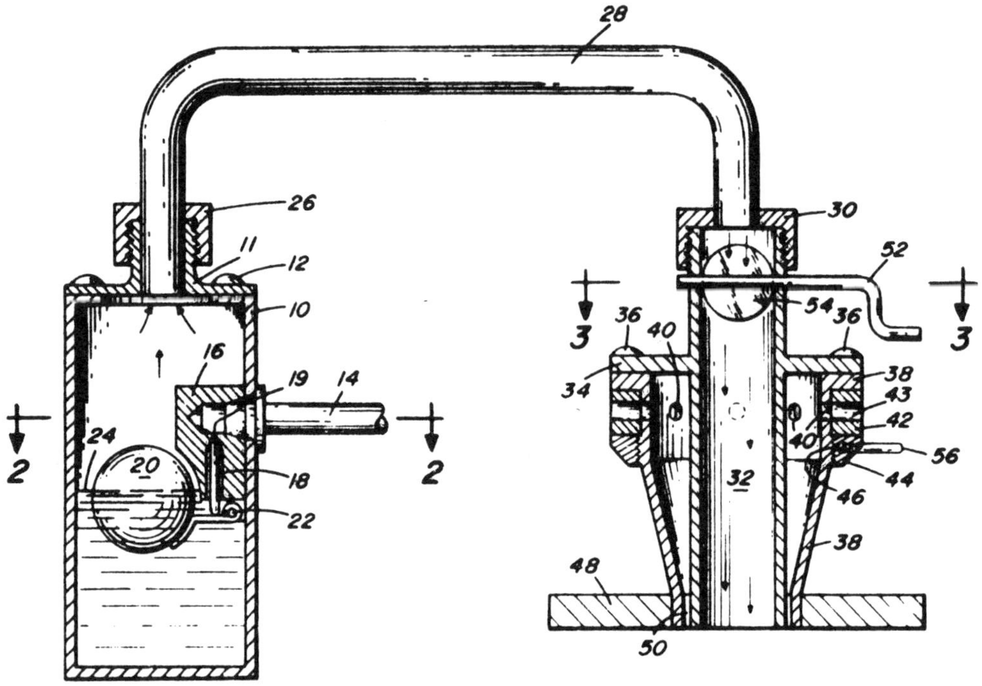

FIG. 1

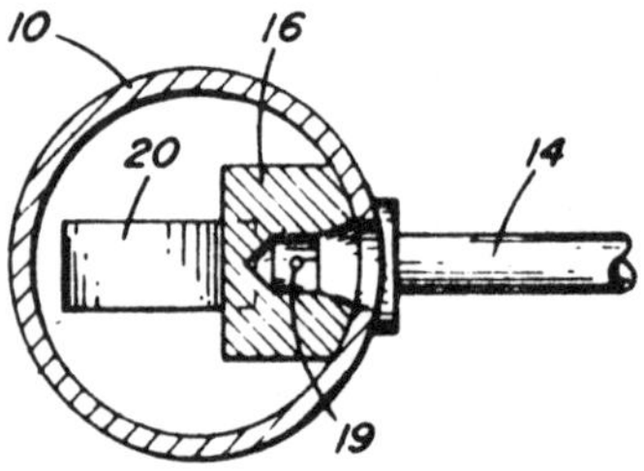

FIG. 2

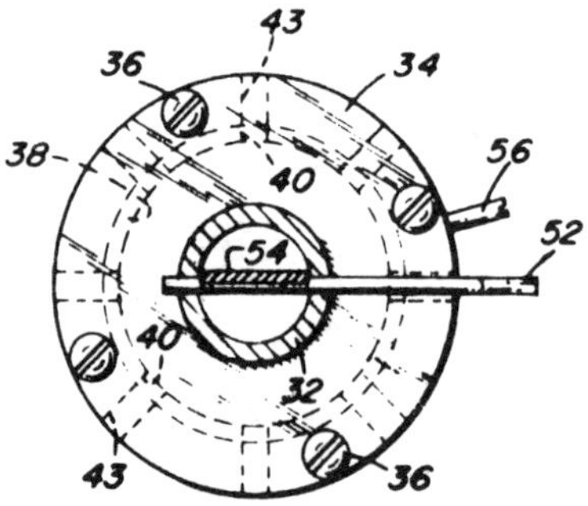

FIG. 3

INVENTOR.
JOSEPH H. BALDWIN
BY

3,630,698

1

FUEL SYSTEM

The present invention relates to fuel systems and, more particularly, to a fuel system for internal combustion engines wherein gaseous fuel rather than liquid fuel is supplied to the engine.

It is widely known among those skilled in the art that while the conventional jet-type carburetor produces fair performance over a wide speed range of the engine, nevertheless, due to inefficient droplets of gasoline which are not converted to vapor, there is the disadvantage that much of the fuel is never converted into enough of a vapor to burn properly. As these larger droplets of fuel are unburned they cause unnecessary wear on the cylinder walls due to their tendency to wash of any oil film which should be present for proper lubrication of the cylinder.

Then, too, in spite of many improvements in jet carburetion during the course of the years, it still remains a most inefficient system. The needle valves and jet openings in the carburetor must be carefully adjusted for even acceptable operation, and often they are out of adjustment with the result that too much gasoline is sprayed into the engine. The engine then will be hard to start, will be easily flooded, and will get very poor mileage to the gallon of gasoline. An internal combustion engine requires an accurately computed mixture of gasoline vapor and air for proper operation, and the process of spraying liquid gasoline into an engine manifold often does not produce the proper mixture.

While others have previously contemplated the concept of feeding gasoline vapor directly to the intake manifold of an internal combustion engine, these prior attempts have not been entirely successful and have not been adopted. Thus the Smith U.S. Pat. No. 3,049,850 requires an outside source of heat, means to force air through the liquid fuel, including agitating means and a fuel pump; besides being very expensive because of the presence of heating means and air baffles, the device will operate efficiently only when the engine has reached its optimum operating temperature. The Williams U.S. Pat. No. 3,338,223 also depends on the passage of air through the liquid fuel, and no provision is made to control the quantity of fuel vapor mixed with incoming air. Epifanio et al. U.S. Pat. No. 2,844,364 require a conventional carburetor or mechanical fuel pump and the system provided is merely supplementary. Hidden U.S. Pat. No. 1,210,169 provides no effective regulation of vapor mixing with the circulating air current Still other devices have other defects in operation and/or expense.

However, the present invention overcomes the above-mentioned shortcomings and weaknesses of the prior art devices and discloses a fuel system which is outstandingly effective and highly efficient. The liquid gasoline is vaporized into a highly combustible vapor which is then mixed with the necessary amount of air, in a venturi section, and this mixture is ultimately fed directly into the internal combustion engine manifold. There is no excess of unvaporized liquid gasoline to cause poor operation and damage to the automobile, the device has a minimum of moving parts, the carburetor and fuel pump have been eliminated, and the system provides greater fuel economy than other known systems.

A general object of the invention is the provision of a fuel system which furnishes gaseous fuel to internal combustion engines and which overcomes the defects of the prior art, such as indicated above. Another object of the invention is the provision of a fuel system which eliminates the carburetor and fuel pump.

Still another object of the invention is the provision of a fuel system which has a minimum of moving parts, and which reduces maintenance.

Yet another object of the invention is the provision of a fuel system which produces a highly combustible vapor whereby there will be no appreciable waste of fuel nor unnecessary wear of the engine due to unburned fuel particles.

Still another object of the invention is the provision of a fuel system which, due to its simplicity of design, can be easily and economically manufactured and maintained.

2

Other objects and many of the attendant advantages of the invention will readily be appreciated as the same becomes better understood by reference to the following detailed description when considered in connection with the accompanying drawings in which like reference numerals designate like parts throughout the figures thereof and wherein:

FIG. 1 shows an elevation view of the invention;

FIG. 2 is a view taken along line 2—2 of FIG. 1;

FIG. 3 is a view taken along line 3—3 of FIG. 1.

Referring now to the drawings there is shown in FIG. 1, which represents a preferred embodiment of the invention, a vacuum chamber 10 having a top cover 11 which is secured to the chamber 10 by means of screws 12 or other suitable connections. A gasket, not shown, is placed between chamber 10 and cover 11, to insure an airtight fit, while chamber 10 may be made of any easily cast metal, such as steel, bronze or aluminum, or other material with sufficient strength to resist the crushing effect of outside atmosphere while the inside is under vacuum. While FIGS. 1 and 2 show the chamber 10 as being cylindrical, it is to be understood that its shape may be modified to fit existing conditions without altering the concept of the invention.

Attached to the side of vacuum chamber 10 by a suitable fitting is a liquid fuel line 14 which is connected to a gasoline supply tank (not shown), for supplying fuel to the vacuum chamber, and ultimately to the internal combustion engine. On the inside of the vacuum chamber 10, and connecting with the fuel line 14, there is a valve 16 having a sliding piston or needle 18 that cooperates with a small circular hole or recess 19 to cut off the flow of fuel through fuel line 14. The sliding piston 18 is actuated by a float 20 which is pivotally suspended at one end by a pin 22 mounted on the sidewall of the vacuum chamber 10. Float 20, piston 18, and recess 19 function together to maintain fuel in the chamber at a level shown at 24.

Passing through the cover 11 of vacuum chamber 10, and held in a tight connection by gland nut 26 or other suitable means, there is a connecting tube 28, which may be of any material that can resist crushing when under a vacuum. Connecting tube 28 passes through a gland nut 30 where it is held in a tight connection with a delivery tube 32. Delivery tube 32, which forms a passage for the gaseous vapors to the internal combustion engine, is fastened to the large end of a venturi 38. This connection between the delivery tube and the venturi is preferably accomplished by means of projections 34 which are an integral part of the delivery tube 32 and which extend outwardly, transversely to the longitudinal axis of the tube. Screws 36 extend through projections 34 and into the venturi, while a gasket (not shown) is placed between the two to insure an airtight fit.

The venturi 38 has elongated, inwardly converging side members or a frustoconical wall, which converge towards the delivery tube 32 until there is formed a narrow passage 50 between the sides of the venturi and the delivery tube, as the venturi and delivery tube pass through an adapter plate 48 which is used to attach these components to the intake manifold of the engine. Passage 50 forms an air passage for admitting an adjustable amount of air to be mixed with the gaseous vapor as the latter enters the engine intake, as will be more fully described below.

Located transversely across the bore of delivery tube 32 there is a butterfly valve 54 which is operated by a crank arm 52 to open and close the valve and thereby control the amount of gaseous vapor being supplied to the engine. Crack arm 52 is connected by suitable mechanical linkage (not shown) to the accelerator pedal of an automobile if the invention is being applied to a car engine, or to any other accelerator control should the invention be applied to any other internal combustion engine.

In order to provide an adjustable air supply to the system there is provided, around the large end of the venturi, an air-mixing valve 42, which takes the configuration of a circular collar that can be rotated relative to, and around the neck of, the venturi. There are provided a plurality of spaced holes 40 drilled around the periphery of the venturi walls, and a like number of holes 43 drilled in the air-mixing valve 42. Rotational adjustment of valve 42 so that holes 40 and 43 line up, or become misaligned, determines the amount of air being admitted by the venturi for mixing with the gaseous vapor. Valve 42 is connected to the engine accelerator by mechanical means (not shown) so that the volume of air being admitted into the engine is constantly varied depending on the speed of the engine. In its closed position the valve acts as a choke by shutting off all of the air supply.

Below the valve 42, and also encircling the neck of the venturi, there is an adjusting collar 44. Collar 44 and the venturi body have matching threads 46 for the correct positioning of valve 42, and a setscrew 56 is provided for locking collar 44 in place and preventing its movement when valve 42 is in correct axial position.

Turning now to the operation of the device, it can be seen that the system disclosed consists of three basic parts; (1) the vacuum chamber 10; (2) the delivery tube 32; and (3) the venturi 38. The operation of the system is based on the fact that a fuel must be converted into a gas before ignition will occur. The present fuel delivery system is designed to change liquid fuel into its gaseous state before introducing it with the mixing air in the venturi. This is accomplished by lowering the pressure on the surface of the confined liquid fuel. Lowering the pressure causes the liquid fuel to boil and changes the top layer of molecules from the liquid to the gaseous state. This action takes place in vacuum chamber 10.

The vacuum chamber 10 is directly connected to the intake manifold by means of the delivery tube 32 and the venturi 38. The venturi 38 has the arrangement of valves 42 covering its ports or openings 40 and this arrangement is used to control the amount of outside air entering the system. Valve 42 is completely closed when starting a cold engine because of the choke; the system is not evacuated at this point, but is at the pressure of the surrounding atmosphere.

As the pistons of the engine begin their intake stroke they act as a vacuum pump by drawing in gases to fill the volume created by their downward movement. At this point the only available gas, mostly air, is that confined in the vacuum chamber 10. As the piston continues its downward movement vapor from the liquid fuel begins to fill the vacuum created by the withdrawn gas. In this way the cycle begins. As the engine accelerates the movement of the pistons is much faster, thus creating a stronger vacuum over the liquid fuel. In order to fill this evacuated space the liquid begins to boil and the gaseous fuel flows from the vacuum chamber 10 through the connecting tube 28 through the delivery tube 32 to the intake manifold and finally into the cylinders themselves. Valve 42 over the venturi ports 40 is now opened to admit outside air into the system in order to obtain the best fuel-to-air mixture. The butterfly valve 54 is installed in delivery tube 32 to regulate the amount of fuel supplied to the engine. The movement of butterfly valve 54 and air-mixing valve 42 must be correlated to obtain optimum engine performance.

The air entering the system through valve holes 43, and venturi ports 40, accelerates as it approaches the throat of the venturi and passes through passage 50. At this point the air passes the open end of delivery tube 32. (The delivery tube extends longitudinally into the venturi with its delivery end in the throat of the venturi, as at the adapter plate 48.) As this high-velocity air passes around the open end of the delivery tube it further reduces the pressure in the connecting tube 28 and vacuum chamber 10 and thereby assists in maintaining a vacuum over the surface of the liquid fuel. Should there be a need for additional heat to increase the rate of evaporation of the fuel, outside sources such as a hot water coil, an electric heating coil, or hot air could be directed around the bottom of the vacuum chamber.

As the engine operates, the fuel level inside the vacuum chamber is lowered due to the boiling of the liquid. In order to maintain the liquid level 24, a float 20 and valve 16 arrangement is installed in the vacuum chamber 10, and as the fuel level lowers, float 20 pivots about its mounting pin 22 causing piston 18 to move downwardly and uncover opening 19 in fuel supply line 14, connecting with a conventional fuel tank. As piston 18 uncovers opening 19 fuel is forced into the vacuum chamber by the atmospheric pressure exerted on the surface of the liquid fuel in the main tank. As the level of the fuel rises, float 20 rotates clockwise about its pivot 22 forcing piston 18 upward to close off opening 19 in the fuel line. Since the fuel is consumed the cycle repeats itself as long as the engine is running.

It should be understood, of course, that as with any fuel system, the sizes and shapes of the individual parts will be dictated by the size and type of equipment on which it is to operate.

From the above description of the structure and operation of the invention it is obvious that there is disclosed a fuel system which provides many improvements over prior known systems. The simplicity of the present system with its minimum of moving parts and greater economy of fuel consumption will replace both the carburetor and fuel pump from a conventionally operated engine. The system will also work with the venturi in any position, and having only three simple valves, the device requires only a minimum of maintenance.

Obviously many modifications and variations of the present invention are possible in the light of the above teachings. It is therefore to be understood that within the scope of the appended claims, the invention may be practiced otherwise than as specifically described.

What is claimed is:

1. A fuel system for supplying gaseous vapor to be burned in an internal combustion engine comprising;

a closed evaporating chamber;

means for maintaining a substantially constant level of vaporizable fuel in said chamber;

a delivery tube means communicating between said evaporating chamber and the intake manifold of the engine for effecting a vacuum in said chamber and for vaporizing fuel therein, said delivery tube means conducting the gaseous vapor to the intake manifold of the engine;

air inlet means associated with said delivery tube means, said air inlet means defining a venturi with said delivery tube means being coupled to said air inlet means at a relatively low-pressure area in the throat of said venturi;

first valve means associated with the large end of said venturi for regulating air input to the large end of said venturi for regulating air input to the system; and

second valve means associated with said delivery tube means for regulating gaseous fuel vapor input to the system, said first and second valve means being operatively interlinked.

2. The fuel system of claim 6 wherein the means to maintain a pool of liquid vaporizable fuel in the evaporating chamber is a float and needle valve connected to a fuel supply line.

3. The fuel system of claim 2 wherein the valve means has a plurality of holes which mate with like holes in the venturi.

4. The fuel system of claim 3 further including a butterfly valve located in the delivery tube for regulating the flow of gaseous vapor through the tube.

5. The fuel system of claim 4 wherein the valve means and the butterfly valve are mechanically coupled to the engine accelerator.

* * * * *

United States Patent

Tucker

[15] **3,653,643**

[45] **Apr. 4, 1972**

[54] **CARBURETOR**

[72] Inventor: **Oliver M. Tucker,** 1613 East Blackthorne Pl., Milwaukee, Wis. 53211

[22] Filed: **Apr. 3, 1970**

[21] Appl. No.: **25,362**

[52] **U.S. Cl.** ....**261/56,** 261/79 R, 261/96, 55/234

[51] **Int. Cl.** **F02m 17/34**

[58] **Field of Search** ....55/234, 255; 261/96, 56, 79 R

[56] **References Cited**

UNITED STATES PATENTS

3,282,033	11/1966	Seppanen	55/234
1,857,070	5/1932	Thomas	261/79 R
2,627,934	2/1953	Martinek	55/234
3,171,820	3/1965	Votz	261/94
2,669,319	2/1954	Inglesby	55/234
3,485,015	12/1969	Vecchio	55/234

Primary Examiner—Tim R. Miles
Attorney—Ronald E. Barry and James E. Nilles

[57] **ABSTRACT**

A carburetor including a housing having a fluid fuel reservoir in the bottom, an air inlet at the top of the housing, a delivery pipe coaxially mounted within the housing and terminating short of the top of the housing, and a porous vaporizing filter substantially filling the reservoir. A baffle is concentrically mounted within said housing and extends partially into the vaporizing filter in the reservoir to deflect the incoming air through the vaporizing filter. The level of liquid fuel in the reservoir is kept above the bottom of the baffle, so that air entering the carburetor through the inlet must pass through the liquid fuel and vaporizing filter in the reservoir before discharge through the outlet. A secondary air inlet is provided in the top of the housing for controlling the fuel air ratio of the vaporized fuel passing into the delivery pipe.

6 Claims, 1 Drawing Figure

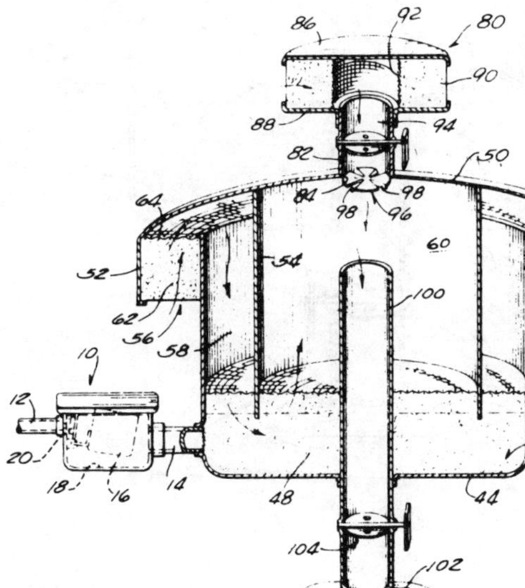

PATENTED APR 4 1972 3,653,643

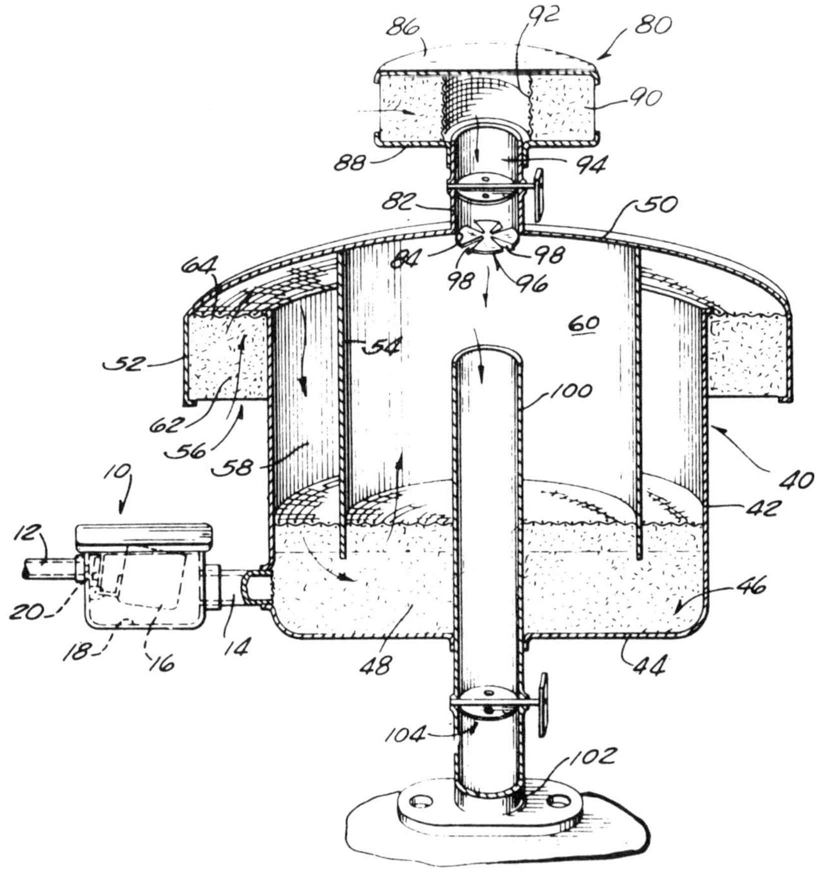

Inventor
Oliver M. Tucker
By Ronald E. Barry
Attorney

CARBURETOR

BACKGROUND OF THE INVENTION

It is generally well known that liquid fuel must be vaporized in order to obtain complete combustion. Incomplete combustion of fuel in internal combustion engines is a major cause of air pollution. In a typical automotive carburetor, the liquid fuel is atomized and injected into the air stream in a manifold of approximately 3.14 sq. in. cross section. In an 8 cylinder 283 cu. in. engine running at approximately 50 miles per hour, the engine operates at approximately 2,400 r.p.m. and requires approximately 340,000 cu. in. of air per minute. The air velocity at this speed in the intake manifold will be approximately 150 feet per second and it will therefore take approximately 0.070 of a second for a particle of fuel to move from the carburetor to the combustion chamber. The fuel will remain in the combustion chamber for approximately 0.0025 seconds when operating at 2,400 r.p.m.

It is conceivable that in this short period of time complete vaporization of the fuel is not achieved and as a consequence, incomplete combustion occurs resulting in further air pollution. The liquid fuel particles if not vaporized can deposit on the cylinder walls and dilute the lubricating oil or oil film on the walls of the cylinders promoting partial burning of the lubricating oil and adding further to the pollution problem. Destruction of the film of lubricating oil by combustion can also increase mechanical wear between the piston and cylinder.

SUMMARY OF THE INVENTION

The carburetor of this invention provides for the complete combustion of liquid fuel in an internal combustion engine with a corresponding decrease of air pollutant in the exhaust. This is achieved by supplying a completely vaporized or dry gas to the combustion chamber. The primary air is initially filtered prior to passing through a vaporizing filter which is immersed in liquid fuel provided in a reservoir in the carburetor. The vaporizing filter continuously breaks the primary air up into small bubbles thereby increasing the surface area available for evaporation of the liquid fuel. Secondary air is added to the enriched fuel-air mixture through a secondary air filter prior to admission of the fuel-air mixture into the combustion chamber of the engine. Initial filtration of both the primary and secondary air removes any foreign particles which may be present in the air which could cause increased wear within the engine. The carburetor also assures delivery of a clean dry gas to the engine due to the gravity separation of any liquid or dirt particles from the fuel enriched primary air.

Other objects and advantages will become apparent from the following detailed description when read in connection with the accompanying drawing, in which the single figure shows a perspective cross sectional view of the carburetor of this invention.

DESCRIPTION OF THE INVENTION

The carburetor **40** disclosed herein is adapted for use in an internal combustion engine wherein air is drawn through the carburetor to vaporize the fuel in the carburetor prior to admission to the combustion chamber.

In this regard, the flow of liquid fuel, gas or oil, to the carburetor **40** is controlled by means of a float valve assembly **10** connected to a source of liquid fuel by a liquid fuel line **12** and to the carburetor **40** by a connecting tube **14**. The flow of liquid fuel through the float valve assembly **10** is controlled by a float **16** pivotally mounted within a float chamber **18** and operatively connected to a float valve **20**.

In accordance with the invention, the liquid fuel admitted to the carburetor **40** through tube **14** is completely evaporated by the primary air for the engine within the carburetor **40** and mixed with secondary air prior to admission into a delivery tube **100** which is connected to the manifold **102** of the engine. More specifically the carburetor **40** includes a cylindrical housing or pan **42** having a bottom wall **44** which forms a liquid fuel and filter reservoir **46**. A vaporizing filter **48** is positioned within the reservoir **46** and extends upward a distance from the bottom wall **44** of the housing **42**. The vaporizing filter **48** is used to continuously break up the primary air into a large number of small bubbles as the primary air passes through the liquid fuel in the reservoir **46**. This increases the surface area per volume of air available for evaporation of the liquid fuel, as more particularly described hereinafter. This filter **48** is formed of a three-dimensional skeletal material that is washable and is not subject to break-down under the operating conditions of the carburetor. A foamed cellular plastic polyurethane filter having approximately 10 to 20 pores per inch has been used successfully in the carburetor **40**.

The housing **42** is closed at the top by a hood or cover **50** which can be secured thereto by an appropriate means. The hood **50** has a larger diameter than the diameter of the housing **42** and includes a depending flange **52** and a depending baffle **54**. The flange **52** is concentrically arranged and spaced outwardly from the outer surface of the housing **42** to form a primary air inlet **56**. The baffle **54** is concentrically arranged within the housing **42** to define a primary air chamber **58** and a central mixing chamber **60** within the housing **42**.

Primary air is drawn into the housing **42** through the air inlet **56** and is filtered by means of a primary air filter **62** which is removably mounted in the space between the flange **52** and the wall of the housing **42**. The primary air filter **62** is prevented from being drawn into the housing **42** by means of a screen **64**. The primary air filter **62** can be made of the same filtering material as the vaporizing filter **48**.

The primary air as it enters the primary air chamber **58** is deflected through the liquid fuel in the reservoir **46** by means of the cylindrical baffle **54**. The baffle **54** extends downwardly from the hood **50** far enough to penetrate the upper portion of the vaporizing filter **48**. The primary air must pass around the bottom of the baffle **54** and through both the liquid fuel and the vaporizing filter **48** prior to entering the mixing chamber **60**.

The level of the liquid fuel in the reservoir **46** is maintained above the bottom edge of the baffle **54** by means of the float valve assembly **10**. The operation of the float valve assembly **10** is well known. The float chamber **18** is located at approximately the same elevation as the reservoir **46** and the float **16** pivots in response to a change in the level of the liquid fuel in the float chamber to open the float valve **20**.

One of the important features of the present invention is the efficiency of evaporation of the liquid fuel by the flow of the large number of bubbles through the reservoir. This is believed to be caused by the continual break up of the bubbles as they pass through the vaporizing filter **48**. It is well known that the rate of evaporation by a bubble of air passing through a liquid unmolested is relatively slow due to the surface tension of the bubble. However, if the bubble is continuously broken, the surface tension of the bubble is reduced and a continual evaporating process occurs. This phenomenon is believed to be the cause of the high evaporation rate of the liquid fuel in the carburetor of this invention.

Another feature of the carburetor of this invention is the ability of the carburetor to supply dry gas to the central mixing chamber **60** in the housing **42**. Since the flow of primary air in the central mixing chamber **60** is vertically upward, the force of gravity will prevent any droplets of liquid fuel from rising high enough in the carburetor to enter the delivery tube **100**. The delivery of dry gas to the delivery tube increases the efficiency of combustion and thereby reduces the amount of unburned gases or pollutants that are exhausted from the engine into the air.

Means are provided for admitting secondary air into the central mixing chamber **60** to achieve the proper fuel-air ratio required for complete combustion. Such means is in the form of a secondary air filter assembly **80** mounted on an inlet tube **82** provided in an opening **84** in the hood **50**. The secondary air filter assembly **80** includes an upper plate **86**, a lower plate **88**, and a secondary air filter **90** positioned between the plates

86 and **88**. The secondary air filter **90** is prevented from being drawn into the inlet tube **82** by means of a cylindrical screen **92** which forms a continuation of the tube **82**. The secondary air passes through the outer periphery of the secondary air filter **90** through the screen **92** and into the tube **82**. The flow of secondary air through the tube **82** is controlled by means of a butterfly valve **94** as is generally understood in the art.

Complete mixing of the dry gas enriched primary air with the incoming secondary air within the housing **42** is achieved by means of a turbulator or deflector **96** positioned at the end of the tube **82**. The turbulator **96** includes a number of vanes **98** which are twisted to provide an outwardly directed circular air flow into the central mixing chamber **60** and thereby creating an increase in the turbulence of the secondary air as it combines with the fuel enriched primary air. The turbulator prevents cavitation from occurring at the upper end of the outlet or delivery tube **100**.

The flow of fuel-air mixture to the engine is controlled by means of a throttle valve **104** provided in the outlet or delivery tube **100**. The operation of the throttle valve **104** and butterfly valve **94** are both controlled in a conventional manner.

THE OPERATION OF THE CARBURETOR

Primary air is drawn into the housing **42** through the primary air inlet **56** and passes upwardly through the primary air filter **62** where substantially all foreign particles are removed from the primary air. The filtered primary air flows downward through the primary air chamber **58**, under the baffle **54**, through the fuel filter reservoir **46**, and upward into the central mixing chamber **60**. All of the primary air passes through the vaporizing filter **48** provided in the reservoir **46**. The vaporizing filter **48** continuously breaks the primary air stream into thousands of small bubbles reducing surface tension and increasing the air surface area available for evaporization of the liquid fuel. Since the outside surface of each bubble is being continuously broken up by the vaporizing filter **48** and is in constant contact with the liquid fuel as the bubble passes through the vaporizing filter **48**, there is a greater opportunity for evaporization of the fuel prior to entering the central mixing chamber **60** in the housing **42**. The vertical upward flow of the fuel enriched primary air in the central mixing chamber assures that no liquid fuel droplets will be carried into the delivery tube **100**.

The fuel enriched primary air is thoroughly mixed with the secondary air entering through tube **82** by means of the turbulator **96** which increases the turbulence of the primary and secondary air within the central mixing chamber and prevents cavitation from occurring in the delivery tube **100**. The completely mixed fuel enriched primary air and the secondary air then passes through the delivery tube **100** into the combustion chamber of the engine.

I claim:

1. A carburetor for an internal combustion engine, said carburetor comprising:
   - a housing having a liquid fuel reservoir, a vaporized fuel outlet projecting upwardly above said reservoir, and a primary inlet around said outlet,
   - an open cell plastic vaporizing filter immersed in said reservoir,
   - baffle means extending downwardly into said filter for deflecting the primary air downwardly through said filter prior to flowing upwardly toward said outlet,
   - means for maintaining a static fuel level above the lower end of said baffle means,
   - a secondary air inlet connected to the top of said housing to admit air to said housing prior to the flow of vaporized fuel into said outlet,
   - means for controlling the flow of secondary air into said housing, and
   - means for deflecting the secondary air away from the vaporized fuel outlet.
2. The carburetor according to claim 1 including a primary air filter in said primary air inlet.
3. The carburetor according to claim 1 wherein said baffle means comprises a cylindrical baffle positioned between said primary air inlet and said vaporized fuel outlet.
4. A carburetor which is adapted to be mounted on the manifold of an internal combustion engine, said carburetor comprising:
   - a cylindrical housing having a bottom wall defining a reservoir in the bottom of said housing,
   - a porous cellular vaporizing filter in said reservoir,
   - a delivery tube mounted in said bottom wall and extending partially upward within said housing through said vaporizing filter,
   - a hood mounted on said housing and spaced therefrom to define an annular primary air inlet around the outside of said housing,
   - a secondary air inlet in said hood, means mounted on said hood for deflecting secondary air away from said delivery tube,
   - a cylindrical baffle depending downwardly from said hood and penetrating said vaporizing filter, and
   - means for maintaining the level of liquid fuel in said reservoir above the bottom of said baffle whereby air entering said housing must pass through said vaporizing filter and liquid prior to entering said delivery tube.
5. The carburetor according to claim 4 including a primary air filter in said primary air inlet.
6. The carburetor according to claim 4 wherein said delivery tube extends upwardly from said vaporizing filter a distance sufficient to prevent liquid fuel droplets from rising far enough to enter said delivery tube.

* * * * *

United States Patent [19] | [11] **3,789,820**
Douglas et al. | [45] **Feb. 5, 1974**

[54] **COMPRESSED GASEOUS FUEL SYSTEM**

[75] Inventors: **Lloyd A. Douglas; Kirk T. Garanflo,** both of Fort Worth, Tex.

[73] Assignee: **Victor Equipment Company,** Denton, Tex.

[22] Filed: **Oct. 19, 1971**

[21] Appl. No.: **190,550**

[52] **U.S. Cl.** .... **123/27 GE,** 123/1 A, 123/121
[51] **Int. Cl.** **F02m 13/06,** F02m 13/68
[58] **Field of Search** .... 123/121, 27 GE, 1 A

[56] **References Cited**

UNITED STATES PATENTS

2,315,881	4/1943	Thomas	123/121
2,358,840	9/1944	Walker	123/121
2,767,691	10/1956	Mengelkamp	123/1
2,558,884	7/1951	Steven	123/127
2,381,304	8/1945	Merrill	123/121
3,650,255	3/1972	McJones	123/120
2,855,759	10/1958	Chaiser	62/51
3,672,341	6/1972	Smith	123/119
3,184,295	5/1965	Bauerstock	48/184

Primary Examiner—Laurence M. Goodridge
Assistant Examiner—Ronald B. Cox
Attorney, Agent, or Firm—James C. Fails; Wofford, Felsman & Fails

[57] **ABSTRACT**

Apparatus for enabling a conventional vehicle to be operated on a compressed gaseous fuel, characterized by a high degree of safety with features such as concentrating the high pressure portion of the apparatus into a section that is located outside of the passenger compartment for safety; employing a remote indicating fuel gauge that does not have a fuel connection inside the passenger compartment; employing double regulators with respective relief valves vented to a safe area; employing special factory-tested high pressure assemblies, and employing an excess flow shut-off valve for blocking flow in the event of a line failure. Also disclosed are specific embodiments such as employing a single stage compensated high pressure regulator for uniform performance until the container of compressed gaseous fuel is substantially empty; and parallel connected containers for increased range.

12 Claims, 7 Drawing Figures

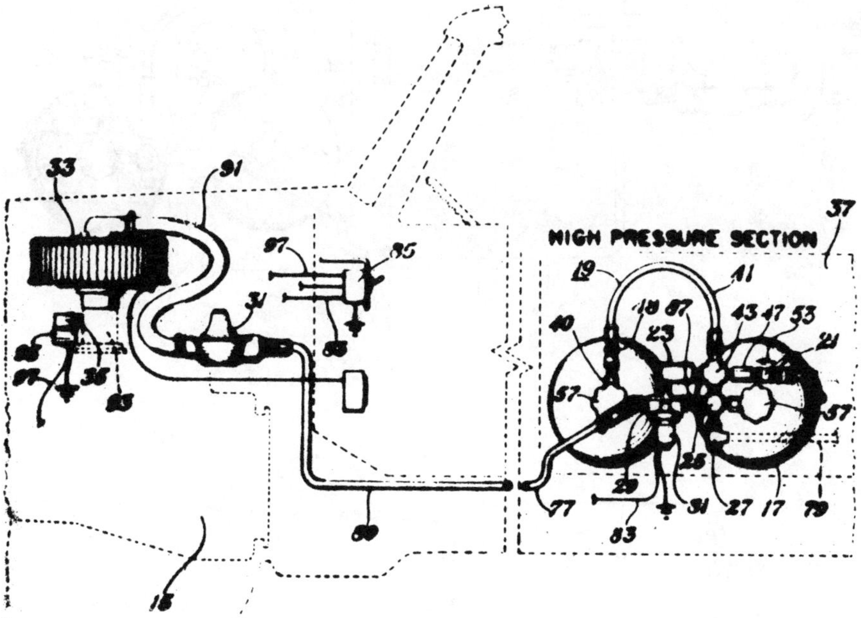

PATENTED FEB 5 1974 **3,789,820**

SHEET 1 OF 2

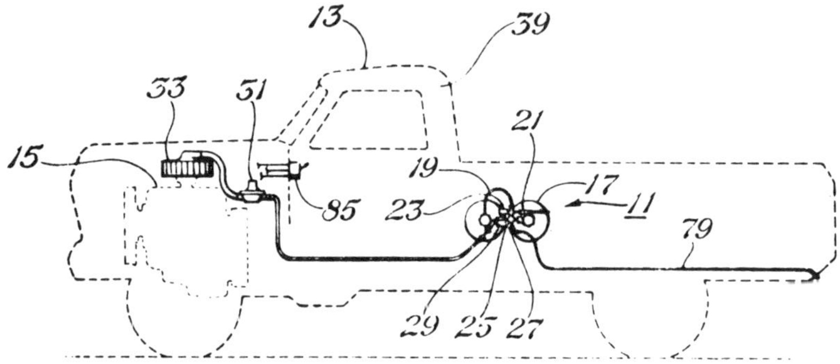

Fig. 1

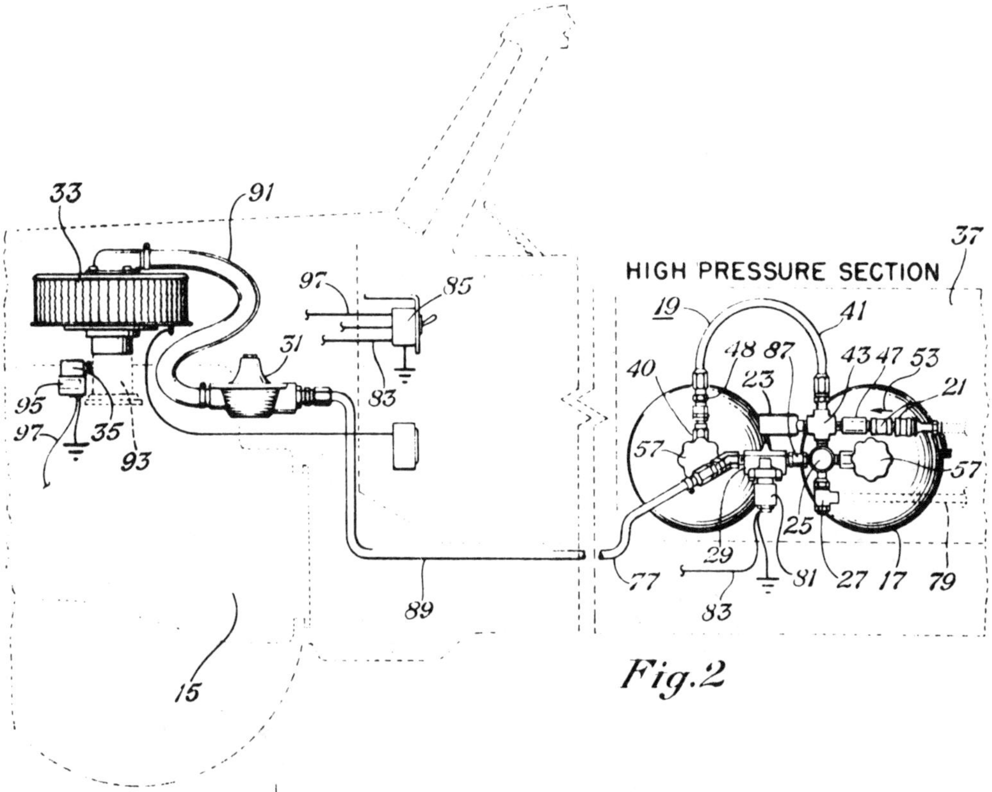

Fig. 2

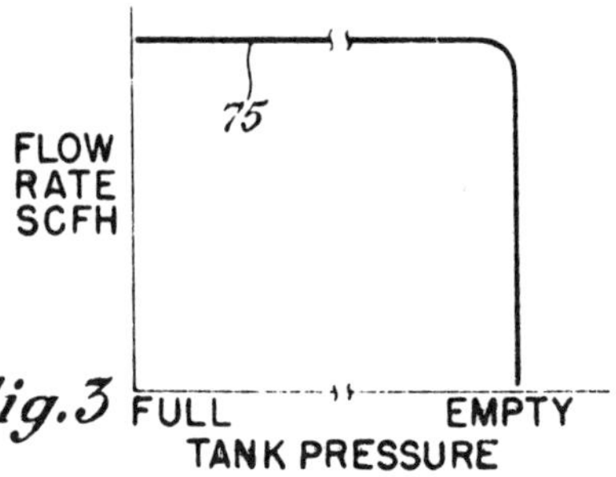

Fig. 3

INVENTORS
Lloyd A. Douglas
Kirk T. Garantlo
BY
Wofford, Felsman & Fails
ATTORNEYS

PATENTED FEB 5 1974 3,789,820

SHEET 2 OF 2

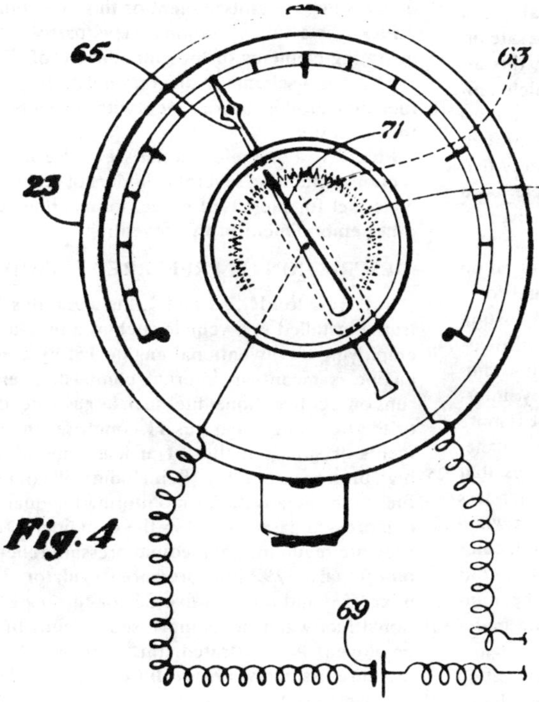

Fig.4

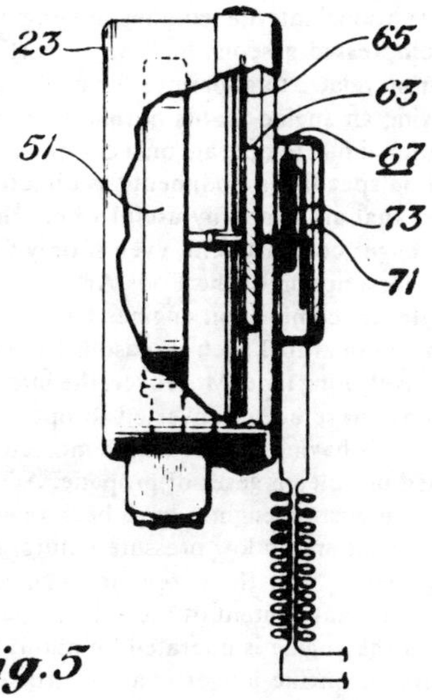

Fig.5

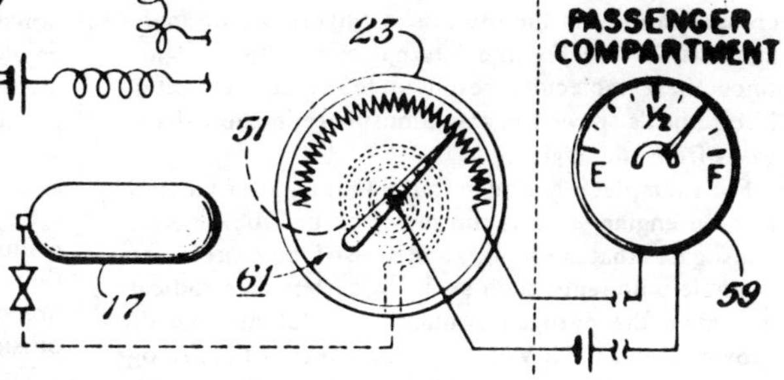

Fig.6

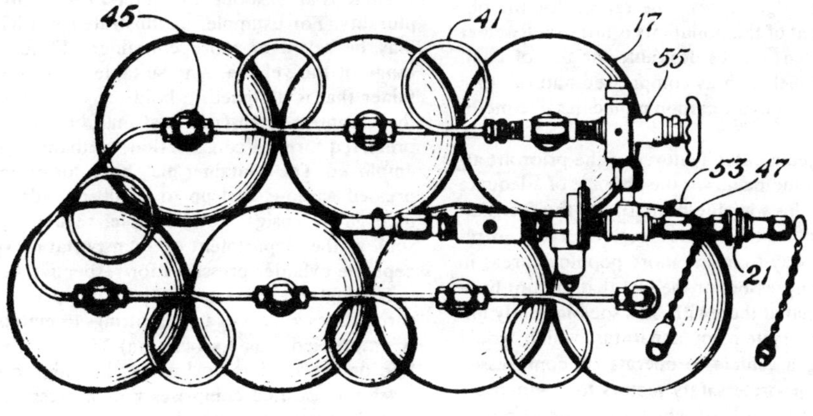

Fig.7

INVENTORS
Lloyd A. Douglas
Kirk T. Carantlo
BY
Wofford, Felsman & Faile
ATTORNEYS

COMPRESSED GASEOUS FUEL SYSTEM

BACKGROUND OF THE INVENTION

1. Field of the Invention

This invention relates to apparatus for enabling a conventional internal combustion engine to operate on a compressed gaseous fuel. More particularly, this invention relates to apparatus for enabling a vehicle employing an engine that is normally operated on a conventional fuel to operate on a compressed gaseous fuel; and, in specific embodiments, is directed at safety features that are primarily useful where the vehicle has a passenger compartment, even if only for the driver.

2. Description of the Prior Art

Internal combustion engines have been operated on conventional fuel such as gasoline or diesel fuel for a relatively long time. Moreover, the internal combustion engines have been converted to operate on hydrocarbon fuels having shorter length molecules; such as, liquified petroleum gases or propane. Additionally, internal combustion engines have been operated at remote field locations on low pressure natural gas or methane. Experience with these remote engines indicates that the pollutant content of the exhaust gases is much less when the engine is operated on natural gas than when operated on the longer chain hydrocarbon molecules such as employed in gasoline. With the advent of concern over polluting our atmosphere, there has been increased interest in the use of short hydrocarbon molecules as the fuel for the internal combustion engine, since these molecules result in fewer contaminating free radicals spewed in the atmosphere in the exhaust gases from the engine.

For example, it has been found that an internal combustion engine such as run in a pick-up truck has exhaust gases that are remarkably free of the more objectionable pollutants; such as, hydrocarbon free radicals and even the nitrogen containing pollutants like the brown nitrous oxide which is a constituent of the smog-type haze in the larger cities. The reason for the observed low content of the pollutants is not exactly clear but the low content makes desirable the use of compressed gaseous fuel such as compressed natural gas, comprising the short hydrocarbon molecules like methane.

One of the objectionable features of the prior art attempts to employ methane was the absence of adequate safety features. This was due primarily to the fact that the internal combustion engines were operated in remote locations away from the more populous areas in powering gas compressors or the like that did not have passengers associated therewith. It is sociologically imperative that any system, or apparatus, which would allow converting a vehicle to operate on compressed gaseous fuel incorporate safety factors to compensate for the increased hazards of which the general public may not be fully aware.

BRIEF DESCRIPTION OF THE DRAWINGS

FIG. 1 is a side elevational view, partly schematic, illustrating one embodiment of this invention employed on a conventional pick-up truck.

FIG. 2 is a more detailed side elevational view, also partly schematic, of the embodiment of FIG. 1.

FIG. 3 is a graph of the flow rate versus pressure to illustrate desirable performance characteristics of an embodiment of this invention.

FIG. 4 is a front elevational view of a fuel gauge means for monitoring the quantity of the fuel in accordance with one embodiment of this invention.

FIG. 5 is a side elevational view, partly cut away, illustrating elements of the embodiment of FIG. 4.

FIG. 6 is a schematic diagram illustrating the remote fuel indicator in accordance with an embodiment of this invention.

FIG. 7 is a side elevational view of a high pressure section containing several cylinders of compressed gaseous fuel for increased range, in accordance with another embodiment of this invention.

DESCRIPTION OF PREFERRED EMBODIMENTS:

Referring to FIGS. 1 and 2, the apparatus 11 is illustrated installed in a vehicle such as a pick-up truck 13 employing a conventional engine 15. By conventional engine is meant an internal combustion engine that runs on a conventional fuel such as gasoline, diesel fuel, or the like. The apparatus 11 comprises the major elements or subassemblies of at least one container 17, high pressure assembly 19, including fill connection 21, fuel gauge means 23 for monitoring the quantity of the compressed gaseous fuel in the container 17, and high pressure regulator 25, medium pressure relief valve 27, on-off valve 29, low pressure regulator 31, gas-air mixer 33, and on-off valve 35 for turning off conventional fuel when the compressed gaseous fuel is being employed. As illustrated, the container 17, the high pressure assembly 19, the fill connection 21, the fuel gauge means 23, and the high pressure regulator 25 are arranged in a high pressure section 37 that is located rearwardly of the passenger comparatment 39 for safety. In FIGS. 1 and 2, the high pressure section 37 is illustrated as being installed in the pick-up bed portion of a pick-up truck 13. In an automobile, the high pressure section is customarily located in the truck, outside of the passenger compartment and out of sight.

There is at least one container 17; and, preferably, a plurality. For example, as illustrated in FIG. 7, there may be seven or more containers 17 for increased range of the vehicle. Any suitable high pressure container that is designed to hold the desired pressure of the compressed gaseous fuel may be employed. In a preferred form of construction, methane cylinders are employed. The container may hold, for example, compressed gaseous fuel up to 2,200 pounds per square inch gauge (psig). For example, regulatory agencies such as the Department of Transportation specify acceptable cylinder pressures for respective vehicles.

The high pressure assembly 19 comprises high pressure fittings 40; for example, fittings to match the CGA (Compressed Gas Association) 350 methane connection on the methane cylinders. The high pressure assembly illustrated comprises a high pressure manifold including respective conduit means such as tubing 41 which may be of stainless steel; and fittings such as cross 43, FIG. 2, at least the interior half of fill connection 21, check valve 47, fuel gauge means 23, high pressure regulator 25, high flow shut-off valve 87, on-off valve 29, and the respective adapter fittings 48 for connecting the tubing 41 to the container-matching fittings 40. Because of the need for safety, the high pressure assembly 19 is preferably factory assembled and tested for leaks before it is released to be sold or used.

The fill connection 21 is connected into the high pressure assembly 19 for refilling the container 17 with the gaseous fuel. The fill connection 21 may comprise any convenient, sealing, connect-disconnect fitting. For additional safety, at least check valve means are disposed on both sides of the connect-disconnect fitting. The check valve means are operable to permit flow into the high pressure assembly when the connect-disconnect fitting is connected; and to prevent flow from the free ends of the connect-disconnect fitting when disconnected. For example, a conventional quick-connect fitting having respective bayonet and receiving connector portions may be employed. In the conventional quick-connect fitting, the check valves therein may be moved to the open position when the bayonet is inserted into the receiving connector portion for connection; but are moved to the closed position when the bayonet is removed so as to prevent flow from the disconnected ends, as might occur when the connection is broken; thereby reducing fire hazard and the like. A check valve 47, FIGS. 2 and 7, is also provided to permit flow in one direction, as indicated by the arrow 53 so as to enable filling the containers, but to prevent flow in the opposite direction. This check valve 47 is includes to prevent reverse flow when the quick-connect halves are coupled and a second vehicle at lower tank pressure is connected to a fill system; this prevents partial emptying of the first vehicle. If desired, a single valve such as valve 55, FIG. 7, may be provided to prevent flow outwardly from the manifold of the high pressure assembly, or, as illustrated in FIG. 2, the respective valves 57 on each respective cylinder may be employed to control the flow from the cylinders.

The fuel gauge means 23 is screwed into cross 43 and includes a pressure responsive element such as bourdon tube 51, FIGS. 5 and 6, in fluid connection with the container for monitoring the pressure therein. Any high pressure gauge mav be employed as fuel gauge means 23 as long as it is commensurate with the range of pressure employed in the containers 17. Preferably, the fuel gauge means 23 includes, also, a remote indicating gauge 59, FIG. 6, that may be emplaced in the passenger compartment to read the quantity of fuel. For safety, no line containing the gaseous fuel should be run into the passenger compartment, so the remote indicating gauge 59 must operate without directly sensing the pressure. It has been recognized that the running of lines with the combustible gaseous fuel into the passenger compartment is undesirable, particularly where the fuel is at super atmospheric pressure. Several states have enacted statutes prohibiting employing any direct reading pressure gauges, with the attendant pressurized fuel lines, in the passenger compartment. As illustrated in FIGS. 1, 2 and 6, the pressure responsive element such as the bourdon tube 51 in the fuel gauge means 23 is disposed adjacent the container 17 and outside the passenger compartment. Preferably, the gauge encompassing the bourdon tube 51 will also have a direct reading dial and pointer thereon to facilitate monitoring the amount of fuel and the quantity that is filled when fuel is added via the fill connection 21, without requiring continually adverting to the remote fuel indicator 59 in the passenger compartment. In the preferred form in which the remote indicating fuel gauge 59 is employed, a transducing means 61 is employed for converting the pressure into a continuous, single-valued transmissible signal which the remote indicating fuel gauge 59 can use to effect a visual display of the quantity of the fuel in response to the signal. As illustrated, the transducing means 61 comprises a magnet 63 that is carried by a pointer 65, FIGS. 4 and 5; a potentiometer 67; and a power source such as battery 69 that is electrically connected serially with the potentiometer 67. As illustrated, the potentiometer 67 is a circular potentiometer. The circular potentiometer 67 is disposed above the magnet 63 and has a potentiometer wiper 71 that is magnetically coupled with the magnet 63 carried by pointer 65. Consequently, the potentiometer wiper 71 will follow the pointer 65. The potentiometer wiper 71 is electrically conductive and has a brush that wipes along the circular resistor 73 of the circular potentiometer 67. As the potentiometer wiper 71 assumes a predetermined position, a unique voltage output is effected. Expressed otherwise, the voltage output is a function of the position of potentiometer wiper 71 on the potentiometer resistor 73. The remote indicating fuel gauge 59 comprises a galvanometer type instrument such as a voltmeter that is responsive to the voltage and is calibrated to read the voltage as a proportion of a full container of the gaseous fuel.

The high pressure regulator 25 is connected into the high pressure assembly 19 and directly with the valve 57 of cylinder 17, FIG. 2, for regulating a dowstream pressure; such as, through on-off valve 29; to a medium pressure. Preferably, the high pressure regulator 25 is a single stage compensated regulator for uniform performance until the container 17 is substantially empty, as illustrated in FIG. 3. Therein, the flow rate, in standard cubic feet per hour (scfh); and, consequently, engine performance; is a flat curve 75 regardless of the pressure of the gaseous fuel in the container 17, FIG. 2, until the container 17 gets below the differential pressure necessary for proper flow of fuel and operation of the low pressure regulator 31. This low pressure denotes a substantially empty container 17. Better maximum speed results from such flat curves. A desirable single stage compensated regulator is described and claimed in copending application Ser. No. 827,711, entitled "Single Stage, Compensated Pressure Regulator," inventor Ray R. Zimmer, filed May 26, 1969, assigned to the assignee of this invention. The medium pressure may range from several psi to as high as 150 psi.

The medium pressure relief vlave 27 is connected with the downstream side of the high pressure regulator 25 for venting the gaseous fuel at pressures greater than a preset maximum. The medium pressure relief valve 27 has a means such as a covered screw (not shown) for altering the preset maximum. Any conventional medium pressure relief valve may be employed as long as its range of preset maximum pressures is commensurate with the medium pressure desired in the conduit means 77 conveying the gaseous fuel to the engine 15. While the medium pressure relief valve 27 may be vented directly to the atmosphere in an open pick-up bed, it is preferred that it be vented, as via conduit means 79, to a safe place such as at the rear of the vehicle. The conduit means 79 is shown in dashed lines in FIG. 2. The conduit means 79 is necessary when the high pressure section 37 is emplaced in the trunk of a vehicle, in order to prevent the build up of explosive gases in the trunk in the event of pressure relief.

The on-off valve 29 is interposed in the conduit means 77 for preventing or permitting flow of the compressed gaseous fuel to the engine 15. Preferably, the on-off valve 29 is a solenoid operated on-off valve. Such on-off valves are conventional. Preferably also, the on-off valve 29 is connected with the downstream side of the high pressure regulator 25 so that it may have a relatively less expensive low pressure body instead of requiring a body that is structurally strong enough to withstand the high pressure of the compressed gaseous fuel in the high pressure assembly 19. As illustrated, the on-off valve 29 is a solenoid operated valve having a solenoid 81 that is electrically connected with ground and via conductor 83 with a switch 85 located inside of of the passenger compartment 39. The switch 85 may comprise a three position toggle switch effecting, in two of the positions, respective on and off operation of the solenoid 81. Conventionally, the switch 85 will be connected through a master switch such as the ignition switch so that the operation of the switch 85 is not necessary each time the engine is started or stopped.

The conduit means 77 effects a closed passageway for the transmission of the gaseous fuel from the high pressure regulator 25 to the gas-air mixer 33 on the engine. The conduit means 77 necessarily incorporates the requisite conventional fittings for connection with the elements interposed therein; for example, the on-off valve 29; a high flow shut-off valve 87, if emplaced therein; and the low pressure regulator 31. The conduit means 77 may comprise pressure tubing 89 and a flexible hose 91 for carrying the respective medium and low pressures. For example, the tubing 89 may be nylon tubing reinforced with steel mesh and covered for protection, or metallic tubing such as steel, copper or aluminum.

The low pressure regulator 31 is interposed into the conduit means 77 downstream of the on-off valve 29. The low pressure regulator 31 reduces the medium pressure to a low pressure for feeding to the gas-air mixer 33. The low pressure may range from a few inches of water to a few psi, depending upon the type of gas-air mixer 33 employed.

To add an additional safety feature, we have resorted to a low pressure relief valve that is connected with the downstream side of the low pressure regulator to vent the gaseous fuel in the event the low pressure becomes greater than a second preset maximum. This protects the gas-air mixer 33 as well as other low pressure components of the system. We employ a low pressure relief valve that is connected into the body of the low pressure regulator 31. Such low pressure relief valves are conventional and need not be described in detail herein.

The gas-air mixer 33 is connected; for example, via a conventional carburetor 93; with the intake manifold of the engine 15, and with the hose 91. The gas-air mixer 33 has means for admixing a combustion supporting gas such as air with the gaseous fuel. Any of the conventional gas-air mixers may be employed. We have found the gas-air mixer supplied commercially by Impco to be satisfactory.

The on-off valve means 35 is interposed in a line supplying the conventional fuel for stopping the flow of the conventional fuel to the engine when the gaseous fuel is being employed. Preferably, the on-off valve means 35 comprises a solenoid operated on-off valve having

solenoid 95 for effecting the on-off operation. The solenoid 95 is connected with ground and via conductor 97 with fuel control switch 85. The solenoid 95 operates conversely to solenoid 81 such that the flow of conventional fuel is stopped when the flow of the gaseous fuel is started or the flow of conventional fuel is started when the flow of gaseous fuel is stopped. Any of the conventionally operable solenoid-operated valves may be employed. The pressure that the body of the valve is designed to withstand will be determined by the type of fuel being employed. Ordinarily, where gasoline and diesel fuel are being employed, a low pressure body is adequate to withstand the pressure of the conventional fuel pump.

We have found it advantageous from a safety standpoint to provide an excess flow shut-off valve, such as high flow valve 87 for blocking the flow in the event there is a failure at some point in the flow path of the compressed gaseous fuel. For example, if tubing 89 were ruptured, the excess flow valve 87 would close and prevent the flow of the gaseous fuel which would otherwise result. The excess flow valves are conventional and are commercially available. Ordinarily, they comprise a check valve means such as a ball or poppet that is held in a predetermined position, as by suitable biasing means, but which is operated into a shut position against a seat to stop the flow as long as the rate of flow tends to be above a predetermined maximum. The predetermined maximum is responsive to pressure differential. The high flow shut-off valve 87 is located downstream of the high pressure regulator 25 for constant inlet pressure. Since the details of the excess flow shut-off valve are not being claimed, per se, and are commercially available, there is no need to supply the well-known details thereof.

In operation, the mechanical and electrical elements are installed to effect the system described hereinbefore and illustrated in the figures. The containers 17 are filled with the compressed natural gas to the desired pressure; for example, 2,200 psig. The high pressure connections are checked to ensure there are no leaks. The fuel control switch 85 is set for the conventional fuel; for example, the gasoline position; and the engine is started. Then, the lines and connections are checked for leaks at the on-off valve means 35. The engine is converted to run on gaseous fuel by positioning the fuel control switch 85 in another position and the engine started and the connections checked for any compressed gaseous fuel leaks. Smooth running of the engine on gaseous fuel is accomplished by adjusting the lean-rich adjusting screw of the gas-air mixer 33 to obtain a smooth idle. On the other hand, if the supply of compressed gaseous fuel becomes low, or it is desirable to run the engine on the conventional fuel for other reasons, the fuel control switch 85 is operated to the opposite position and transistion to the conventional fuel is smoothly made. If desired, the low pressure regulator may be preset to maintain from five to ten inches of water pressure on the compressed gaseous fuel. If natural gas is to be the only fuel used, the engine ignition timing may be set to 7° to 10° more advanced than the manufacturer's recommendations for the conventional fuel. For increased economy and performance, if natural gas is to be the prime fuel the spark plug gap setting should be decreased to half of the normal gap setting. The engine timing is set at the maximum recommended by the vehicle manufacturer if satisfactory operation is desired on both fuels.

The compressed gaseous fuel should be dried to have a dew point of at last −65° Fahrenheit at 2,250 psig, because, otherwise, the large pressure drop between a filled container 17 and the medium pressure conduit means 77 lowers the temperature sufficiently to cause operational difficulties.

The materials of construction ordinarily employed in this art are adequate for use herein and no new exotic materials need be employed.

Although the invention has been described with a certain degree of particularity, it is understood that the present disclosure has been made only by way of example and that numerous changes in the details of construction and the combination and arrangement of parts may be resorted to without departing from the spirit and the scope of the invention.

What is claimed is:

1. In apparatus for enabling a vehicle engine normally employing a conventional fuel to operate on a second fuel, and normally including:

a fuel-air mixer connected with the intake manifold of said engine and with a low pressure regulator and having means for admixing a combustion supporting gas with said second fuel;

on-off valve interposed in a conduit means for carrying said second fuel to said vehicle engine; said conduit means being connected with said fuel-air mixer and with a pressure regulator;

low pressure regulator interposed in said conduit means downstream of said on-off valve for reducing upstream pressure to a low pressure in the range of from a few inches of water to a few pounds per square inch guage (psig) for supplying to said fuel-air mixer; and

on-off valve interposed in a fuel line supplying said conventional fuel for stopping flow of said conventional fuel to said engine when said second fuel is being employed; the improvement comprising:

a. a high pressure section located outside of the passenger compartment of said vehicle for safety, said high pressure section including:

i. at least one container of a compressed gaseous fuel under superatmospheric pressure; said compressed gaseous fuel serving as said second fuel; said container being capable of withstanding pressure in excess of 1,000 pounds per square inch gauge (psig); and

ii. a high pressure assembly connected with said container of compressed gaseous fuel; said high pressure assembly including a first high pressure fitting for said container of said compressed gaseous fuel; high pressure tubing connected to said first high pressure fitting; a second high pressure fitting connected to said tubing;

a fill connection connected with said second high pressure fitting and into the remainder of said high pressure assembly for refilling said container with said compressed gaseous fuel;

fuel gauge means connected responsively with said second high pressure fitting and into the remainder of said high pressure assembly for monitoring the quantity of said compressed gaseous fuel in said container;

and high pressure regulator connected with said second high pressure fitting and into the remainder of said high pressure assembly for regulating a downstream pressure to a medium pressure in the range of from several psig to 150 psig; and

b. medium pressure relief valve connected with the downstream side of said high pressure regulator for venting said gaseous fuel at pressures greater than a preset maximum; said medium pressure relief valve having means for altering said preset maximum; and

c. a gas-air mixer serving as said fuel-air mixer.

2. The apparatus of claim 1 wherein said high pressure regulator comprises a single stage compensated regulator.

3. The apparatus of claim 1 wherein said high pressure assembly is leak-free.

4. The apparatus of claim 1 wherein said fill connection includes a convenient, sealing connect-disconnect fitting and has check valve means disposed on both sides of the joint of said fill connection, said check valve means being operable to permit flow when said fitting is connected for filling said container and operable to prevent flow from either of the free ends in the event of an inadvertent disconnection during filling.

5. The apparatus of claim 1 wherein a low pressure relief valve is connected with the downstream side of said low pressure regulator for venting said gaseous fuel at pressures above a second preset maximum low pressure; said low pressure relief valve having means for altering said second preset maximum low pressure.

6. The apparatus of claim 1 wherein said apparatus is installed in an automobile, said high pressure section is installed in the trunk of said automobile; and said medium pressure relief valve has a vent pipe that traverses to the exterior of said trunk and to a safe place for venting any gaseous fuel from said medium pressure relief valve.

7. The apparatus of claim 1 wherein an excess flow shut-off valve is incorporated into said apparatus for blocking flow from said high pressure assembly in the event there is a failure at some point that would otherwise vent said gaseous fuel at a rate greater than a predete. mined rate.

8. The apparatus of claim 1 wherein there are a plurality of containers of said compressed gaseous fuel and said high pressure assembly comprises a manifold interconnected with said plurality of containers; said containers comprising cylinders of compressed natural gas.

9. The apparatus of claim 1 wherein said fill connection has check valve means and a redundant check valve is interposed in the flow line to prevent back flow when said fill connection is connected.

10. In apparatus for enabling a vehicle engine normally employing a conventional fuel to operate on a second fuel, and normally including:

a fuel-air mixer connected with the intake manifold of said engine and with a low pressure regulator and having means for admixing a combustion supporting gas with said second fuel;

on-off valve interposed in a conduit means for carrying said second fuel to said vehicle engine; said conduit means being connected with said fuel-air mixer and with a pressure regulator;

low pressure regulator interposed in said conduit means downstream of said on-off valve for reducing upstream pressure to a low pressure in the range of from a few inches of water to a few pounds

United States Patent [19] [11] **3,817,233**

Kihn [45] **June 18, 1974**

[54] **VAPOR FUEL CARBURETION SYSTEM**

[75] Inventor: **Harold E. Kihn**, Pacific, Wash.

[73] Assignee: **The Raymond Lee Organization, Inc.**, New York, N.Y. ; a part interest

[22] Filed: **Sept. 18, 1972**

[21] Appl. No.: **289,802**

[52] U.S. Cl. .... **123/133**, 123/122 H

[51] Int. Cl. **F02m 17/18**

[58] Field of Search .... 123/133, 122 R, 122 F, 123/122 E, 122 H, 34, 34 A, 35, 180 E; 261/140, 142

[56] **References Cited**

UNITED STATES PATENTS

2,884,917 5/1959 Quinby 123/133

Primary Examiner—Manuel A. Antonakas
Assistant Examiner—R. H. Lazarus
Attorney, Agent, or Firm—Daniel Jay Tick

[57] **ABSTRACT**

A vapor fuel carburetion system for an internal combustion engine comprises a vaporizing device for vaporizing the fuel in the fuel reservoir of the engine. A directing system directs air and vaporized fuel to the compression chamber.

3 Claims, 1 Drawing Figure

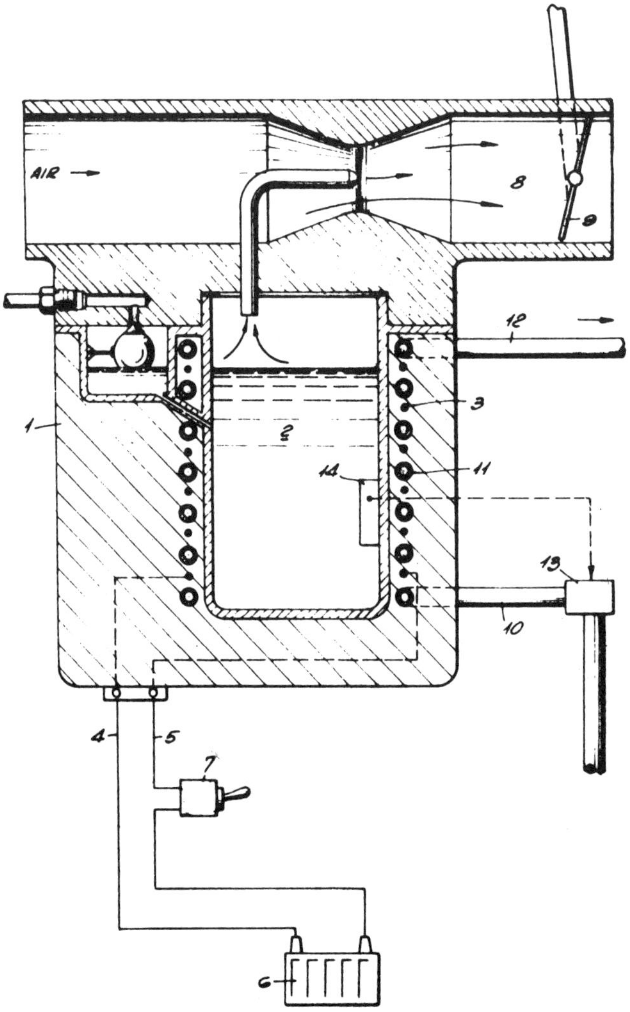

PATENTED JUN 18 1974 3,817,233

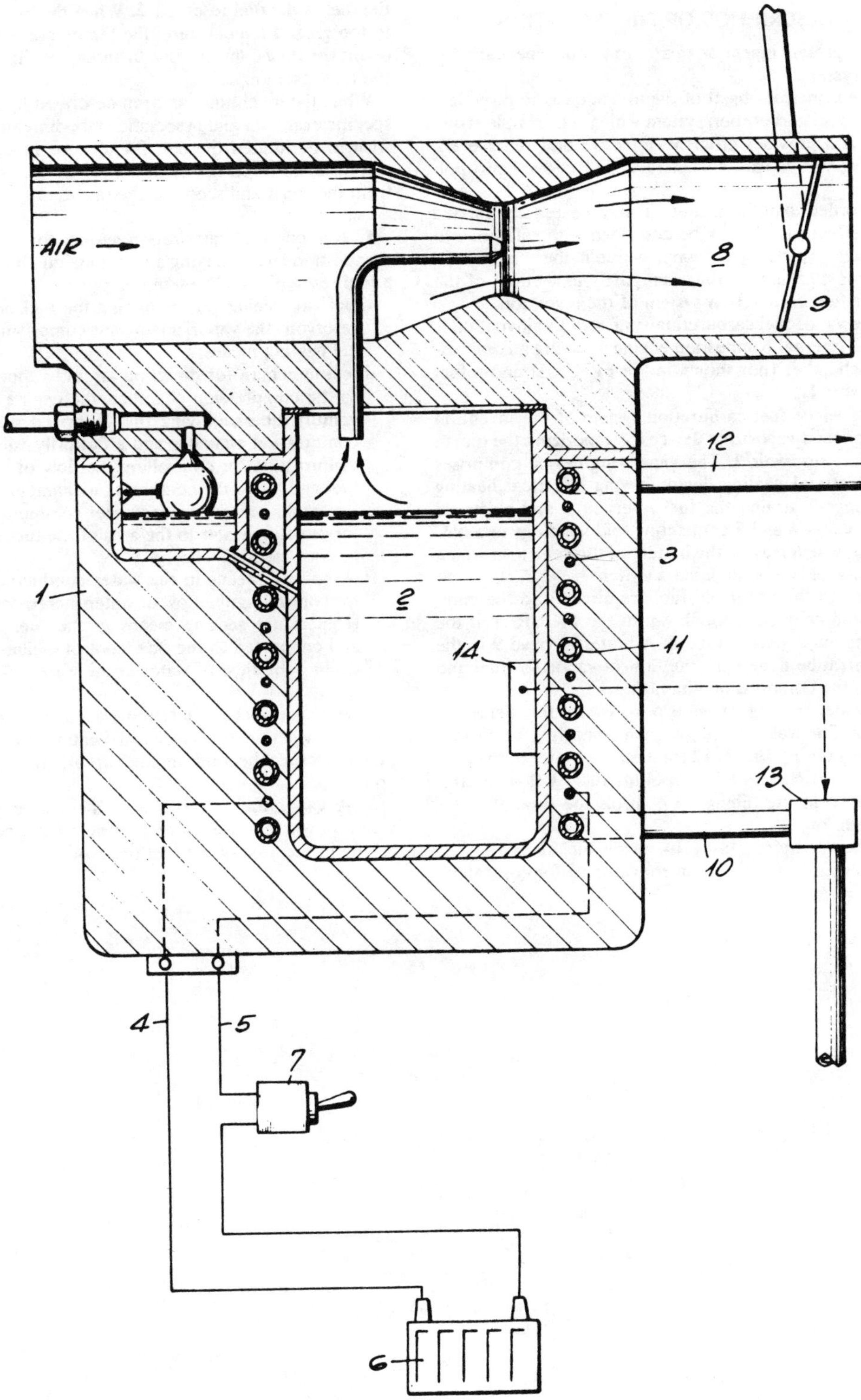

3,817,233

1

VAPOR FUEL CARBURETION SYSTEM

DESCRIPTION OF THE INVENTION

The present invention relates to a vapor fuel carburetion system.

The principal object of the invention is to provide a vapor fuel carburetion system which is of simple structure and provides efficient, effective and reliable engine operation with considerably reduced exhaust pollution.

In order that the invention may be readily carried into effect, it will now be described with reference to the accompanying drawing, wherein the single FIGURE is a schematic diagram, partly in section, of the vapor fuel carburetion system of the invention.

The vapor fuel carburetion system of the invention is for an internal combustion engine 1 having a compression chamber (not shown in the FIGURE) and a fuel reservoir 2.

The vapor fuel carburetion system of the invention comprises a vaporizing device for vaporizing the fuel in the fuel reservoir 2. The vaporizing device comprises an electrical heating device having electrical heating windings 3 around the fuel reservoir 2 and electrical conductors 4 and 5 connecting said windings to a battery 6, which may be the battery of the vehicle in which the engine is mounted, via a switch 7.

Air and the vaporized fuel are directed to the compression chamber (not shown in the FIGURE) of the engine via a venturi tube 8. A butterfly valve 9 in the venturi tube 8 controls the flow of vaporized fuel and air to the compression chamber.

A water cooling system is provided for the fuel reservoir 2. The water cooling system comprises water conducting tubing 10, 11, 12 for directing water to the area of the fuel reservoir 2 to cool the fuel. A flow control valve 13 in the tubing 10 controls the flow of water therethrough.

A temperature sensor 14 in the fuel reservoir 2 is connected to the flow control valve 13 for controlling

2

the flow of water in the water cooling system. The water cooling system thus controls the temperature of the fuel in the fuel reservoir 2. When the fuel is heated to too great a temperature, the temperature sensor 14 opens the flow control valve to supply cooling water to the fuel reservoir 2.

While the invention has been described by means of specific examples and in specific embodiments, I do not wish to be limited thereto, for obvious modifications will occur to those skilled in the art without departing from the spirit and scope of the invention.

I claim:

1. A vapor fuel carburetion system for an internal combustion engine having a compression chamber and a fuel reservoir, said system comprising
   - vaporizing means for vaporizing the fuel in the fuel reservoir, the vaporizing means comprising electrical heating means;
   - directing means for directing air and vaporized fuel to the compression chamber, the engine having a venturi tube supplying the vaporized fuel to the compression chamber and a butterfly valve in the venturi tube for controlling the flow of vaporized fuel and air to the compression chamber;
   - water cooling means having water conducting means for directing water to the area of the fuel reservoir to cool the fuel;
   - flow control means in the water conducting means for controlling the flow of water therethrough; and
   - temperature sensing means in the fuel reservoir and connected to the flow control means for controlling the flow of water in the water conducting means.

2. A vapor fuel carburetion system as claimed in claim 1, wherein the electrical heating means comprises electrical heating windings around the fuel reservoir.

3. A vapor fuel carburetion system as claimed in claim 1, wherein the water cooling means comprises tubing in the area of the fuel reservoir.

* * * * *

United States Patent [19] [11] **3,851,633**

Shih [45] **Dec. 3, 1974**

[54] **FUEL SYSTEM FOR AN INTERNAL COMBUSTION ENGINE**

[75] Inventor: **Kelvin Shih,** Detroit, Mich.

[73] Assignee: **General Motors Corporation,** Detroit, Mich.

[22] Filed: **Oct. 27, 1972**

[21] Appl. No.: **301,652**

[52] **U.S. Cl.**.............. **123/127,** 123/133, 123/179 G
[51] **Int. Cl.**........................ **F02d 19/06,** F02b 43/04
[58] **Field of Search**........ 123/3, 133, 179 G, 122 E, 123/122 F, 180 E, 180 EH, 34 R, 34 A, 35, 127

[56] **References Cited**

UNITED STATES PATENTS

1,559,214	10/1925	Woolson	123/3
1,744,953	1/1930	Dienner	123/3
2,625,920	1/1953	Farrell	123/133
3,688,755	9/1972	Grayson et al.	123/3

FOREIGN PATENTS OR APPLICATIONS

576,404	8/1924	France	123/180 E

Primary Examiner—Charles J. Myhre
Assistant Examiner—W. Rutledge, Jr.
Attorney, Agent, or Firm—Arthur N. Krein

[57] **ABSTRACT**

A fuel supply system for an internal combustion engine having a vapor carburetor wherein liquid fuel is vaporized in a fuel vaporizing chamber to supply fuel vapor to the engine with a portion of the lighter end of the fuel thus vaporized being stored in a holding fuel vaporizing chamber for a fuel system of an internal combustion engine used in starting the engine.

6 Claims, 2 Drawing Figures

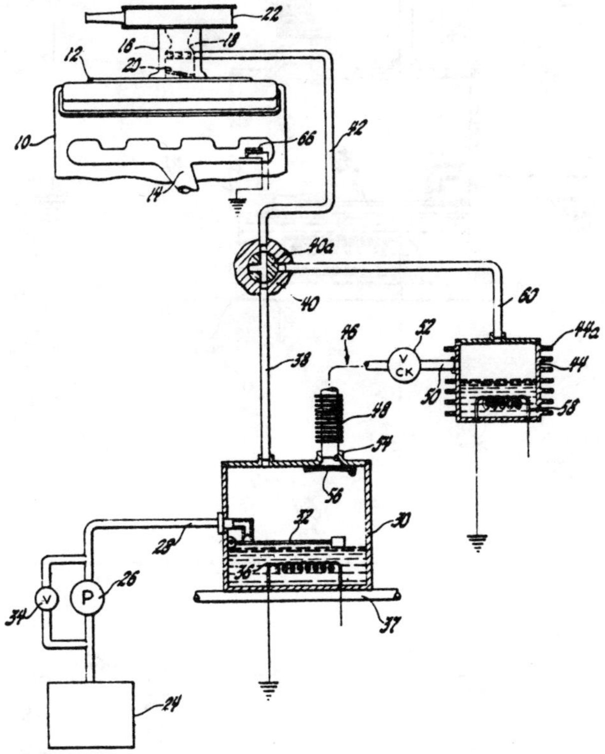

PATENTED DEC 3 1974 3,851,633

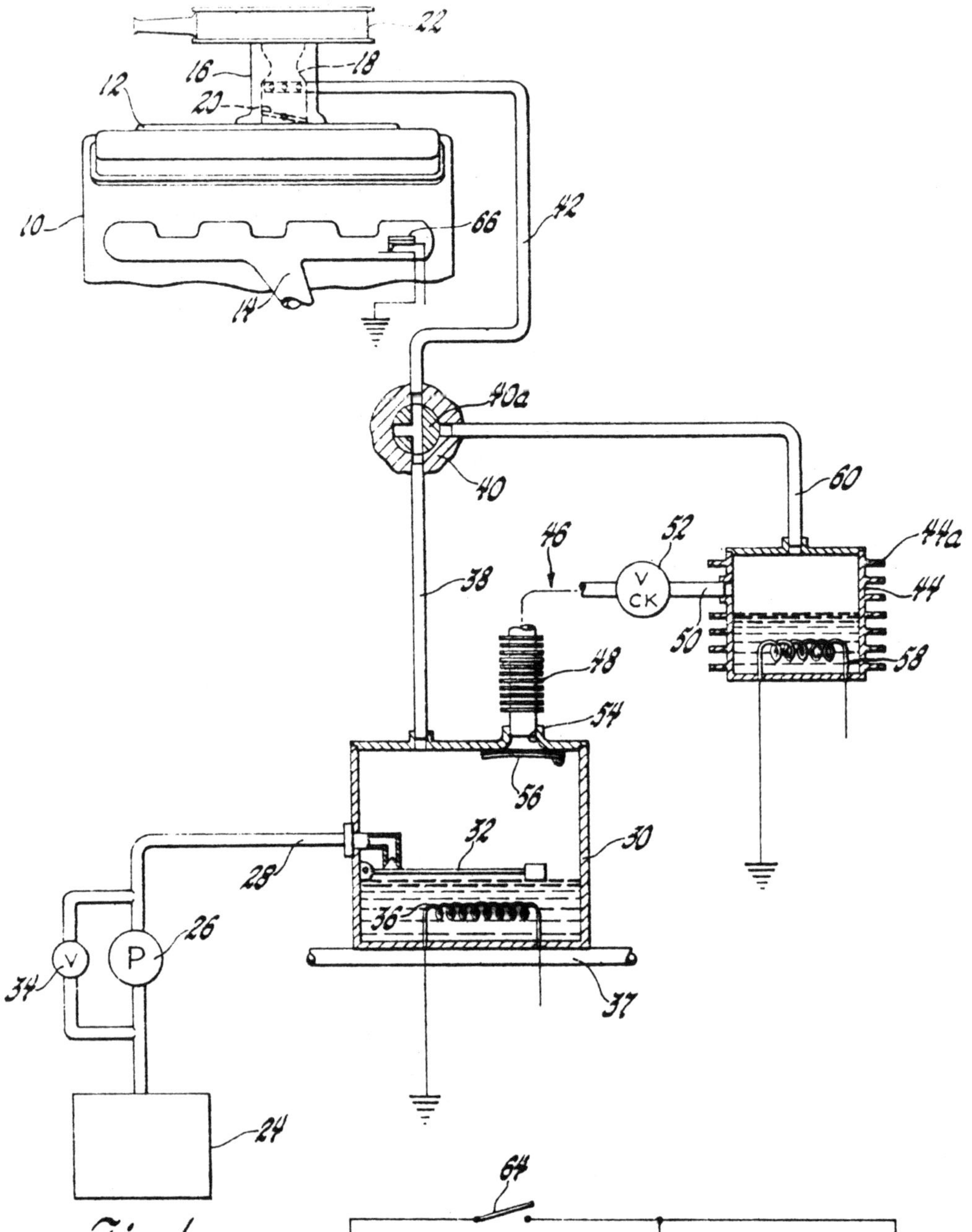

Fig.1

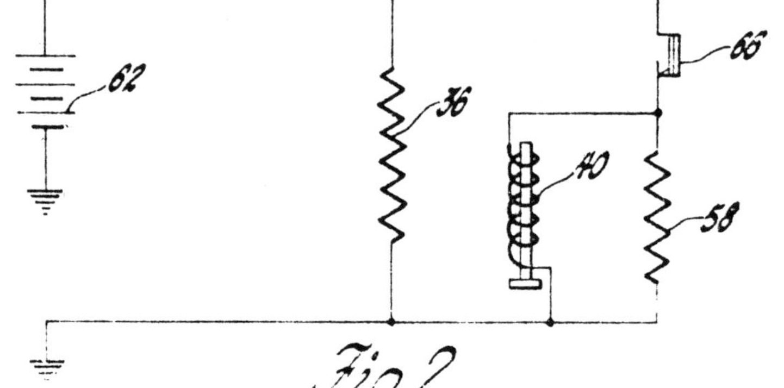

Fig.2

FUEL SYSTEM FOR AN INTERNAL COMBUSTION ENGINE

This invention relates to a fuel system for internal combustion engines and, in particular, to a fuel system for supplying dual fuel vapors to an automotive vehicle engine.

It is well known to use a fuel vapor mixed with air by means of a vapor carburetor, such as a conventional liquid petroleum gas (LP gas) carburetor to effect operation of an internal combustion engine. If the fuel used is a liquid fuel, such as gasoline, means are provided to heat the liquid fuel to generate the necessary fuel vapors for engine operation before the fuel is injected into the induction passage of the engine. With gasoline, both light end and heavy end fuel vapors are generated since gasoline vaporizes within the approximate range of 95°F to 450°F because of the wide variation in components making up gasoline, such as Pentane, Hexane, Heptane, Octane, Nonane, Decane and Hendecane.

However, in these prior art systems in which gasoline is used as the fuel, electrical heat elements must normally be used when starting the vehicle engine to vaporize the liquid gasoline in a suitable fuel vaporizing chamber which has been supplied with gasoline from a fuel reservoir. The vehicle's electrical power is normally used to vaporize the liquid fuel, both light and heavy end fuel vapors being generated for delivery to the engine induction system to effect engine starting and subsequent engine operation. Since the raw liquid gasoline is being vaporized, considerable electrical energy is used to effect this vaporization during engine starting.

It is therefore the principal object of this invention to provide a fuel system for an internal combustion engine in which liquid fuel is partially distilled to provide a dual fuel supply of vapors for engine starting and operation.

Another object of this invention is to provide a fuel system for an internal combustion engine wherein the more volatile fuel vapors are stored and then used during engine starting and initial operation of the engine until less volatile fuel vapors are generated for use in subsequent engine operation.

These and other objects of the invention are attained by means of a fuel supply system for an internal combustion engine wherein liquid fuel is vaporized in a first fuel vaporizing chamber to supply fuel vapor to the engine during engine operation with the light end of the fuel vapors thus generated being stored in a second fuel vaporizing chamber for use during initial engine starting.

For a better understanding of the invention, as well as other objects and further features thereof, reference is had to the following detailed description of the invention to be read in connection with the accompanying drawings, wherein:

FIG. **1** is a schematic illustration of a fuel system for an internal combustion engine in accordance with the invention and,

FIG. **2** is a schematic electrical wiring diagram of the fuel system of FIG. **1**.

Referring now to FIG. **1**, a conventional internal combustion engine **10** includes an intake manifold **12** and an exhaust manifold **14**, with a conventional vapor carburetor **16** mounted over the inlet of the intake

manifold. The vapor carburetor **16** is provided with an induction passage including a venturi **18**, flow through which is controlled by a throttle valve **20**, air being supplied through an air cleaner **22** mounted on top of the vapor carburetor.

The fuel system for the internal combustion engine **10**, in accordance with the invention, includes a fuel reservoir or tank **24** with a fuel pump **26** supplying liquid fuel, such as gasoline, therefrom via a conduit **28** to a vaporizing chamber **30**. As shown, the pump delivers liquid fuel to the vaporizing chamber **30** which contains a float controlled valve **32** so as to maintain a constant level of fuel available for vaporizing within the vaporizing chamber **30**. As is common practice, a pressure valve controlled fuel bypass **34** is provided around pump **26**.

Liquid fuel entering the vaporizing chamber **30** is vaporized by a heat source, such as electrical heating coil **36**, suitably mounted in heat transfer relationship to the fuel within the vaporizing chamber and may also include additional heating means receiving heat as an incident to engine operation, such as, for example, a conduit **37** receiving hot exhaust gases from the exhaust manifold **14** and also being positioned to be in heat transfer relationship to the vaporizing chamber **30**.

Fuel vapors generated within the vaporizing chamber **30** are delivered via a conduit **38**, a solenoid controlled valve **40** and conduit **42** to the vapor carburetor **16** for induction into the combustion chambers, not shown, of the internal combustion engine **10**.

The solenoid controlled valve **40** is provided with a rotatable valve element **40*a***, normally positioned as shown in FIG. **1** when de-energized, to place conduit **38** in fluid communication with conduit **42** but which, when energized, will rotate clockwise to block the flow from conduit **38**.

In addition, part of the vapors generated within the vaporizing chamber **30** are supplied to a fuel vaporizing and holding chamber **44** through a vapor separator, generally designated **46**. The vapor separator **46** includes a finned condenser conduit **48** and conduit **50** joined together, with the latter having a one-way check valve **52** therein so that fuel vapors can flow from the vaporizing chamber **30** to the chamber **44** but reverse flow of fuel vapors is prevented.

The flow of fuel vapors from the vaporizing chamber **30** through the discharge port **54** therein into the finned condenser conduit **48** is controlled by a suitable thermostat valve, such as bimetallic valve **56**, which at low temperature uncovers the port **54** but at a predetermined temperature covers this port. Bimetallic valve **56** is used to block the flow of fuel vapors into vapor separator **46** at a predetermined elevated temperature, depending on the cooling capability of the finned condenser conduit **48**. Otherwise, if heavy end fuel vapors above this predetermined temperature were permitted to enter the vapor separator **46**, they would not condense in the finned condenser conduit **48** for return to the vaporizing chamber **30** as a liquid but instead would enter the chamber **44** to condense therein and mix with the more volatile fuel stored in chamber **44**.

As fuel vapors are generated in the vaporizing chamber **30**, with the bimetallic valve **56** open, the vapors will travel upward through the finned condenser conduit **48** toward chamber **44**. Since the temperature at the top of this condenser conduit **48** will be lower because of heat loss, as compared to the bottom portion of this condenser conduit directly adjacent to the vaporizing chamber **30**, the heavy end fuel vapors of the gasoline will condense along the walls of this conduit and this liquid gas will then flow back into the vaporizing chamber **30**. Only the light end fuel vapors will continue to travel through the vapor separator into the fuel vaporizing and holding chamber **44**. Since the fuel vaporizing and holding chamber **44** is provided with suitable fins **44**a on the exterior thereof for heat dissipation, these light end fuel vapors will also condense in the chamber **44**.

Liquid fuel, consisting only of the light ends of the gasoline, thus stored within the chamber **44** are then vaporized, when desired, by a heat source, such as electrical heating coil **58** suitably mounted in heat transfer relationship to the fuel within the fuel vaporizing and holding chamber **44**.

Fuel vapors generated within the fuel vaporizing and holding chamber **44** are delivered via a conduit **60** and the previously described solenoid controlled valve **40** and conduit **42** to the vapor carburetor **16** during engine starting in a manner to be described.

As seen in FIG. **2**, the heating coils **36** and **58** are connectable to a source of electrical power, such as the vehicle battery **62**, through the ignition switch **64**, the heating coil **58** also being connected through a normally closed thermostatic control switch **66** positioned in thermal contact with, for example, a catalytic converter (not shown), or as shown to the engine exhaust manifold **14**. In addition, the solenoid controlled valve **40** is connected in parallel with the heating coil **58**.

With light end fuel of high volatility stored in the vaporizing and holding chamber **44** and with a cold engine so that the switch **66** is in its normally closed position, as the ignition switch **64** is closed to start the engine, the heating coils **36** and **58** and the solenoid controlled valve **40** are energized. As the solenoid controlled valve **40** is energized, the valve element **40a** thereof is rotated in a clockwise direction from the position shown in FIG. **1** to place the conduit **60** in communication with conduit **42**. Heat from the heating coil **58** will vaporize the light end fuel in chamber **44** in a few seconds and the thus generated vapors will then flow through conduit **60**, valve **40** and conduit **42** to the vapor carburetor **16** for induction into the combustion chambers of the engine. With this arrangement, both the starting and initial operation of the engine are effected by the use of only the more volatile components of the original gasoline stored in the fuel tank, that is, the light end components of the gasoline. This permits the engine to be started with a leaner air-fuel mixture than is normal using regular gasoline and, the resulting combustion of this fuel is much cleaner during initial engine operation. Depending on the size of the engine, sufficient fuel is stored in the chamber **44** to permit engine operation for a predetermined period of time so as to allow the engine to reach a normal operating temperature.

When the temperature of the engine, as determined at the exhaust manifold **14**, for example, reaches a predetermined temperature as sensed by the thermostatic control switch **66**, the contact of this switch will open to de-energize the heating coil **58** and the solenoid controlled valve **40**. As the solenoid controlled valve **40** is de-energized, the valve element **40a** will then rotate counterclockwise back to the position shown in FIG. **1** to place conduit **38** in fluid communication with the

conduit **42** so that fuel vapors generated within the vaporizing chamber **30** are then used for further operation of the engine. During subsequent engine operation, the vaporizing and holding chamber **44** will then start to recharge until it is once again full of the low end fuel components of the gasoline being vaporized in the vaporizing chamber **30** in the manner previously described.

Although in the embodiment disclosed a thermostatic control switch **66** is used to determine when the engine operation should be switched from the fuel in chamber **44** to the fuel supply from the vaporizing chamber **30**, it is to be realized that in lieu of this thermostatic control switch **66**, a suitable timer, such as a thermal timer, not shown, can be used to effect deenergizing of the heating coil **58** and the solenoid controlled valve **40** a predetermined time interval after engine start-up.

It is also to be realized that the fuel tank **24**, vaporizing chamber **30** and the vaporizing and holding chamber **44** can also be connected in a suitable manner to an evaporztive loss control device of the type disclosed, for example, in U.S. Pat. No. 3,093,124 issued on June 11, 1963 to Joseph T. Wentworth which device would also serve as a venting device for the subject fuel system. However, such an evaporative loss control device is not shown since it forms no part of the subject invention.

What is claimed is:

1. A fuel supply system for an internal combuation engine having an air induction passage provided with a venturi, a fuel reservoir for liquid fuel, a first fuel vaporizing chamber having a first heat source associated therewith, pump means connected to said fuel reservoir and to said first fuel vaporizing chamber for supplying fuel from said fuel reservoir to said first fuel vaporizing chamber, a second fuel vaporizing and storage chamber for light end fuel having a second heat source associated therewith, a vapor separator and valve controlled conduit means connecting said second fuel vaporizing chamber to said first fuel vaporizing chamber whereby only the light end fuel portion of the fuel vapor generated in said first fuel vaporizing chamber flows to said second fuel vaporizing and storage chamber and, valve controlled conduit means selectively connecting said first fuel vaporizing chamber and said second fuel vaporizing chamber to said venturi.

2. A fuel supply system according to claim **1** wherein said vapor separator and valve controlled conduit means includes a heat exchanger conduit connected to said first fuel vaporizing chamber for the flow of fuel vapors therefrom, a thermostat valve at one end thereof for controlling the flow of fuel vapor from said first fuel vaporizing chamber into said heat exchanger conduit and a one-way check valve at the opposite end of said heat exchanger conduit to prevent the flow of fuel vapor from said second fuel vaporizing and storage chamber to said first fuel vaporizing chamber.

3. A fuel supply system according to claim **2** wherein said first heat source includes an electrical heating coil connectable through a first switch to a source of electrical power and wherein said second heat source includes a second electrical heating coil connectable through said first switch and an engine operating temperature sensitive switch to said source of electrical power and wherein said valve controlled conduit means includes a solenoid controlled valve connected in parallel with said second electrical heating coil.

4. A fuel supply system for an internal combustion engine having an induction passage provided with a venturi, a fuel reservoir for liquid fuel, a first fuel vaporizing chamber having a first heat source associated therewith, energizing means connected to said first heat source for energizing said first heat source during engine operation, pump means connected to said fuel reservoir and to said first fuel vaporizing chamber for supplying fuel from said fuel reservoir to said first fuel vaporizing chamber, a second fuel vaporizing and storage chamber having a second heat source associated therewith, said energizing means including means connected to said second heat source for energizing said second heat source as a function of an engine operating condition, a vapor separator and valve controlled conduit means connecting said second fuel vaporizing chamber with said first fuel vaporizing chamber whereby fuel vapors flowing therein from said first fuel vaporizing chamber are separated so that only the light end fuel vapors flow to said second fuel vaporizing and storage chamber and, valve controlled conduit means operatively connecting said second fuel vaporizing chamber to said venturi when said second heat source is being energized and to connect said first fuel vaporizing chamber to said venturi when said second heat source is not being energized.

5. A fuel supply system according to claim **4** wherein said vapor separator and valve controlled conduit means includes a heat exchanger conduit connected to said first fuel vaporizing chamber for the flow of fuel vapors therefrom, a thermostat valve at one end thereof for controlling the flow of fuel vapor from said first fuel vaporizing chamber into said heat exchanger conduit and a one-way check valve at the opposite end of said heat exchanger conduit to prevent the flow of fuel vapor from said second fuel vaporizing and storage chamber to said first fuel vaporizing chamber.

6. A fuel supply system according to claim **4** wherein said first heat source is a first electrical heating coil and said second heat source is a second electrical heating coil and wherein said energizing means includes a source of electrical power connectable through a first switch to said first electrical heating coil and to said second electrical heating coil and an engine operating temperature sensitive switch interposed between said first switch and said second electrical heating coil, and wherein said valve controlled conduit means includes a solenoid controlled valve connected in parallel with said second electrical heating coil.

* * * * *

United States Patent [19] [11] **3,872,849**
Chester et al. [45] **Mar. 25, 1975**

[54] **ENGINE FUEL VAPORIZER**

[75] Inventors: **Frederic J. Chester**, Oxford; **Richard P. Krygowski**, Livonia, both of Mich.

[73] Assignee: **Ford Motor Company**, Dearborn, Mich.

[22] Filed: **Mar. 7, 1973**

[21] Appl. No.: **338,796**

[52] **U.S. Cl.** .... **123/122 AA**, 123/122 AC
[51] **Int. Cl.** **F02m 31/00**
[58] **Field of Search**.. 123/122 AB, 122 AH, 122 AC, 123/141, 122 B, 52 MW

[56] **References Cited**

UNITED STATES PATENTS

1,128,998	2/1915	Mulvaney	123/122 AA
1,200,608	10/1916	Hall	123/52 MV
1,373,290	3/1921	Berruet	123/122 AA
1,526,963	2/1925	Chandler	123/141
1,777,472	10/1930	Mock	123/122 AB
3,310,045	3/1967	Bartholomew	123/52 MV
3,648,674	3/1972	Proctor	123/141

FOREIGN PATENTS OR APPLICATIONS

870,460	3/1942	France	123/122 AA
1,190,803	5/1970	United Kingdom	123/141

Primary Examiner—Charles J. Myhre
Assistant Examiner—R. H. Lazarus
Attorney, Agent, or Firm—Keith L. Zerschling; Robert E. McCollum

[57] **ABSTRACT**

An engine exhaust gas crossover passage located in the intake manifold directly beneath the carburetor throttle riser bores is defined by a thin sheet metal tube that quickly radiates and transfers its heat to liquid fuel and fuel droplets in the carburetor air/fuel mixture splashed against it, the riser bores having guide members to direct the fuel against the tube, the tube quickly dissipating the heat upon engine shutdown so as not to add to the conventional engine hot soak.

6 Claims, 6 Drawing Figures

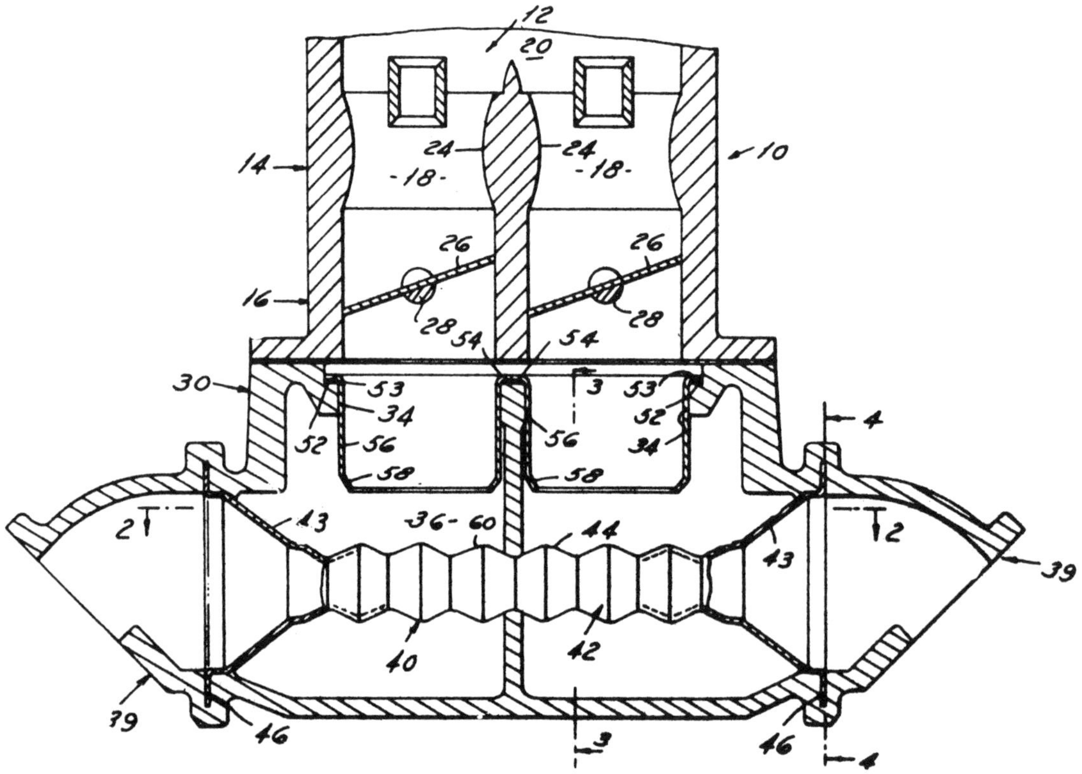

PATENTED MAR 25 1975 3.872.849

SHEET 1 OF 3

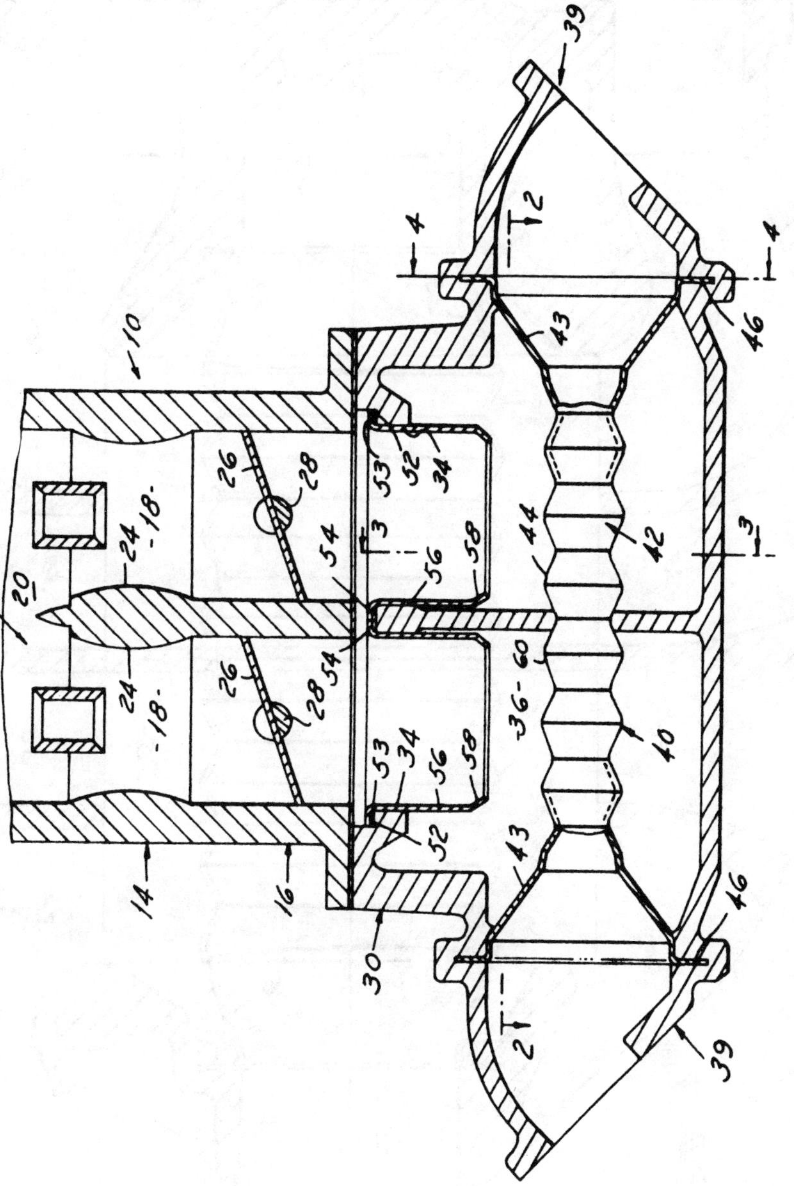

FIG. 1

PATENTED MAR 25 1975 3.872,849

SHEET 2 OF 3

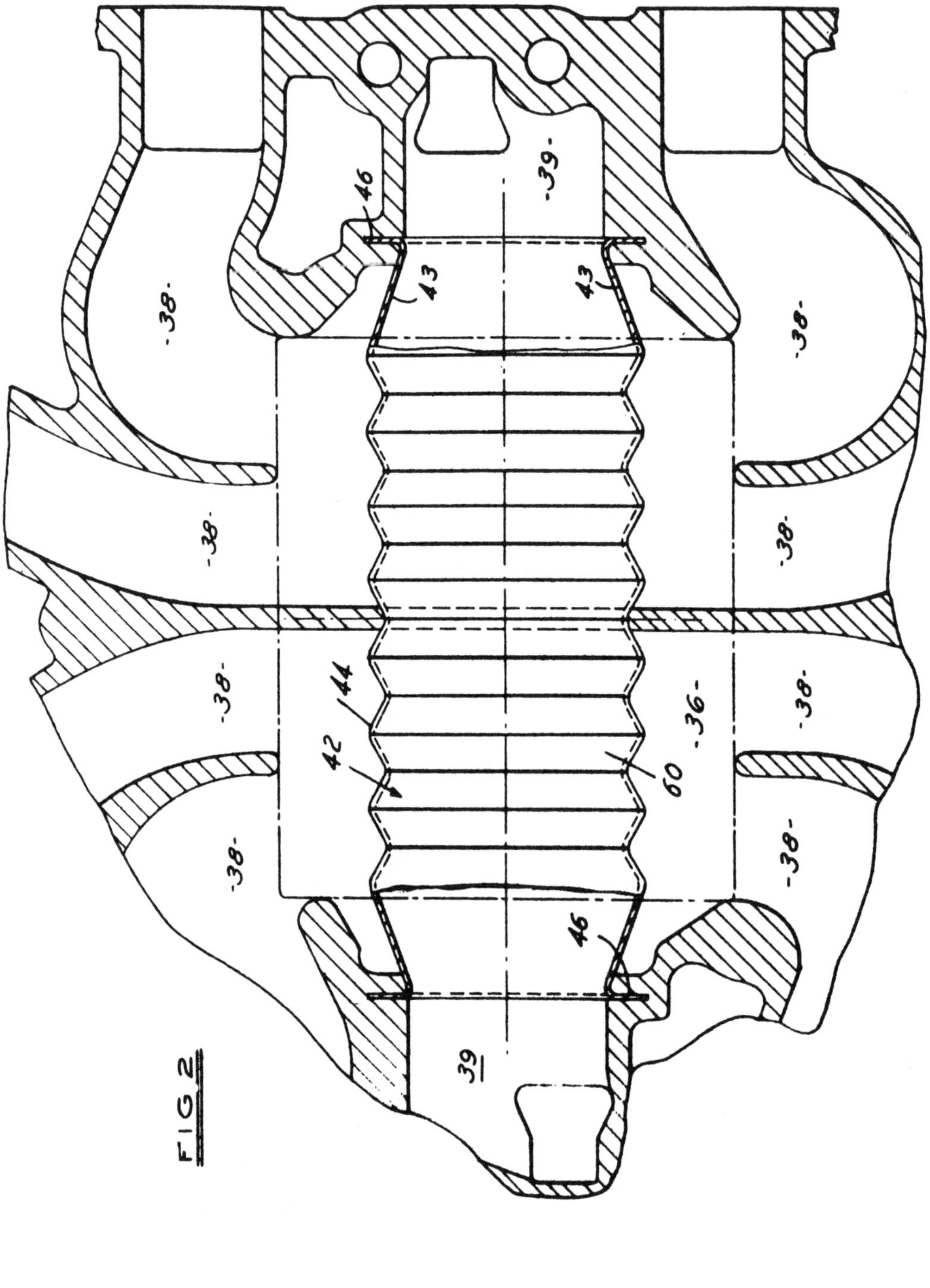

FIG 2

PATENTED MAR 25 1975 3,872,849

SHEET 3 OF 3

FIG. 3

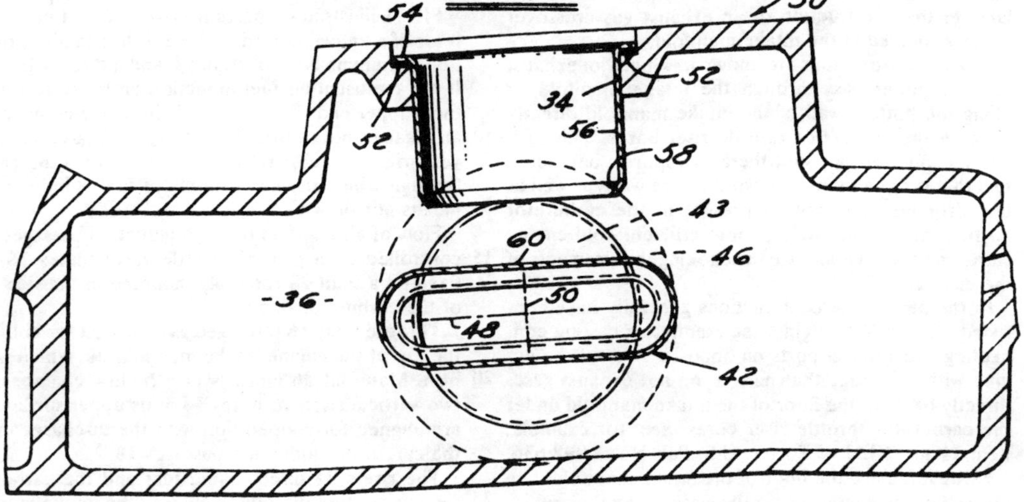

FIG. 5

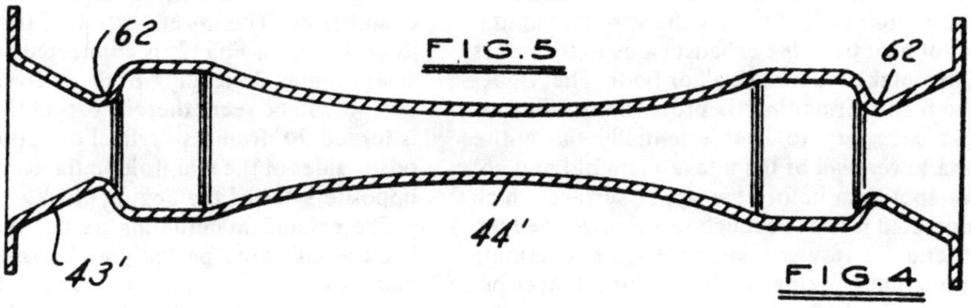

FIG. 4

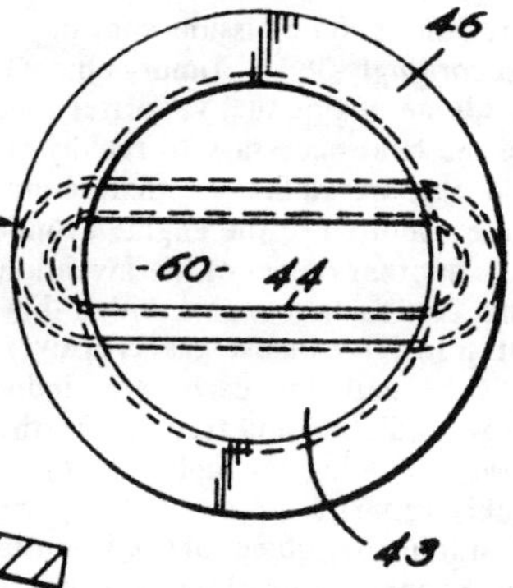

FIG. 6

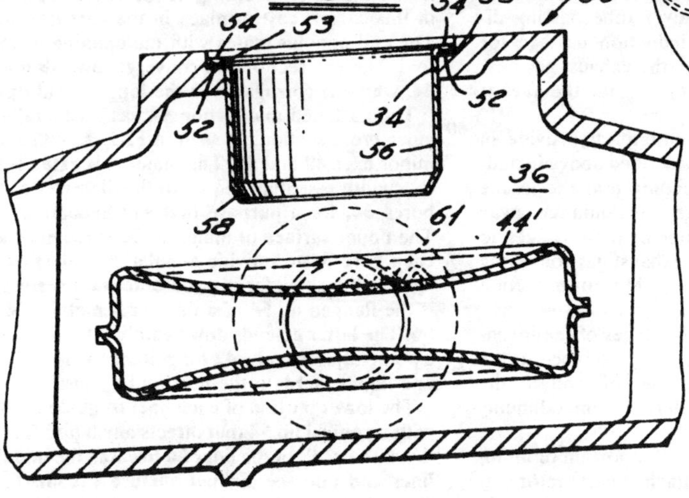

3,872,849

1

ENGINE FUEL VAPORIZER

This invention relates in general to an internal combustion engine fuel vaporizer. More particularly, it relates to the construction of an exhaust gas crossover passage located in the intake manifold.

Engine constructions are known in which hot exhaust gases at times pass through the intake manifold or along the bottom wall of the intake manifold directly beneath the carburetor throttle riser bores. This provides a so-called hot spot there to vaporize any liquid fuel that may be running down the bore walls as well as fuel droplets that may be present in the carburetor air/fuel mixture to provide more efficient cold engine starts and also to minimize the choking requirements of the engine.

In the past, these constructions generally have consisted of, in a V–8 engine, for example, a casting connecting two exhaust ports on opposite sides of the engine with a passage that passed the hot exhaust gases directly beneath the floor of the intake manifold under the carburetor throttle riser bores. See, for example, item 21 in FIG. 2 of Porter, U.S. Pat. No. 1,998,636.

However, since the floor of the intake manifold, like the rest of the engine, generally was of high thermal inertia cast iron, this resulted in only the slow transmittal or radiation of heat from the exhaust gases to the upper surface of the intake manifold wall or floor. That is, because of the high thermal inertia properties of the cast iron, it was necessary to heat essentially the entire length of the lower wall of the intake manifold and not just the hot spot area before the upper surface which the fuel contacted was hot enough to vaporize the fuel.

This, of course, resulted in the engine operating under rich mixture choking conditions for a longer period than is desirable for both engine operating efficiency and good emission control.

Accordingly, it is a primary object of the invention to provide an engine fuel vaporizer that will quickly provide the heat necessary to rapidly vaporize liquid fuel in the carbureted fuel/air mixture inducted through the intake manifold to the engine cylinders.

It is another object of the invention to provide an internal combustion engine with a thin sheet metal, high heat transfer exhaust gas crossover tube passing directly beneath the carburetor induction passage for quickly radiating and transferring the exhaust gas heat to the incoming air/fuel mixture so that the fuel is quickly vaporized.

A still further object of the invention is to provide an engine construction of the type described above including thin sheet metal liners extending into the intake manifold and having portions directing liquid fuel away from the walls towards the center of the passages to splash against a heat conducting exhaust gas conveying tube located directly beneath the carburetor induction passages.

Other objects, features and advantages of the invention will become more apparent upon reference to the succeeding detailed description thereof, and to the drawings illustrating the preferred embodiments thereof; wherein,

FIG. 1 is a cross-sectional view of a portion of an internal combustion engine with attached carburetor embodying the invention;

FIGS. 2, 3 and 4 are cross-sectional views taken on planes indicated by and viewed in the direction of the

2

arrows 2—2, 3—3, and 4—4, respectively, of FIG. 1; and,

FIGS. 5 and 6 are cross-sectional views of modifications of details shown in FIGS. 1 and 3, respectively.

FIG. 1 illustrates a portion 10 of one-half of a carburetor of a known downdraft type. It has an air horn section 12, a main body portion 14, and a throttle body 16. It has the usual air/fuel induction passages 18 open at their upper ends 20 to fresh air from the conventional air cleaner, not shown. The passages 18 have fixed area venturiis 22 cooperating with booster venturiis 24 through which the main supply of fuel is inducted, by means not shown.

Flow of air and fuel through induction passages 18 is controlled by a pair of throttle valve plates 26 each fixed on a shaft 28 rotatably mounted in the side walls of the carburetor body.

Throttle body 16 is flanged as indicated for bolting to the top of the engine intake manifold 30, which is cast iron. Manifold 30 consists of a hollow chamber with two vertical risers or bores 34 in its upper surface that are aligned for cooperation with the discharge end of the carburetor induction passages 18.

The risers 34 direct liquid fuel and the carbureted fuel/air mixture directly down into the intake manifold chamber 36. The lower portion of the intake manifold, as best seen in FIG. 2, is connected to eight, four on a side, runners 38 leading to the individual engine cylinders. It will be seen, therefore, that the fuel/air mixture is turned 90° from its vertical direction to pass out opposite sides of the manifold to the banks of cylinders on opposite sides of the engine block.

The exhaust manifolding part of the engine cylinder head is indicated partially at 39, and includes an exhaust gas crossover passage 40. The hot exhaust gases pass between the exhaust manifolds, not shown, on each side of the engine through the intake manifold chamber 36 directly beneath the riser bores 34 to vaporize the air/fuel mixture.

More specifically, passage 40 is defined by a thin sheet metal, high heat transfer tube 42. The tube has conical or bell mouth shaped end sections 43 joined to a flattened midportion 44. The end sections 43 each have an annular mounting flange 46 (see FIG. 3) which in this case is cast in place in the cast iron manifold. The end annulus mates with the opening of the manifold. The end sections 43 converge towards midportion 44, which is corrugated in the longitudinal direction.

The flattened midportion 44 has a generally rectangular cross section, as seen in FIG. 3, with major and minor axes 48 and 50. The major axis extends laterally to a width essentially equal to the diameter of the riser bores 34, for a purpose that will become clear later. The upper surface of manifold 30 is recessed adjacent each bore 34 to provide annular shoulders 52. Seated against the shoulder in each bore on a sealing ring 53 is the flanged lip 54 of a thin sheet metal tube or liner 56. The latter extends downwardly into the manifold to a point adjacent tube 42 for guiding the fuel/air mixture and liquid fuel into the manifold against tube 42.

The lower portion of each liner or guide 56 is formed with a conical lip 58 that directs any liquid fuel running down the walls of the liner 56 towards the middle of the liner and into the air/fuel mixture stream. The liquid fuel splashes against tube 42 and also passes around the tube with the air/fuel mixture into the dead air space below the tube to absorb the heat of the tube and pass into the manifold trunks or runners 38 at right angles.

It will be seen, therefore, that the hot exhaust gases will rapidly radiate their heat to the outer surfaces 60 of tube 42; that the flattened corrugated midportion of the tube provides a heat transfer surface area for the entire width of the air/fuel mixture stream; that both top, bottom and side surfaces of tube 42 provide heat; and that, therefore, the liquid fuel particles will be quickly vaporized because of a maximum rate of heat transfer between the exhaust gas the the intake charge. The necessity for a conventional choke fuel enrichment system, therefore, is essentially eliminated, with a resultant reduction in the emission of unburned hydrocarbons. The raw fuel flowing against the tube is vaporized and mixed with inlet air to provide the engine with a combustible charge without additional enrichment only seconds after a start.

An added advantage of the sheet metal tube heater over the conventional cast iron plate or lower manifold wall is heat dissipation upon engine shutdown. The conventional cast iron intake manifold with cast iron floor at the hot spot acts as a heat sink during engine shutdown since the heat only slowly dissipates from it. Accordingly, hot engine restart problems may be encountered by too much fuel being vaporized during engine shutdown; that is, the heat previously radiated from the crossover passage to the entire manifold radiates to other parts of the engine and up through the carburetor walls to the fuel bowl. The fuel then vaporizes and can provide too rich a mixture for engine restart.

The thin sheet metal tube 42 of the invention, on the contrary, has no ability to retain heat once the exhaust gas flow is cut off, and excess evaporation of fuel in the carburetor float bowl on engine shutdown, therefore, does not occur.

FIGS. 5 and 6 show modifications of the shape of the crossover tube 42. The midportion 44' of the two piece (FIG. 6) tube is dished both longitudinally and transversely so that liquid fuel splashed against the upper surface will puddle and be retained until vaporized instead of some escaping into the engine cylinders in a liquid state. That is, the tube midportion is shaped to form a dam along its outer edge, with a depressed central portion. Also, the surface can be corrugated, as indicated in dotted lines 61 in FIG. 6, if desired.

Additionally, these figures show the end sections 43' as having pinched annular portions 62. These permit longitudinal thermal expansion and contraction of the tube.

While the invention has been shown and described in its preferred embodiments, it will be clear to those skilled in the arts to which it pertains, that many changes and modifications may be made thereto without departing from the scope of the invention.

We claim:

1. An engine fuel vaporizer for use on an internal combustion engine having an intake manifold, a carburetor mounted on the intake manifold and having one or more air/fuel induction passages aligned so as to discharge an air/fuel mixture into the intake manifold, exhaust manifolding including an exhaust gas crossover passage connecting spaced parts of the exhaust manifolding and passing through the intake manifold directly beneath the carburetor induction passages, the intake and exhaust manifolding being constructed of high thermal inertia slow heat transfer material, the improvement comprising,

the crossover passage being defined by a thin walled sheet metal tube of high heat transfer potential for quickly radiating and transferring the heat of exhaust gases flowing through the tube through the tube walls in all directions to the fuel in the air/fuel mixture adjacent and contiguous to the tube,

the tube having a flattened midportion of a generally rectangular-like cross section providing major and minor axes and being spaced from the walls of the intake manifold so as to be encompassed by the air/fuel mixture for maximum heat transfer therebetween,

the tube extending laterally to a width comparable to the diameter of an induction passage to constitute a splash shield for the vaporization of liquid fuel and fuel vapor coming into contact therewith,

the tube having bell mouth-like shaped end sections mating with essentially annular exhaust manifolding portions, the end sections converging towards the flattened tube midportion for spreading the exhaust gas flow laterally of the tube longitudinal axis for better heat transfer through the tube walls.

2. A fuel vaporizer as in claim 1, the tube being longitudinally corrugated to provide a greater effective heat transfer surface.

3. A fuel vaporizer as in claim 1, the tube having at least the upper portion thereof of a dished shape in longitudinal cross-section for containment and subsequent vaporization of liquid fuel coming into contact with the upper portion thereof.

4. A fuel vaporizer as in claim 1, the tube having concave longitudinal and end cross-sections for the puddling of liquid fuel deposited on the tube upper surface adjacent the liner.

5. A fuel vaporizer as in claim 1, the tube having pinched portions formed therein providing thermal expansion and contraction portions.

6. A fuel vaporizer as in claim 1, the liner comprising a removable tubular insert having an annular lip at its upper end seating against the intake manifold.

* * * * *